T0203014

The Foundations of Computability Theory

Borut Robič

The Foundations
of Computability Theory

Borut Robič
Faculty of Computer and Information
 Science
University of Ljubljana
Ljubljana
Slovenia

ISBN 978-3-662-51601-0   ISBN 978-3-662-44808-3   (eBook)
DOI 10.1007/978-3-662-44808-3

Springer Heidelberg New York Dordrecht London
© Springer-Verlag Berlin Heidelberg 2015
Softcover reprint of the hardcover 1st edition 2015

Printed on acid-free paper

Springer-Verlag GmbH Berlin Heidelberg is part of Springer Science+Business Media
(www.springer.com)

To Marinka, Gregor and Rebeka Hana

I still think that the best way to learn a new idea is to see its history, to see why someone was forced to go through the painful and wonderful process of giving birth to a new idea. ... Otherwise, it is impossible to guess how anyone could have discovered or invented it.

— *Gregory J. Chaitin,* Meta Maths, The Quest for Omega

Preface

Context

The paradoxes discovered in Cantor's set theory sometime around 1900 began a crisis that shook the foundations of mathematics. In order to reconstruct mathematics, freed from all paradoxes, Hilbert introduced a promising program with formal systems as the central idea. Though the program was unexpectedly brought to a close in 1931 by Gödel's famous theorems, it bequeathed burning questions: "What is computing? What is computable? What is an algorithm? Can every problem be algorithmically solved?" This led to *Computability Theory*, which was born in the mid-1930s, when these questions were resolved by the seminal works of Church, Gödel, Kleene, Post, and Turing. In addition to contributing to some of the greatest advances of twentieth century mathematics, their ideas laid the foundations for the practical development of a universal computer in the 1940s as well as the discovery of a number of algorithmically unsolvable problems in different areas of science. New questions, such as "Are unsolvable problems equally difficult? If not, how can we compare their difficulty?" initiated new research topics of *Computability Theory*, which in turn delivered many important concepts and theorems. The application of these is central to the multidisciplinary research of *Computability Theory*.

Aims

Monographs in *Theoretical Computer Science* usually strive to present as much of the subject as possible. To achieve this, they present the subject in a definition–theorem–proof style and, when appropriate, merge and intertwine different related themes, such as computability, computational complexity, automata theory, and formal language theory. This approach, however, often blurs historical circumstances, reasons, and the motivation that led to important goals, concepts, methods, and theorems of the subject.

My aim is to compensate for this. Since the fundamental ideas of theoretical computer science were either motivated by historical circumstances in the field, or were developed by pure logical reasoning, I describe *Computability Theory*, a part of *Theoretical Computer Science*, from this point of view. Specifically, I describe difficulties that arose in mathematical logic, the attempts to recover from them, and how these attempts led to the birth of *Computability Theory* and later influenced it. Although some of these attempts fell short of their primary goals, they put forward crucial questions about computation and led to the fundamental concepts of *Computability Theory*. These in turn logically led to still new questions, and so on. By describing this evolution I want to give the reader a deeper understanding of the foundations of this beautiful theory. The challenge in writing this book was therefore to keep it accessible by describing the historical and logical development while at the same time introducing as many modern topics as needed to start the research. Thus, I will be happy if the book makes good reading before one tackles more advanced literature on *Computability Theory*.

Contents

There are three parts in this book.

Part I (Chaps. 1–4) *Chapter 1* is introductory: it discusses the *intuitive* comprehension of the concept of the algorithm. This comprehension was already provided by Euclid and sufficed since 300 B.C.E. or so. In the next three chapters I explain how the need for a rigorous, *mathematical definition* of the concepts of the algorithm, computation, and computability was born. *Chapter 2* describes the events taking place in mathematics around 1900, when *paradoxes* were discovered. The circumstances that led to the paradoxes and consequently to the foundational crisis in mathematics are explained. The ideas of the three main schools of recovery—the *intuitionism*, *logicism*, and *formalism*—that attempted to reconstruct mathematics, are described. *Chapter 3* delves into formalism. This school gathered the ideas and results of other schools in the concept of the *formal axiomatic system*. Three particular systems that played crucial roles in the future events are described; these are the formal axiomatic systems of *logic*, *arithmetic*, and *set theory*. *Chapter 4* presents *Hilbert's Program*, a promising formalistic attempt that would use formal axiomatic systems to eliminate all the paradoxes from mathematics. It is explained how *Hilbert's Program* was unexpectedly shattered by Gödel's *Incompleteness Theorems*, which state, in effect, that not every truth can be proved (in a formal system).

Part II (Chaps. 5–9) *Hilbert's Program* left open a question about the existence of a particular algorithm, the algorithm that would solve the *Entscheidungsproblem*. Since this algorithm might not exist, it was necessary to formalize the concept of the algorithm—only then would a proof of the non-existence of the algorithm be possible. Therefore, *Chapter 5* discusses the fundamental questions: "What is an algorithm? What is computation? What does it mean when we say that a function or problem is computable?" It is explained how these intuitive, informal concepts were formally defined in the form of the *Computability* (*Church–Turing*) *Thesis* by the different yet equivalent *models of computation*, such as partial and general recursive functions, λ-calculus, the Turing machine, the Post machine, and the Markov algorithms. *Chapter 6* focuses on the *Turing machine* that most convincingly formalized the intuitive concepts of computation. Several equivalent variants of the Turing machine are described. Three basic uses of the Turing machines are presented: function computation, set generation, and set recognition. The existence of the *universal Turing machine* is proved and its impact on the creation and development of *general-purpose computers* is described. The equivalence of the Turing machine and the RAM model of computation is proved. In *Chapter 7*, the first basic yet important theorems are deduced. These include the relations between *decidable*, *semi-decidable*, and *undecidable sets* (i.e., decision problems), the *Padding Lemma*, the *Parametrization* (i.e., *s-m-n*) *Theorem*, and the *Recursion Theorem*. The latter two are also discussed in view of the recursive procedure calls in the modern general-purpose computer. *Chapter 8* is devoted to *incomputability*. It uncovers a surprising fact that, in effect, not everything that is defined can be computed (on a usual model of computation). Specifically, the chapter shows that not every computational problem can be solved by a computer. First, the incomputability of the *Halting Problem* is proved. To show that this is not just a unique event, a list of selected incomputable problems from various fields of science is given. Then, in *Chapter 9*, methods of proving the incomputability of problems are explained; in particular, proving methods by *diagonalization*, *reduction*, the *Recursion Theorem*, and *Rice's Theorem* are explained.

Part III (Chaps. 10–15)   In this part attention turns to *relative computability*. I tried to keep the chapters "bite-sized" by focusing in each on a single issue only. *Chapter 10* introduces the concepts of the oracle and the *oracle Turing machine*, describes how computation with such an external help would run, and discusses how oracles could be replaced in the real world by actual databases or networks of computers. *Chapter 11* formalizes the intuitive notion of the "*degree of unsolvability*" of a problem. To do this, it first introduces the concept of *Turing reduction*, the most general reduction between computational problems, and then the concept of *Turing degree*, which formalizes the notion of the degree of unsolvability. This formalization makes it possible to define, in *Chapter 12*, an operator called *Turing jump* and, by applying it, to construct a *hierarchy* of infinitely many Turing degrees. Thus, a surprising fact is discovered that for every unsolvable problem there is a more difficult unsolvable problem; there is no most difficult unsolvable problem. *Chapter 13* embarks on this intriguing fact. It first introduces a view of the *class of all Turing degrees* as a mathematical structure. This eases expression of relations between the degrees. Then several properties of this class are proved, revealing a highly complex structure of the class. *Chapter 14* introduces *computably enumerable* (*c.e.*) *Turing degrees*. It then presents *Post's Problem*, posing whether there exist c.e. degrees other than $\mathbf{0}$ below the degree $\mathbf{0}'$. Then the *priority method*, discovered and used by Friedberg and Muchnik to solve *Post's Problem*, is described. *Chapter 15* introduces the *arithmetical hierarchy*, which gives another, arithmetical view of the degrees of unsolvability. Finally, *Chapter 16* lists some suggestions for further reading.

Approach

The main lines of the approach are:

- **Presentation levels.** I use two levels of presentation, the fast track and detours. The *fast track* is a *fil rouge* through the book and gives a bird's-eye view of *Computability Theory*. It can be read independently of *detours*. These contain detailed proofs, more demanding themes, additional historical facts, and further details, all of which can safely be skipped while reading on the fast track. The two levels differ visually: detours are written in small font and are put into *Boxes* (between gray bars, with broken lower bar), so they can easily be skipped or skimmed on first reading. Proofs are given on both levels whenever they are difficult or long.

- **Clarity.** Whenever possible I give the motivation and an explanation of the circumstances that led to new goals, concepts, methods, or theorems. For example, I explicitly point out with **NB** (nota bene) marks those situations and achievements that had important impact on further development in the field. Sometimes **NB** marks introduce conventions that are used in the rest of the book. New notions are introduced when they are naturally needed. Although I rigorously deduce theorems, I try to make proofs as intelligible as possible; this I do by commenting on tricky inferences and avoiding excessive formalism. I give intuitive, informal explanations of the concepts, methods, and theorems. Figures are given whenever this can add to the clarity of the text.

- **Contemporary terminology.** I use the recently suggested terminology and describe the reasons for it in the Bibliographic Notes; thus, I use partial computable (p.c.) functions (instead of partial recursive (p.r.) functions); computable functions (instead of recursive functions); computably enumerable (c.e.) sets (instead of recursively enumerable (r.e.) sets); and computable sets (instead of recursive sets).

- **Historical account.** I give an extended historical account of the mathematical and logical roots of *Computability Theory*.

- **Turing machine.** After describing different competing models of computation, I adopt the Turing machine as *the* model of computation and build on it. I neither formally prove the equivalence of these models, nor do I teach how to program Turing machines; I believe that all of this would take excessive space and add little to the understanding of *Computability Theory*. I do, however, rigorously prove the equivalence of the Turing machine and the RAM model, as the latter so closely abstracts real-life, general-purpose computers.

- **Unrestricted computing resources.** I decouple *Automata Theory* and *Formal Language Theory* from *Computability Theory*. This enables me to consider general models of computation (i.e., models with unlimited resources) and hence focus freely on the question "What can be computed?" In this way, I believe, *Computability Theory* can be seen more clearly and it can serve as a natural basis for the development of *Computational Complexity Theory* in its study of "What can be computed efficiently?" Although I don't delve into *Computational Complexity Theory*, I do indicate the points where *Computational Complexity Theory* would take over.

- **Short-cuts to relative computability.** I introduce oracles in the usual way, after explaining classical computability. Readers eager to enter relative computability might want to start with Part II and continue on the fast track.

- **Practical consequences and applications.** I describe the applications of concepts and theorems, whenever I am aware of them.

Finally, in describing *Computability Theory* I do not try to be comprehensive. Rather, I view the book as a first step towards more advanced texts on *Computability Theory*, or as an introductory text to *Computational Complexity Theory*.

Audience

This book is written at a level appropriate for undergraduate or beginning graduate students in computer science or mathematics. It can also be used by anyone pursuing research at the intersection of theoretical computer science on the one hand and physics, biology, linguistics, or analytic philosophy on the other.

The only necessary prerequisite is some exposure to elementary logic. However, it would be helpful if the reader has had undergraduate-level courses in set theory and introductory modern algebra. All that is needed for the book is presented in App. A, which the reader can use to fill in the gaps in his or her knowledge.

Teaching

There are several courses one can teach from this book. A course offering the *minimum* of *Computability Theory* might cover (omitting boxes) Chaps. 5, 6, 7; Sects. 8.1, 8.2, 8.4; and Chap. 9. Such a course might be continued with a course on *Complexity Theory*. An *introductory* course on *Computability Theory* might cover Parts I and II (omitting most boxes of Part I). A beginning *graduate* level course on *Computability Theory* might cover all three parts (with all the details in boxes). A course offering a *shortcut* (some 60 pages) to *Relative Computability* (Chaps. 10 to 15) might cover Sect. 5.3; Sects. 6.1.1, 6.2.1, 6.2.2; Sects. 6.3, 7.1, 7.2, 7.3; Sects. 7.4.1, 7.4.2, 7.4.3; Sects. 8.1, 8.2, 9.1, 9.2; and then Chaps. 10 through 15.

PowerPoint slides covering all three parts of the text are maintained and available at:

$$\text{http://lalg.fri.uni-lj.si/fct}$$

Origin

This book grew out of two activities: (1) the courses in *Computability and Compu-tational Complexity Theory* that I have been teaching at the University of Ljubljana, and (2) my intention to write a textbook for a course on algorithms that I also teach.

When I started working on (2) I wanted to explain the $\mathcal{O}$-notation in a satisfactory way, so I planned an introductory chapter that would cover the basics of *Computational Complexity Theory*. But to explain the latter in a satisfactory way, the basics of *Computability Theory* had to be given first. So, I started writing on computability. But the story repeated once again and I found myself describing the *Mathematical Logic* of the twentieth century. This regression was due to (i) my awareness that, in the development of mathematical sciences, there was always some *reason* for in-troducing a new notion, concept, method, or goal, and (ii) my belief that the text should describe such reasons in order to present the subject as clearly as possible. Of course, many historical events and logical facts were important in this respect, so the chapter on *Computability Theory* continued to grow.

At the same time, I was aware that students of *Computability and Computational Complexity Theory* often have difficulty in grasping the meaning and importance of certain themes, as well as in linking up the concepts and theorems to a whole. It was obvious that before introducing a new concept, method, or goal, a student should be given a historical or purely logical *motivation* for such a step. In addition, giving a *bird's eye view* of the theory developed up to the last milestone also proved to be extremely advantageous.

These observations coincided with my wishes about the chapter on *Computabil-ity Theory*. So the project continued in this direction until the "chapter" grew into a text on *The Foundations of Computability Theory*, which is in front of you.

Acknowledgements

I would like to express my sincere thanks to all the people who read all or parts of the manuscript and suggested improvements, or helped me in any other way. I benefited from the comments of my colleagues *Uroš Čibej* and *Jurij Mihelič*. In particular, *Marko Petkovšek* (University of Ljubljana, Faculty of Mathematics and Physics, Department of Mathematics), and *Danilo Šuster* (University of Maribor, Faculty of Arts, Department of Philosophy) meticulously read the manuscript and suggested many improvements. The text has benefited enormously from their assis-tance. Although errors may remain, these are entirely my responsibility.

Many thanks go to my colleague *Boštjan Slivnik* who skilfully helped me on several occasions in getting over TeX and its fonts. I have used drafts of this text in courses on *Computability and Computational Complexity Theory* that are given to students of computer science by our faculty, and to students of computer science and mathematics in courses organized in collaboration with the Faculty of Mathematics and Physics, University of Ljubljana. For their comments I particularly thank the students *Žiga Emeršič, Urša Krevs, Danijel Mišanović, Rok Resnik, Blaž Sovdat, Tadej Vodopivec,* and *Marko Živec.* For helpful linguistic suggestions, discussions on the pitfalls of English, and careful proofreading I thank *Paul McGuiness.*

I have made every reasonable effort to get permissions for including photos of the scientists whose contributions to the development of *Computability Theory* are described in the book. It turned out that most of the photos are already in the public domain. Here, *Wikimedia* makes praiseworthy efforts in collecting them; so does the online *MacTutor History of Mathematics Archive* at the University of St Andrews, Scotland. They were both very helpful and I am thankful to them. For the other photos I owe substantial thanks to the *Archives of the Mathematisches Forschungsinstitut Oberwolfach*, Germany, *King's College Library*, Cambridge, UK, and the *Los Alamos National Laboratory Archives*, USA. I have no doubt that photos make this serious text more pleasant. The following figures are courtesy of Wikimedia: Figs. 1.3, 1.4, 1.5, 1.7, 2.5, 2.6, 2.7, 2.9, 3.5, 3.8, 5.4, and 5.10. The following figures are courtesy of the MacTutor History of Mathematics archive: Figs. 1.6, 2.4, 2.8, 4.8, 5.5, and 5.8. The following figures are courtesy of the King's College Library, Cambridge: Figs. 5.6 (AMT/K/7/9), 6.1 (AMT/K/7/14). The following figures are courtesy of the Archives of the Mathematisches Forschungsinstitut Oberwolfach: Figs. 2.2, 2.10, 3.4, 3.7, 4.5, 5.2, and 5.3. Figure 3.6 is courtesy of *Los Alamos National Laboratory Archives,* US. (Unless otherwise indicated, this information has been authored by an employee or employees of the Los Alamos National Security, LLC (LANS), operator of the Los Alamos National Laboratory under Contract No. DE-AC52-06NA25396 with the U.S. Department of Energy. The U.S. Government has rights to use, reproduce, and distribute this information. The public may copy and use this information without charge, provided that this Notice and any statement of authorship are reproduced on all copies. Neither the Government nor LANS makes any warranty, express or implied, or assumes any liability or responsibility for the use of this information.)

I also thank the staff at Springer for all their help with the preparation of this book. In particular, I thank *Ronan Nugent*, my editor at Springer in Heidelberg, for his advice and kind support over the past few years. Finally, I thank the anonymous reviewers for their many valuable suggestions.

Ljubljana, January 2015   *Borut Robič*

Contents

In this part we will describe the events taking place in mathematics at the beginning of the twentieth century. Paradoxes, discovered in Cantor's set theory around 1900, started a crisis in the foundations of mathematics. To reconstruct the mathematics freed from all paradoxes, Hilbert introduced a promising program with formal systems as the central idea. Though the program was unexpectedly put to an end in 1931 by Gödel's famous theorems, it bequeathed burning questions: What is computing? What is computable? What is an algorithm? Can every problem be algorithmically solved? These were the questions that led to the birth of *Computability Theory*.

Chapter 1
Introduction

A recipe is a set of instructions describing how to prepare something. A culinary recipe for a dish consists of the required ingredients and their quantities, equipment and environment needed to prepare the dish, an ordered list of preparation steps, and the texture and flavor of the dish.

Abstract The central notions in this book are those of the algorithm and computation, not a particular algorithm for a particular problem or a particular computation, but the algorithm and computation in general. The first algorithms were discovered by the ancient Greeks. Faced with a specific problem, they asked for a set of instructions, whose execution in the prescribed order would eventually provide the solution to the problem. This view of the algorithm sufficed since the fourth century B.C.E., which meant there was no need to ask questions about algorithms and computation in general.

1.1 Algorithms and Computation

In this section we will describe how the concept of the algorithm was traditionally intuitively understood. We will briefly review the historical landmarks connected with the concept of the algorithm up to the beginning of the twentieth century.

1.1.1 The Intuitive Concept of the Algorithm and Computation

Every computational problem is associated with two sets: a set $\mathcal{A}$, which consists of all the possible input data to the problem, and a set $\mathcal{B}$, which consists of all the possible solutions. For example, consider the problem

"Find the greatest common divisor of two positive natural numbers."

After we have chosen the input data (e.g., 420 and 252), the solution to the problem is defined (84). Thus, we can think of the problem as a function $f : \mathcal{A} \to \mathcal{B}$, which maps the input data to the corresponding solution to the problem (see Fig. 1.1).

© Springer-Verlag Berlin Heidelberg 2015
B. Robič, *The Foundations of Computability Theory*,
DOI 10.1007/978-3-662-44808-3_1

Fig. 1.1 A problem is viewed
as a function mapping the input data to the corresponding
solution

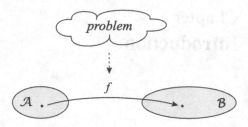

What does f look like? How do we compute its value $f(x)$ for a given x? The
way we do this is described and governed by the associated *algorithm* (see Fig. 1.2).

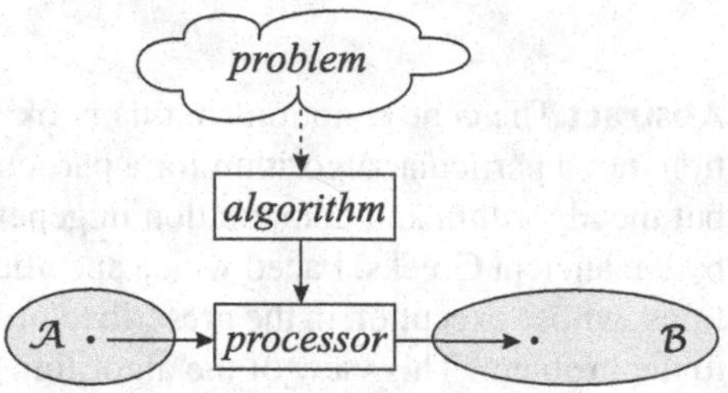

Fig. 1.2 The algorithm directs the processor in order to
compute the solution to the
problem

Definition 1.1. (Algorithm Intuitively) An **algorithm** for solving a problem is
a finite set of instructions that lead the processor, in a finite number of steps,
from the input data of the problem to the corresponding solution.

This was an informal definition of the algorithm. As such, it may raise questions,
so we give some additional explanations. An algorithm is a *recipe*, a finite list of
instructions that tell how to solve a problem. The *processor* of an algorithm may be
a human, or a mechanical, electronic, or any other device, capable of mechanically
following, interpreting, and executing instructions, while using no self-reflection or
external help. The *instructions* must be simple enough, so that the processor can
execute them, and they have to be unambiguous, so that every next step of the execution is precisely defined. A *computation* is a sequence of such steps. The *input
data* must be reasonable, in the sense that they are associated with solutions, so that
the algorithm can bring the processor to a solution.

In general, there are many algorithms for solving a computational problem. However, they differ because they are based on different ideas or different design methods.

Example 1.1. (Euclid's Algorithm) An algorithm for finding the greatest common divisor of two numbers was described around 300 B.C.E. by Euclid[1] in his work *Elements*. The algorithm is:

> Divide the larger number by the other.
> Then keep dividing the last divisor by the last remainder—unless the last remainder is 0.
> In that case the solution is the last nonzero remainder.[2]

For the input data 420 and 252, the computation is:

$$
\begin{aligned}
420 : 252 \quad &= 1 + \text{remainder } 168; \\
252 : 168 \quad &= 1 + \text{remainder } 84; \\
168 : 84 \quad &= 2 + \text{remainder } 0; \quad \text{and the solution is 84.}
\end{aligned}
$$

Fig. 1.3 Euclid
(Courtesy: See Preface)

Observe that there exists another algorithm for this problem that is based on a different idea:

> Factorize both numbers
> and multiply the common factors.
> The solution is this product.

For the input data 420 and 252, the computation is:

$$
\begin{aligned}
420 &= 2^2 \times 3 \times 5 \times 7 \\
252 &= 2^2 \times 3^2 \times 7 \\
& 2^2 \times 3 \times 7 = 84; \quad \text{and the solution is 84.}
\end{aligned}
$$

□

1.1.2 Algorithms and Computations Before the Twentieth Century

Euclid's algorithm is one of the first known nontrivial algorithms designed by a human. Unfortunately, the clumsy Roman number system, which was used in ancient

---

[1] Euclid, 325–265 B.C.E., Greek mathematician, lived in Alexandria, now in Egypt.

[2] This algorithm was probably not discovered by Euclid. The algorithm was probably known by Eudoxus of Cnidus (408–355 B.C.E.), but it may even pre-date Eudoxus.

Europe, hindered computation with large numbers and, consequently, the application of such algorithms. But, after the positional decimal number system was discovered between the first and fourth centuries in India, large numbers could be written succinctly. This enabled the Persian mathematician al-Khwarizmi[3] to describe in the year 825 algorithms for computing with such numbers, and in 830 algorithms for solving linear and quadratic equations. His name is the origin of the word *algorithm*.

Fig. 1.4 al-Khwarizmi
(Courtesy: See Preface)

In the seventeenth century, the first attempts were made to *mechanize* the algorithmic solving of *particular* computational problems of interest. In 1623, Schickard[4] tried to construct a machine capable of executing the operations $+$ and $-$ on natural numbers. Ten years later, a similar machine was successfully constructed by Pascal.[5] But Leibniz[6] saw further into the future. From 1666 he was considering a universal language (Latin *lingua characteristica universalis*) that would be capable of describing any notion from mathematics or the other sciences. His intention was to associate basic notions with natural numbers in such a way that the application of arithmetic operations on these numbers would return more complex numbers that would represent new, more complex notions. Leibniz also considered a universal computing machine (Latin *calculus ratiocinator*) capable of computing with such numbers. In 1671 he even constructed a machine that was able to carry out the operations $+, -, \times, \div$. In short, Leibniz's intention was to replace certain forms of human reflection (such as thinking, inferring, proving) with mechanical and mechanized arithmetic.

In 1834, a century and a half later, new and brilliant ideas led Babbage[7] to his design of a new, conceptually different computing machine. This machine, called the *analytical engine*, would be capable of executing *arbitrary programs* (i.e., algorithms written in the appropriate way) and hence of solving *arbitrary computational problems*. This led Babbage to believe that every computational problem is mechanically, and hence algorithmically, solvable.

---

[3] Muhammad ibn Musa al-Khwarizmi, 780–850, Persian mathematician, astronomer, geographer; lived in Baghdad.

[4] Wilhelm Schickard, 1592–1635, German astronomer and mathematician.

[5] Blaise Pascal, 1623–1662, French mathematician, physicist and philosopher.

[6] Gottfried Wilhelm Leibniz, 1646–1716, German philosopher and mathematician.

[7] Charles Babbage, 1792–1871, British mathematician.

Fig. 1.5 Blaise Pascal
(Courtesy: See Preface)

Fig. 1.6 Gottfried Leibniz
(Courtesy: See Preface)

Fig. 1.7 Charles Babbage
(Courtesy: See Preface)

Nowadays, Leibniz's and Babbage's ideas remind us of several concepts of contemporary *Computability Theory* that we will look at in the following chapters. Such concepts are Gödel's arithmetization of formal axiomatic systems, the universal computing machine, computability, and the *Computability Thesis*. So Leibniz's and Babbage's ideas were much before their time. But they were too early to have any practical impact on mankind's comprehension of computation.

As a result, the concept of the algorithm remained firmly at the intuitive level, defined only by common sense as in the Definition 1.1. It took many events for the concept of the algorithm to be rigorously defined and, as a result, for *Computability Theory* to be born. These events are described in the next chapter.

1.2 Chapter Summary

The algorithm was traditionally intuitively understood as a recipe, i.e., a finite list of directives written in some language that tells us how to solve a problem mechanically. In other words, the algorithm is a precisely described routine procedure that can be applied and systematically followed through to a solution of a problem. Because there was no need to define formally the concept of the algorithm, it remained firmly at the intuitive, informal level.

Chapter 2
The Foundational Crisis of Mathematics

A paradox is a situation that involves two or more facts or qualities which contradict each other.

Abstract The need for a formal definition of the concept of algorithm was made clear during the first few decades of the twentieth century as a result of events taking place in mathematics. At the beginning of the century, Cantor's naive set theory was born. This theory was very promising because it offered a common foundation to all the fields of mathematics. However, it treated infinity incautiously and boldly. This called for a response, which soon came in the form of logical paradoxes. Because Cantor's set theory was unable to eliminate them—or at least bring them under control—formal logic was engaged. As a result, three schools of mathematical thought—intuitionism, logicism, and formalism—contributed important ideas and tools that enabled an exact and concise mathematical expression and brought rigor to mathematical research.

2.1 Crisis in Set Theory

In this section we will describe the axiomatic method that was used to develop mathematics since its beginnings. We will also describe how Cantor applied the axiomatic method to develop his set theory. Finally, we will explain how the paradoxes revealed themselves in this theory.

2.1.1 Axiomatic Systems

The basic method used to acquire new knowledge in mathematics and similar disciplines is the *axiomatic method*. Euclid was probably the first to use it when he was developing his geometry.

© Springer-Verlag Berlin Heidelberg 2015
B. Robič, *The Foundations of Computability Theory*,
DOI 10.1007/978-3-662-44808-3_2

9

Axiomatic Method

When using the axiomatic method, we start our treatment of the field of interest by carefully selecting a few *basic notions* and making a few basic statements,[1] called *axioms* (see Fig. 2.1). An axiom is a statement that asserts either that a basic notion has a certain property, or that a certain relation holds between certain basic notions. We do not try to define the basic notions, nor do we try to prove the axioms. The basic notions and axioms form our *initial theory* of the field.

We then start *developing* the theory, i.e., extending the initial knowledge of the field. We do this systematically. This means that we must *define* every new notion in a clear and precise way, using only basic or previously defined notions. It also means that we must try to *prove* every new proposition,[2] i.e., *deduce*[3] it only from axioms or previously proven statements. Here, a *proof* is a finite sequence of mental steps, i.e., inferences,[4] that end in the realization that the proposition is a logical consequence of the axioms and previously proven statements. A provable proposition is called a *theorem* of the theory. The process of proving is informal, *content-dependent* in the sense that each conclusion must undoubtedly follow from the *meaning* of its premises.

Informally, the development of a theory is a process of discovering (i.e., deducing) new theorems—in Nagel and Newman's words, as Columbus discovered America—and defining new notions in order to facilitate this process. (This is the *Platonic* view of mathematics; see Box 2.1.) We say that axioms and basic notions constitute an *axiomatic system*.

Fig. 2.1 A theory has axioms, theorems, and unprovable statements

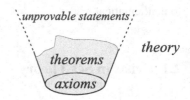

Box 2.1 (Platonism).

According to the *Platonic view* mathematics *does not create* new objects but, instead, *discovers already existing* objects. These exist in the non-material world of abstract *Ideas*, which is accessible only to our intellect. For example, the idea of the number 2 exists per se, capturing the state of "twoness," i.e., the state of any gathering of anything and something else—and nothing else. In the

---

[1] A *statement* is something that we say or write that makes sense and is either true or false.

[2] A *proposition* is a statement for which a proof is either required or provided.

[3] A *deduction* is the process of reaching a conclusion about something because of other things (called premises) that we know to be true.

[4] An *inference* is a conclusion that we draw about something by using information that we already have about it. It is also the process of coming to a conclusion.

material world, *Ideas* present themselves in terms of imperfect copies. For example, the *Idea* of a triangle is presented by various copies, such as the figures △, ▽, ◁, ▷ (and love triangles too). It can take considerable instinct to discover an *Idea*, the comprehension of which is consistent with the sensation of its copies in the material world. The agreement of these two is the criterion for deciding as to whether the *Idea* really exists.

Evident Axiomatic Systems

From the time of Euclid to the mid-nineteenth century it was required that axioms be statements that are in perfect agreement with human experience in the particular field of interest. The validity of such axioms was beyond any doubt, because they were clearly confirmed by the reality. Thus, no proofs of axioms were required. Such axioms are *evident*. Euclidean elementary geometry is an example of an evident axiomatic system, because it talks of points, lines and planes, which are evident idealizations of the corresponding real-world objects.

However, in the nineteenth century serious doubts arose as to whether evident axiomatic systems are always appropriate. This is because it was found that experience and intuition may be misleading. (For an example of such a situation see Box 2.2.) This led to the concept of the *hypothetical* axiomatic system.

Box 2.2 (Euclid's Fifth Axiom).

In two-dimensional geometry a line parallel to a given line L is a line that does not intersect with L. Euclid's fifth axiom, also called the *Parallel Postulate*, states that at most one parallel can be drawn through any point not on L. (In fact, Euclid postulated this axiom in a different but equivalent form.)

To Euclid and other ancients the fifth axiom seemed less obvious than the other four of Euclid's axioms. This is because a parallel line can be viewed as a line segment that never intersects with L, even if it is extended indefinitely. The fifth axiom thus implicitly speaks about a certain occurrence in arbitrarily removed regions of the plane, that is, that the segment and L will never meet. However, since Aristotle the ancients were well aware that one has to be careful when dealing with infinity. For example, they were already familiar with the notion of asymptote, a line that "approaches a given curve but it meets the curve only in the infinity."

To avoid the vagueness and controversy of Euclid's fifth axiom, they undertook to deduce it from Euclid's other four axioms; these caused no disputes. However, all attempts were unsuccessful until 1868, when Beltrami proved that Euclid's fifth axiom *cannot be deduced* from the other four axioms. In other words, Euclid's fifth axiom is *independent* of Euclid's other four axioms.

NB *The importance of Beltrami's discovery is that it does not belong to geometry, but rather to the science about geometry, and, more generally, to metamathematics, the science about mathematics. About fifty years later, metamathematics would come to the fore more explicitly and play a key role in the events that led to a rigorous definition of the notion of the algorithm.*

Since the eleventh century, Persian and Italian mathematicians had tried to prove Euclid's fifth axiom indirectly. They tried to refute all of its alternatives. These stated either that there are no parallels, or that there are several different parallels through a given point. When they considered these alternatives they unknowingly discovered *non-Euclidean geometries*, such as elliptic and hyperbolic geometry. But they cast them away as having no evidence in reality. According to the usual experience they viewed reality as a space where only Euclidean geometry can rule.

In the nineteenth century, Lobachevsky, Bolyai, and Riemann thought of these geometries as true alternatives. They showed that if Euclid's fifth axiom is replaced by a different axiom, then a different non-Euclidean geometry is obtained. In addition, there exist in reality examples, also called *models*, of such non-Euclidean geometries. For instance, Riemann replaced Euclid's fifth axiom with the axiom that states that there is no parallel to a given line L through a given point. The resulting geometry is called *elliptic* and is modelled by a sphere. In contrast to this, Bolyai and Lobachevsky selected the axiom that allows several parallels to L to pass through a given point. The resulting *hyperbolic* geometry holds, for example, on the surface of a saddle.

NB *These discoveries shook the traditional standpoint that axioms should be obvious and clearly agree with reality. It became clear that intuition and experience may be misleading.*

Hypothetical Axiomatic Systems

After the realization that instinct and experience can be delusive, mathematics gradually took a more abstract view of its research subjects. No longer was it interested in the (potentially slippery) *nature* of the basic notions used in an axiomatic system. For example, arithmetic was no longer concerned with the question of *what* a natural number really is, and geometry was no longer interested in *what* a point, a line, and a plane really are. Instead, mathematics focused on the *properties* of and *relations* between the basic notions, which could be defined without specifying what the basic notions are in reality. A basic notion can be *any* object if it fulfills all the conditions given by the axioms.

Thus the role of the axioms has changed; now an axiom is only a *hypothesis*, a speculative statement about the basic notions taken to hold, although nothing is said about the true nature and existence of such basic notions. Such an axiomatic system is called *hypothetical*. For example, by using nine axioms Peano in 1889 described properties and relations typical of natural numbers without explicitly defining a natural number. Similarly, in 1899 Hilbert developed elementary geometry, where no explicit definition of a point, line and plane is given; instead, these are defined implicitly, only as possible objects that satisfy the postulated axioms.

Because the nature of basic notions lost its importance, also the requirement for the evidence of axioms as well as their verifiability in reality was abandoned. The obvious link between the subject of mathematical treatment and reality vanished. Instead of axiomatic evidence the fertility of axioms came to the fore, i.e., the number of theorems deduced, their expressiveness, and their influence. The reasonableness and applicability of the theory developed was evaluated by the importance of successful *interpretations*, i.e., applications of the theory on various domains of reality. Depending on this, the theory was either accepted, corrected, or cast off.

NB *This freedom, which arose from the hypothetical axiomatic system, enabled scientists to make attempts that eventually bred important new areas of mathematics. Set theory is such an example.*[5]

2.1.2 Cantor's Naive Set Theory

A theory with a hypothetical axiomatic system, which will play a special role in what follows, was the *naive set theory* founded by Cantor.[6] Let us take a quick look at this theory.

Fig. 2.2 Georg Cantor
(Courtesy: See Preface)

Basic Notions, Concepts and Axioms

In 1895 Cantor defined the concept of a set as follows:

Definition 2.1. (Cantor's Set) A **set** is any collection of definite, distinguishable objects of our intuition or of our intellect to be conceived as a whole (i.e., regarded as a single unity).

Thus, an object can be any thing or any notion, such as a number, a pizza, or even another set. If an object x is in a set S, we say that x is a *member* of S and write $x \in S$. When x is not in S, it is not a member of S, so $x \notin S$. Given an object x and a set S, either $x \in S$ or $x \notin S$ — there is no third choice. This is known as the *Law of Excluded Middle*.[7]

---

[5] Another example of a theory with a hypothetical axiomatic system is group theory.

[6] Georg Cantor, 1845–1918, German mathematician.

[7] The law states that for any logical statement, either that statement is true, or its negation is — there is no third possibility (Latin *tertium non datur*).

Cantor did not develop his theory from explicitly written axioms. However, later analyses of his work revealed that he used three principles in the same fashion as axioms. For this reason we call these principles **Axioms of Extensionality**, **Abstraction**, and **Choice**. Let us describe them.

Axiom 2.1 (Extensionality). *A set is completely determined by its members.*

Thus a set is completely described if we list all of its members (by convention between braces "{" and "}"). For instance, $\{\diamond, \triangleleft, \circ\}$ is a set whose members are $\diamond, \triangleleft, \circ$, while one of the three members of the set $\{\diamond, \{\triangleleft, \triangleright\}, \circ\}$ is itself a set. When a set has many members, say a thousand, it may be impractical to list all of them; instead, we may describe the set perfectly by stating the *characteristic* property of its members. Thus a set of objects with the property P is written as $\{x \mid x$ has the property $P\}$ or as $\{x \mid P(x)\}$. For instance, $\{x \mid x$ is a natural number $\wedge\ 1 \leqslant x \leqslant 1000\}$.

What property can P be? Cantor's liberal-minded standpoint in this matter is summed up in the second axiom:

Axiom 2.2 (Abstraction). *Every property determines a set.*

If there is no object with a given property, the set is *empty*, that is, $\{\}$. Due to the *Axiom of Extensionality* there is only one empty set; we denote it by $\emptyset$.

Cantor's third principle is summed up in the third axiom:

Axiom 2.3 (Choice). *Given any set $\mathcal{F}$ of nonempty pairwise disjoint sets, there is a set that contains exactly one member of each set in $\mathcal{F}$.*

We see that the set and the membership relation $\in$ are such basic notions that Cantor defined them informally, in a descriptive way. Having done this he used them to define rigorously other notions in a true axiomatic manner. For example, he defined the relations $=$ and $\subseteq$ on sets. Specifically, two sets $\mathcal{A}$ and $\mathcal{B}$ are *equal* (i.e., $\mathcal{A} = \mathcal{B}$) if they have the same objects as members. A set $\mathcal{A}$ is a *subset* of a set $\mathcal{B}$ (i.e., $\mathcal{A} \subseteq \mathcal{B}$) if every member of $\mathcal{A}$ is also a member of $\mathcal{B}$. Cantor also defined the operations $\bar{\ }, \cup, \cap, -, 2^{\cdot}$ that construct new sets from existing ones. For example, if $\mathcal{A}$ and $\mathcal{B}$ are two sets, then also the *complement* $\overline{\mathcal{A}}$, the *union* $\mathcal{A} \cup \mathcal{B}$, the *intersection* $\mathcal{A} \cap \mathcal{B}$, the *difference* $\mathcal{A} - \mathcal{B}$ and the *power set* $2^{\mathcal{A}}$ are sets.

Application

Cantor's set theory very quickly found applications in different fields of mathematics. For example, Kuratowski used sets to define the *ordered pair* (x, y), i.e., a set of two elements with one being the first and the other the second in some order. The definition is $(x, y) \overset{\text{def}}{=} \{\{x\}, \{x, y\}\}$. (The ordering of $\{x\}$ and $\{x, y\}$ is implicitly imposed by the relation $\subseteq$, since $\{x\} \subseteq \{x, y\}$, but not vice versa.) Two ordered pairs are equal if they have equal first elements and equal second elements. Now the *Cartesian product* $\mathcal{A} \times \mathcal{B}$ could be defined as the set of all ordered pairs (a, b), where $a \in \mathcal{A}$ and $b \in \mathcal{B}$. The sets $\mathcal{A}$ and $\mathcal{B}$ need not be distinct. In this case, $\mathcal{A}^2$ was used to denote $\mathcal{A} \times \mathcal{A}$ and, in general, $\mathcal{A}^n \overset{\text{def}}{=} \mathcal{A}^{n-1} \times \mathcal{A}$, where $\mathcal{A}^1 = \mathcal{A}$.

Many other important notions and concepts, which were in common use although informally defined, were at last rigorously defined in terms of set theory, e.g., the concepts of function and natural number. For example, a *function* $f : \mathcal{A} \to \mathcal{B}$ is a set of ordered pairs (a,b), where $a \in \mathcal{A}$ and $b = f(a) \in \mathcal{B}$, and there are no two ordered pairs with equal first components and different second components. Based on this, set-theoretic definitions of injective, surjective, and bijective functions were easily made. For example, a bijective function is a function $f : \mathcal{A} \to \mathcal{B}$ whose set of ordered pairs contains, for each $b \in \mathcal{B}$, at least one ordered pair with the second component b (surjectivity), and there are no two ordered pairs having different first components and equal second components (injectivity).

Von Neumann used sets to construct *natural numbers* as follows. Consider number 2. We may imagine that it represents the state of "twoness," i.e., the gathering of one element and one more different element—and nothing else. Since the set $\{0,1\}$ is an example of such a gathering, we may define $2 \stackrel{\text{def}}{=} \{0,1\}$. Similarly, if we imagine 3 to represent "threeness," we may define $3 \stackrel{\text{def}}{=} \{0,1,2\}$. Continuing in this way, we arrive at the general definition $n \stackrel{\text{def}}{=} \{0,1,2,\ldots,n-1\}$. So a natural number can be defined as a *set* of all of its predecessors. What about the number 0? Since 0 has no natural predecessors, the corresponding set is empty. Hence the definition $0 \stackrel{\text{def}}{=} \emptyset$. We can now see that natural numbers can be constructed from $\emptyset$ as follows: $0 \stackrel{\text{def}}{=} \emptyset$; $1 \stackrel{\text{def}}{=} \{\emptyset\}$; $2 \stackrel{\text{def}}{=} \{\emptyset, \{\emptyset\}\}$; $3 \stackrel{\text{def}}{=} \{\emptyset, \{\emptyset\}, \{\emptyset, \{\emptyset\}\}\}$; $\ldots$; $n+1 \stackrel{\text{def}}{=} n \cup \{n\}$; $\ldots$ Based on this, other definitions and constructions followed (e.g., of rational and real numbers).

NB *Cantor's set theory offered a simple and unified approach to all fields of mathematics. As such it promised to become the foundation of all mathematics.*

But Cantor's set theory also brought new, quite surprising discoveries about the so-called cardinal and ordinal numbers. As we will see, these discoveries resulted from Cantor's *Axiom of Abstraction* and his view of infinity. Let us go into details.

Cardinal Numbers

Intuitively, two sets have the same "size" if they contain the same number of elements. Without any counting of their members we can assert that two sets are *equinumerous*, i.e., of the same "size," *if* there is a bijective function mapping one set onto the other. This function pairs every member of one set with exactly one member of the other set, and vice versa. Such sets are said to have the same *cardinality*. For example, the sets $\{\diamond, \triangleleft, \circ\}$ and $\{a,b,c\}$ have the same cardinality because $\{(\diamond,a), (\triangleleft,b), (\circ,c)\}$ is a bijective function. In this example, each of the sets has cardinality ("size") 3, a *natural* number. We denote the cardinality of a set $\mathcal{S}$ by $|\mathcal{S}|$.

Is cardinality always a natural number? Cantor's *Axiom of Abstraction* guarantees that the set $\mathcal{S}_P = \{x \mid P(x)\}$ exists for *any* given property P. Hence, it exists also for a P which is shared by infinitely many objects. For example, if we put

$P \equiv$ "is natural number," we get a set of *all* natural numbers. This set is not only an interesting and useful mathematical object, but (according to Cantor) it also *exists* as a perfectly defined and accomplished unity. Usually, it is denoted by $\mathbb{N}$. It is obvious that the cardinality of $\mathbb{N}$ cannot be a natural number because any such number would be too small. Thus Cantor was forced to introduce a new kind of number and designate it with some new symbol not used for natural numbers. He denoted this number by $\aleph_0$ (read *aleph zero* [8]). Cardinality of sets can thus be described by the numbers that Cantor called *cardinal numbers*. A *cardinal number* (or *cardinal* for short) can either be *finite* (in that case it is natural) or *transfinite*, depending on whether it measures the size of a finite or infinite set. For example, $\aleph_0$ is a transfinite cardinal which describes the size of the set $\mathbb{N}$ as well as the size of any other infinite set whose members can all be listed in a sequence.

Does every infinite set have the cardinality $\aleph_0$? Cantor discovered that this is not so. He proved (see Box 2.3) that the cardinality of a set S is strictly less than the cardinality of its power set 2^S — *even when S is infinite!* Consequently, there are larger and larger infinite sets whose cardinalities are larger and larger transfinite cardinals—and this never ends. He denoted these transfinite cardinals by $\aleph_1, \aleph_2, \ldots$. Thus, there is no largest cardinal.

Cantor also discovered (using diagonalization, a method he invented; see Sect. 9.1) that there are more real numbers than natural ones, i.e., $\aleph_0 < c$, where c denotes the cardinality of $\mathbb{R}$, the set of real numbers. (For the proof see Example 9.1 on p. 193.) He also proved that $c = 2^{\aleph_0}$. But where is c relative to $\aleph_0, \aleph_1, \aleph_2, \ldots$? Cantor conjectured that $c = \aleph_1$, that is, $2^{\aleph_0} = \aleph_1$. This would mean that there is no other transfinite cardinal between $\aleph_0$ and c and consequently there is no infinite set larger than $\mathbb{N}$ and smaller than $\mathbb{R}$. Yet, no one succeeded in proving or disproving this conjecture, until Gödel and Cohen finally proved that *neither* can be done (see Box 4.3 on p. 59). Cantor's conjecture is now known as the *Continuum Hypothesis*.

Box 2.3 (Proof of Cantor's Theorem).

Cantor's Theorem states: $|S| < |2^S|$, *for every set S.*

Proof. (a) First, we prove that $|S| \leq |2^S|$. To do this, we show that S is equinumerous to a subset of 2^S. Consider the function $f : S \to 2^S$ defined by $f : x \mapsto \{x\}$. This is a bijection from S onto $\{\{x\} | x \in S\}$, which is a subset of 2^S. (b) Second, we prove that $|S| \neq |2^S|$. To do this, we show that there is no bijection from S onto 2^S. So let $g : S \to 2^S$ be an *arbitrary* function. Then g cannot be surjective (and hence, neither is it bijective). To see this, let $\mathcal{N}$ be a subset of S defined by $\mathcal{N} = \{x \in S | x \notin g(x)\}$. Of course, $\mathcal{N} \in 2^S$. But $\mathcal{N}$ is not a g-image of any member of S. *Suppose* it were. Then there would be an $m \in S$, such that $g(m) = \mathcal{N}$. Where would be m relative to $\mathcal{N}$? If $m \in \mathcal{N}$, then $m \notin g(m)$ (by definition of $\mathcal{N}$), and hence $m \notin \mathcal{N}$ (as $g(m) = \mathcal{N}$)! Conversely, if $m \notin \mathcal{N}$, then $m \notin g(m)$ (as $g(m) = \mathcal{N}$), and hence $m \in \mathcal{N}$ (by definition of $\mathcal{N}$)! This is a contradiction. We conclude that g is not a surjection, and therefore neither is it a bijection. Since g was an arbitrary function, we conclude that there is no bijection from S onto 2^S. □

---

[8] $\aleph$ is the first symbol of the Hebrew alphabet.

Ordinal Numbers

We have seen that one can introduce order into a set of *two* elements. This can easily be done with other finite and infinite sets, and it can be done in many different ways. Of special importance to Cantor was the so-called *well-ordering*, because this is the way natural numbers are ordered with the usual relation $\leqslant$. For example, each of the sets $\{0,1,2\}$ and $\mathbb{N}$ is well-ordered with $\leqslant$, that is, $0 < 1 < 2$ and $0 < 1 < 2 < 3 < \cdots$. (Here $<$ is the strict order corresponding to $\leqslant$.) But well-ordering can also be found in other sets and for relations other than the usual $\leqslant$. When two well-ordered sets differ only in the naming of their elements or relations, we say that they are *similar*.

Cantor's aim was to *classify* all the well-ordered sets according to their similarity. In doing so he first noticed that the usual well-ordering of the set $\{0,1,2,\ldots,n\}$, $n \in \mathbb{N}$, can be represented by a single natural number $n + 1$. (We can see this if we construct $n+1$ from $\emptyset$, as von Neumann did.) For example, the number 3 represents the ordering $0 < 1 < 2$ of the set $\{0,1,2\}$. But the usual well-ordering of the set $\mathbb{N}$ cannot be described by a natural number, as any such number is too small. Once again a new kind of a "number" was required and a new symbol for it was needed. Cantor denoted this number by ω and called it the *ordinal number*.

Well-ordering of a set can thus be described by the *ordinal number*, or *ordinal* for short. An ordinal is either *finite* (in which case it is natural) or *transfinite*, depending on whether it represents the well-ordering of a finite or infinite set. For example, ω is the transfinite ordinal that describes the usual well-ordering in $\mathbb{N}$. Of course, in order to use ordinals in classifying well-ordered sets, Cantor required that two well-ordered sets have the same ordinal *iff* they are similar. (See details in Box 2.4.) Then, once again, he proved that there are larger and larger transfinite ordinals describing larger and larger well-ordered infinite sets—and this never ends. There is no largest ordinal.

NB *With his set theory, Cantor boldly entered a curious and wild world of infinities.*

2.1.3 Logical Paradoxes

Unfortunately, the great leaps forward made by Cantor's set theory called for a response. This came around 1900 when logical paradoxes were suddenly discovered in this theory. A *paradox* (or *contradiction*) is an unacceptable conclusion derived by apparently acceptable reasoning from apparently acceptable premises (see Fig. 2.3).

Burali-Forti's Paradox. The first logical paradox was discovered in 1897 by Burali-Forti.[9] He showed that in Cantor's set theory there exists a well-ordered set Ω whose ordinal number is *larger than itself*. But this is a contradiction. (See details in Box 2.4.)

---

[9] Cesare Burali-Forti, 1861–1931, Italian mathematician.

Cantor's Paradox. A similar paradox was discovered by Cantor himself in 1899. Although he proved that, for any set $\mathcal{S}$, the cardinality of the power set $2^{\mathcal{S}}$ is strictly larger than the cardinality of $\mathcal{S}$, he was forced to admit that this cannot be true of *the set $\mathcal{U}$ of all sets*. Namely, the existence of $\mathcal{U}$ was guaranteed by the *Axiom of Abstraction*, just by defining $\mathcal{U} = \{x \mid x = x\}$. But if the cardinality of $\mathcal{U}$ is less than the cardinality of $2^{\mathcal{U}}$, which also exists, then $\mathcal{U}$ is not the largest set (which $\mathcal{U}$ is supposed to be since it is the set of *all* sets). This is a contradiction.

Russell's Paradox. The third paradox was found in 1901 by Russell.[10] He found that in Cantor's set theory there exists a set $\mathcal{R}$ that *both is and is not* a member of itself. How? Firstly, the set $\mathcal{R}$ defined by

$$\mathcal{R} = \{\mathcal{S} \mid \mathcal{S} \text{ is a set} \wedge \mathcal{S} \text{ does not contain itself as a member}\}$$

must exist because of the *Axiom of Abstraction*. Secondly, the *Law of Excluded Middle* guarantees that $\mathcal{R}$ either contains itself as a member (i.e., $\mathcal{R} \in \mathcal{R}$), or does not contain itself as a member (i.e., $\mathcal{R} \notin \mathcal{R}$). But then, using the definition of $\mathcal{R}$, each of the two alternatives implies the other, that is, $\mathcal{R} \in \mathcal{R} \Longleftrightarrow \mathcal{R} \notin \mathcal{R}$. Hence, each of the two is both a true and a false statement in Cantor's set theory.

Fig. 2.3 A paradox is an unacceptable statement or situation because it defies reason; for example, because it is (or at least seems to be) both true and false

Why Do We Fear Paradoxes?

Suppose that a theory contains a logical statement such that both the statement and its negation can be deduced. Then it can be shown (see Sect. 4.1.1) that *any* other statement of the theory can be deduced as well. So in this theory everything is deducible! This, however, is not as good as it may seem at first glance. Since deduction is a means of discovering truth (i.e., what is deduced is accepted as true) we see that in such a theory every statement is true. But a theory in which everything is true has no cognitive value and is of no use. Such a theory must be cast off.

---

[10] Bertrand Russell, 1872–1970, British mathematician, logician, and philosopher.

Box 2.4 (Burali-Forti's Paradox).

A set S is *well-ordered* by a relation $\prec$ if the following hold: 1) $a \not\prec a$; 2) $a \neq b \Rightarrow a \prec b \vee b \prec a$; and 3) every nonempty $\mathcal{X} \subseteq S$ has $m \in \mathcal{X}$, such that $m \prec x$ for every other $x \in \mathcal{X}$. For example, $\mathbb{N}$ is well-ordered with the usual relation $<$ on natural numbers. Well-ordering is a special case of the so-called *linear ordering*, i.e., a well-ordered set is also linearly ordered. For example, $\mathbb{Z}$, the set of integers, is linearly ordered by the usual relation $<$.

Suppose we do not want to distinguish between two linearly ordered sets that differ only in the naming of their elements or relations. We want to consider such sets as being similar, because they obviously share the same "type of order."

Let us define the notion "type of order" precisely. Let two sets $\mathcal{A}$ and $\mathcal{B}$ be linearly ordered with relations $\prec_A$ and $\prec_B$, respectively. We say that $\mathcal{A}$ and $\mathcal{B}$ are *similar* if there is a bijection $f : \mathcal{A} \to \mathcal{B}$ such that $a \prec_A b \Longleftrightarrow f(a) \prec_B f(b)$. The function f renames the elements of $\mathcal{A}$ to the elements of $\mathcal{B}$ while respecting both relations. We can easily prove that similarity is an equivalence relation between linearly ordered sets. So we can define the *order type* to be an equivalence class of similar, linearly ordered sets. Informally, an order type is the feature shared by all linearly ordered sets that differ only in the naming of their elements and relations.

Having defined the order types we might want to compare them. Unfortunately, they may not be comparable. It can be shown, however, that *order types of well-ordered sets* are themselves *linearly* ordered by some relation $\prec_o$. (Actually, $\prec_o$ is the usual set-membership relation $\in$.) Because such order types are ordered in a similar way to integers, we call them *ordinal numbers* (or *ordinals* for short). Hence, the definition: An *ordinal* is an equivalence class of similar well-ordered sets. For example, sets similar to $\{0, 1, \ldots, n\}$ have the same ordinal; we denote it by the natural number $n + 1$. This cannot be done with sets similar to $\mathbb{N}$, so we use ω to denote their ordinal.

For each ordinal α there is exactly one ordinal $\alpha' \stackrel{\text{def}}{=} \alpha \cup \{\alpha\}$ that is the $\prec_o$-successor of α. (We also denote α' by $\alpha + 1$.) It follows that there is no $\prec_o$-largest ordinal.

This is where Burali-Forti entered. He proved that Cantor's set theory allows the construction of a set Ω of *all* the ordinals. He also showed that such an Ω leads to a paradox. Namely, Ω would not only be linearly ordered by $\prec_o$, but also well-ordered by $\prec_o$. As such, Ω would be associated with the corresponding ordinal, say α_Ω. But where would α_Ω be relative to Ω? Since Ω is the set of *all* the ordinals, it must be that $\alpha_\Omega \in \Omega$. On the other hand, α_Ω must be $\prec_o$-larger than any member of Ω, and therefore larger than itself.

2.2 Schools of Recovery

In this section we will describe the three main schools of mathematical thought that significantly contributed to the struggle against paradoxes in mathematics. These are *intuitionism*, *logicism*, and *formalism*. We will show how their discoveries synthesized in the concept of a formal axiomatic system and then in a clear awareness that a higher, metamathematical language is needed to investigate such systems.

2.2.1 Slowdown and Revision

The discovery of the paradoxical sets Ω, $\mathcal{U}$, and $\mathcal{R}$ was shocking, because Cantor's set theory was supposed to become a firm foundation for all other fields of mathematics and should, therefore, have been free of paradoxes. But the simplicity of Russell's Paradox, which used only two basic notions of set and membership relation, revealed that paradoxes originated deep in Cantor's theory, in the very definition of the concept of a set. It was this definition of a set and the unrestricted use of the *Axiom of Abstraction* that allowed the existence of the sets Ω, $\mathcal{U}$, and $\mathcal{R}$ that, in the end, caused and revealed paradoxical situations. So it was clear that objects like Ω, $\mathcal{U}$, and $\mathcal{R}$ should *not* be recognized as existing sets.

Therefore, Cantor's *naive* definition of the concept of a set (Sect. 2.1.2) should be restricted somehow. But this was easier said than done. Namely, Cantor's definition of a set was so natural and of such common sense that it was far from clear how to restrict it and, at the same time, retain all the sound parts of the theory. If a set is not what Cantor thought about, then what was it? And what was it not?

This once again triggered a critical reflection about the basic concepts, notions, principles, methods, and tools of set theory and logic, which might be sources of paradoxes. The aim was to make the necessary corrections to them, so that they, as a whole, would again act as a foundation for the development of mathematics and other axiomatic areas of science, but this time *safe from all paradoxes*. It turned out that no universally accepted resolutions could be expected. The critiques and proposals went in several directions, of which the three mainstream directions were called *intuitionism*, *logicism*, and *formalism*. Because they all contributed to future events, we briefly review them.

2.2.2 Intuitionism

Intuitionism argued for greater mathematical rigor in the process of proving and it advocated a non-Platonic view that the existence of a mathematical object is closely connected to the existence of its mental construction.

Fig. 2.4 Jan Brouwer
(Courtesy: See Preface)

The school was initiated by Brouwer[11] and then further developed by his student Heyting.[12] Brouwer was critical of the way in which Cantor's mathematics viewed the existence of *infinite* sets, and of the way in which mathematics was using the *Law of Excluded Middle*. He proposed a thorough change of this view as well as severe restrictions on the use of the law. Specifically, unlike Cantor, who considered infinite sets as *actualities*, i.e., accomplished objects, intuitionism advocated the classical point of view that infinite sets are no more than *potentialities*, i.e., objects that are *always* under construction, making it possible to construct as many members as needed, but *never all*. This view called for a change in the way that the *existence* of objects in infinite sets should be proven: an object is recognized as a member of an infinite set *if and only if* the object has been *constructed* or the existence of such a construction is beyond doubt. We give the details in Box 2.5.

Using these principles, intuitionism reconstructed several parts of classical mathematics and showed that such intuitionistic mathematics is free of all *known* paradoxes. Unfortunately, the price for this was rather high: large parts of mathematics had to be cast off, because it seemed impossible to reconstruct them according to intuitionistic principles. In addition, in the reconstructed mathematics, surprising changes occurred; for example, every (constructed) function is continuous.

It turned out that only a few researchers were willing to make such radical sacrifices.

NB *Nevertheless, the intuitionistic demand for mathematical rigor survived and was partially taken into account in the events to follow.*

Box 2.5 (Intuitionism).

This school argued for greater mathematical rigor in several ways.

View of Infinity. Since Aristotle, mathematics understood infinity only as the potentiality (i.e., possibility), never as the actuality (i.e., accomplishment). For instance, it is true that natural numbers $0, 1, 2, \ldots$ continue endlessly, yet up to any natural number there are only finitely many of them, and when we say that what remains is infinite we only mean that the rest, although growing ever larger, remains never accomplished. So, in the classical view infinity is by nature never accomplished, never actual. In contrast, Cantor's view of infinity was different, indeed radically Platonic: *"Any set, regardless of its size, is as much real as its members are real,"* he boldly advocated. To Cantor the set $\{0, 1, 2, \ldots\}$ was an actual, accomplished mathematical object.

Intuitionism returned to the classical view of infinity as potentiality. According to this view, using an appropriate procedure, we can find in an infinite set as many members as we wish, but never all of them. To treat infinite sets as actual, accomplished unities, is wrong, said intuitionists, and may lead to paradoxes (as shown by Russell and others).

But there are also differences between classical mathematics and intuitionism.

---

[11] Luitzen Egbertus Jan Brouwer, 1881–1966, Dutch mathematician and philosopher.

[12] Arend Heyting, 1898–1980, Dutch mathematician and logician.

Existence of Objects. Intuitionism treats the existence of mathematical objects differently from classical mathematics. In classical mathematics, mathematical objects exist *per se*, as Platonic ideas (see p. 10). Consequently, statements about mathematical objects are either true or false. Intuitionism does not accept this view. Instead, it advocates that the only things that exist *per se* are mental, mathematical constructions, while the existence of an object that has *not* been constructed remains *dubious*. For intuitionism, *to exist is the same as to be constructed.*

For instance, in classical mathematics, given a set S and a property P sensible for the members of S, we are always allowed to *indirectly* prove that there exists a member of S with the property P, i.e., that the statement $\exists x \in S : P(x)$ is true. To do this, we first make the hypothesis $H \equiv \neg \exists x \in S : P(x)$, stating that such a member *does not* exist. Then we try to deduce from H a contradiction. If we succeed in this, we conclude that H is false. Now comes the critical step: since classical mathematics fully accepts the *Law of Excluded Middle*, there can be no other alternative but to conclude that $\neg H \equiv \exists x \in S : P(x)$ is true, i.e., that such a member of S *exists*. But note that, generally, we have no idea about this member, or how to find it.

Intuitionism does not accept such an indirect proof of existence when the set S is *infinite*. Indeed, it rejects any proof of existence that neither constructs the alleged object, nor describes how to construct it at least in principle.

Use of Logic. The intuitionistic point of view was also reflected in the use of logic. For example, because of the *Law of Excluded Middle*, the classical mathematics takes for granted that, for *any* statement F, either F or $\neg F$ is true. Hence, the statement $F \vee \neg F$ is *a priori* true, even though we may never determine the truth-values of F and $\neg F$. Intuitionism, in contrast, treats the truth-values of F and $\neg F$ as *dubious*, until they are actually determined in some indisputable way.

To explain the reasons for such caution, let S be a set, P a property sensible for the members of S, and F the statement $\forall x \in S : P(x)$. So F conjectures that every member of S has the property P. How can we indisputably determine whether or not F is true? Can we always do this?

First, we can try to prove in one sweep that *all* the members of S have the property P. (We can use various techniques, such as mathematical induction.) If it turns out that we are unable to construct a proof that works for every member of S, it might be that F is *false*. However, it might also be that F is *true*, where $P(x)$ holds for every $x \in S$, but for a *different* reason in each case. That is, our inability to construct a one-sweep proof might be due to the lack of a recognizable pattern, i.e., a common reason for which different members of S share the property P. In this case, neither can we prove F (because the "proof" would be infinitely long) nor refute it (because F is true).

We must therefore resort to some other method to settle the conjecture F. If S is *finite*, we can in principle check, for each $x \in S$ individually, whether or not $P(x)$ holds. When the checking is finished, we know either that F is true, or that it is false. But what if S is *infinite*? We can still do the checking, but we must be aware of the following. We may check as many members of S as we like, say 10^{18}, and find that each of them has the property P — but, generally, there is *no* way of knowing whether, for a member yet to be checked, P holds or not. So we keep checking in the hope that such a member will be reached soon. But, if in truth F is true, the checking will continue indefinitely, and we will never find out whether F is true or false. (By the way, this is the present situation with *Goldbach's Conjecture*; see Box 5.4 on p. 91.)

Finally, we may try to prove $F \equiv \forall x \in S : P(x)$ by contradiction. As usually, we assume the converse, i.e., that $\neg \forall x \in S : P(x)$ is true. In classical mathematics, where the equivalence $\neg \forall x \in S : P(x) \Longleftrightarrow \exists x \in S : \neg P(x)$ holds for arbitrary S, we would try to deduce a contradiction from the more promising right-hand side of the equivalence. In intuitionism, however, the equivalence does not *a priori* hold; namely, if S is infinite, the statement $\exists x \in S : \neg P(x)$ is *dubious* until we have *constructed* an $x \in S$ for which $\neg P(x)$ holds. (As we have seen above, this may not be easy.) As long as $\exists x \in S : \neg P(x)$ is dubious, it cannot be used to deduce a contradiction, and our proving by contradiction is stalled.

So in some situations the truth-value of a statement F cannot be indisputably determined.

2.2.3 Logicism

Logicism aimed to found mathematics on pure logic. As a side-effect it developed the notation by which mathematics was at last given concise and precise expression. The main contributions to this school were made by Boole, Frege, Peano, Russell, and Whitehead.

Fig. 2.5 George Boole (Courtesy: See Preface)

Fig. 2.6 Gottlob Frege (Courtesy: See Preface)

Fig. 2.7 Giuseppe Peano (Courtesy: See Preface)

Boole

In the middle of the nineteenth century scientists noticed that, from Aristotle onward, logical deduction had been using various *self-evident* rules of inference that, surprisingly, had never been rigorously analyzed and written down.

Boole[13] was among the first to become aware of the pitfalls of this. He embarked on the question of how to express logical statements by means of algebraic expressions (containing the operations "and," "or" and "not"), and then algebraically manipulate these expressions to pursue logical deduction. He described his discoveries in the book *The Laws of Thought* (1854) and thus founded *algebraic logic*. His logic was further developed by Peirce and others in the early twentieth century to become what is now known as *Propositional Calculus* **P** (see Appendix A, p. 297). Since then a more precise and clear expression of logical statements has been possible.

Frege and Peano

Frege[14] was aiming even higher. His goal was to show that arithmetic can be deduced from pure logic. In particular, he planned to define number-theoretic notions (i.e., numbers, relations, and operations on numbers) by pure logical notions, and to deduce arithmetical axioms from logical axioms.

---

[13] George Boole, 1815–1864, English mathematician and philosopher.

[14] Friedrich Ludwig Gottlob Frege, 1848–1925, German mathematician, logician, and philosopher.

Like Boole, Frege was well aware that a natural language, such as German, has structural, rhetorical, psychological, and other characteristics that often blur the meaning of its own statements and, consequently, the argumentation of the deductions. This required the introduction of a new, formal notation by which mathematics and logic could be given concise and precise expression. In particular, such a notation should be able to isolate all the important logical principles of inference while throwing off the lumber of natural language. In other words, the notation should be able to support purely logical deduction. So, in 1879, Frege proposed his *Begriffsschrift*, a "conceptual notation," capable of giving mathematics and logic better expression. Begriffsschrift was based on an alphabet of *symbols*, from which mathematical and logical expressions were constructed using *rules of construction*. An important innovation of Frege was that these rules directed the purely *mechanical manipulation* of symbols, without appealing to intuition or to the (possible) meaning of symbols. In addition, Frege introduced *quantified variables* and thus laid the foundations of the *First-Order Logic* (which we will describe later). The inferences were described diagrammatically, so they were in this respect somewhat unusual. Nevertheless, Begriffsschrift was capable of precisely and concisely representing the inferences that involved arbitrary mathematical statements.

At the same time, Peano[15] developed another symbolic language for expressing mathematical statements. He used innovative logical symbols (e.g., $\in, \Rightarrow$) in order to distinguish between logical and other operations. In 1895, he published a book *Fomulario Mathematico* where he expressed fundamental theorems of mathematics in his symbolic language. Peano's notation proved to be more practical than Frege's notation and is after having gone through further development in common use today.

In short, among Frege's and Peano's contributions to logic were the analysis of logical concepts, the foundation of the *First-Order Logic* **L** (see Appendix A, p. 298), and the introduction of a standard formal notation.

Russell and Whitehead

Russell's goal was even more ambitious than Frege's. He wanted to deduce *all* mathematics from logic. Namely, at the end of the nineteenth century it had already been shown that many concepts of algebra and analysis can be defined by means of number-theoretic notions, which, in turn, can be defined with purely logical notions.

To avoid his own paradox, Russell invented the *Theory of Types*. There are three requirements in this theory: 1) A *hierarchy of types* must be established. A *type* can be a member of any well-ordered set, e.g., a natural number. 2) Each mathematical object must be assigned to a type. 3) Each mathematical object must be constructed exclusively from objects of lower types in the hierarchy. As a result, the set $\mathcal{U}$ of all sets cannot exist in this theory (because $\mathcal{U} \in \mathcal{U}$), and neither does Russell's Paradox (for if $\mathcal{R}$ existed, we would have $\mathcal{R} \notin \mathcal{R}$ because of its type, and consequently $\mathcal{R} \in \mathcal{R}$ due to its definition: a contradiction). Similarly, Ω would not exist (as $\Omega \in \Omega$).

---

[15] Giuseppe Peano, 1858–1932, Italian mathematician.

These ideas were described in the 1910–13 book *Principia Mathematica* (*PM*) by Whitehead[16] and Russell. Using symbolic notation based on Peano's work, they developed from logical and three additional axioms the theory of sets and cardinal, ordinal, and real numbers, while avoiding all *known* paradoxes. The deductions were long, even cumbersome, yet many shared the opinion that the remaining fields of mathematics could also be deduced (at least in principle).

Fig. 2.8 Alfred Whitehead
(Courtesy: See Preface)

Fig. 2.9 Bertrand Russell
(Courtesy: See Preface)

Did *Principia Mathematica* put an end to the crisis in mathematics? Not really. There were imperfections in *PM*. First of all, there was a kind of aesthetic flaw in the set of *PM*'s axioms, because in addition to logical axioms there were three axioms not recognized as purely logical. One of these was Cantor's *Axiom of Choice*. More importantly, it remained unclear as to whether *PM* is *consistent*, i.e., it avoids, besides all *known* paradoxes, also *all the other* paradoxes that may still be hidden in various fields of mathematics, patiently awaiting their discovery. This question became known as the *Consistency Problem* of *PM*. In addition, it was not clear whether *PM* was *complete*, i.e., whether exactly true statements are provable within *PM*. This was the *Completeness Problem* of *PM*. Consequently, *PM* was not widely accepted.[17]

NB *Nevertheless, PM was all important for future events, because it finally developed 1) a symbolic language for the concise and precise expression of mathematical statements from an arbitrary field of mathematics; and 2) a concise formulation of all the rules of inference used in the deduction of mathematical theorems. In addition, PM led to a clear formulation of the problems of consistency and completeness of a particular axiom system, the PM.[18]*

---

[16] Alfred North Whitehead, 1861–1947, British mathematician and philosopher.

[17] In addition, it would soon turn out that Russell's Paradox, as well as other paradoxes stemming from Cantor's liberal *Axiom of Abstraction*, can be eliminated just by a *two-level hierarchy of sets and classes*, instead of the complicated infinite hierarchy of types. (See Box 3.5 on p. 45.)

[18] As we will see in Chap. 4, the two problems were later solved in general by Gödel.

The concepts and tools developed by intuitionism and logicism were used by *formalism*, the third of the schools that attempted to resolve the crisis in mathematics.

2.2.4 Formalism

Formalism could not accept the radical measures suggested by intuitionism. It wished to keep all classical mathematics. After all, classical mathematics had been proving its immense usefulness from the very beginning.

To achieve this, formalism focused on a radically different aspect of human mathematical activity. Instead of being the *meaning* (i.e., semantics, contents) of mathematical expressions and inferences, the subject of the formalist's research was their *structure* (i.e., syntax, form). Formalism focused on the formal-language formulation of human mathematical activities and their results, as well as on the relations between these formulations.

This school was initiated by Hilbert[19] and then developed in close collaboration with Ackermann,[20] Bernays,[21] and others.

Fig. 2.10 David Hilbert
(Courtesy: See Preface)

Syntax vs. Semantics

Hilbert became fully aware that it is sensible to draw a distinction between syntactic notions (i.e., notions referring to the structure of mathematical expressions) and semantic notions (i.e., notions referring to the meaning of mathematical expressions). For instance, the interpretation of a theory is a semantic notion. Recall that interpretation gives a meaning to a theory developed in a hypothetical axiomatic system (see p. 12). To describe the interpretation one needs to describe its domain, which is, mathematically, a set. But the concept of a set was not clear at that time. So Hilbert advocated a focus on syntactic notions, as the research of these seemed to require

---

[19] David Hilbert, 1862–1943, German mathematician.
[20] Wilhelm Friedrich Ackermann, 1896–1962, German mathematician.
[21] Paul Isaac Bernays, 1888–1977, Swiss mathematician.

only the non-problematic parts of mathematics, that is, basic logic and some basic arithmetic.

Let us describe these ideas in greater detail. Because it proved that mathematical concepts, such as that of the set, may be vague, also inference incorporating such concepts may be false, and may, eventually, lead to paradoxes. On the other hand, mathematical notions are always expressed in the *words* of some language, either natural, such as English, or symbolic, designed just for this purpose. A word is a finite sequence of *symbols* from some finite alphabet. Formalism noticed that every symbol is perfectly clear *per se*, that is, a symbol is comprehended as soon as it is recognized as a discrete part of the reality, without any further intuitive or logical analysis. This comprehension of symbols is independent of their intended meaning, which might previously be associated with them (such as the operation of addition with the symbol "+"). So why not comprehend words in that manner as well? One should only ignore the intended meaning of the word at hand and comprehend and treat it simply as a finite sequence of symbols. *Expressions*, i.e., sequences of words, could also be treated in the same fashion and, finally, the *sequences of expressions* too.

After the banishment of meaning from language constructs, one would be free to focus on their structure (syntax). But why do that? The reason is that one could found mathematical inference on a clear and precise structure (syntax) of language constructs, instead of on their (sometimes) unclear meaning (semantics). The syntax is always clear, provided it is rigorously and precisely defined (as was the case with logicism). As a result, a proof (deduction) would simply be a finite sequence of language constructs (expressions), built according to a finite number of rules. The gain would be improved control over the process of deduction and, finally, the elimination of paradoxes.

Formal Axiomatic Systems

In order to implement these ideas, formalism invented *formal axiomatic systems*. Each such system offers 1) a rigorously defined *symbolic language*; 2) a set of *rules of construction*, i.e., syntactic rules that are used to build well-formed expressions, called *formulas*, of the language; and 3) a set of *rules of inference* that are used to build well-formed sequences of formulas, called *derivations* or *formal proofs*. Each formula or derivation is viewed and treated exclusively as a finite sequence of symbols of the language. Hence, though each formula has a definite structure, no meaning is to be seen or searched for in it. Some of the formulas are proclaimed as *axioms*. Given a finite set of formulas, one may *infer* a new formula by applying a rule of inference. Formulas that can be derived by a finite sequence of inferences from axioms only are called *theorems*. Axioms, theorems, and other formulas make up the *theory* belonging to the formal axiomatic system at hand. A detailed discussion of formal axiomatic systems and their theories will appear in the next chapter.

Interpretation

Let us stress that formalists were aware of the fact that there was a limit to neglecting the semantics. After all, their ultimate goal was to establish conditions for the development of sound, safely *applicable* theories. They were aware that each theory, developed in a formal axiomatic system, should eventually be given some meaning; otherwise it would be of no use. In other words, the theory should be *interpreted*. Informally, an interpretation of a theory in a field of interest maps formulas of the theory into statements about (some) objects of the field. We will discuss interpretation again shortly.

NB *Formalism cast out the issues of meaning from the development of a theory, and shifted them to a later interpretation. What were the expected benefits of this? Such a theory could clearly show the syntactic properties of its expressions and expose various relations between these properties. Laid bare, the whole theory could be examined by* metamathematics *and subjected to its judgement.*

Metamathematics

When a theory is developed in a formal axiomatic system, the only things that can be examined within or about it are its *expressions*, the *syntactic properties* of expressions, and the *relations* between them. All these are unambiguously determined by the formal system (i.e., its language and rules of construction and inference). Thus, syntactic aspects of the theory can be systematically analyzed without the interference of semantic issues. Only now can one raise well-defined questions *about the theory* and propose answers to such questions.

Fig. 2.11 A statement about the theory belongs to its metatheory

But questions and statements about the theory are no longer part of the theory. Instead, they belong to the higher "theory about the theory," which is called *metatheory*, or, more generally, *metamathematics*.[22] Thus, the subject matter of a metatheory is some other theory.

---

[22] meta- (Greek $\mu\varepsilon\tau\acute{\alpha}$) = after, beyond, about

Metamathematical statements are formulated in a natural language that can be augmented with additional symbols. (We will explain two such symbols, i.e., ⊢ and ⊨, shortly.) The proving of metamathematical statements is still necessary, but it is not formal, in contrast to proving within the formal system. Instead, the usual (i.e., semantic, informal) proving is used, where each inference in a proof must be grounded in the *meaning* of its premises. Of course, premises are metamathematical statements so that they can refer only to syntactic aspects of the theory. In addition, to avoid any doubts that might arise because of the use of infinity, only *finite* objects and techniques are allowed in metamathematical proofs. Such a cautious and indisputable way of reasoning is called *finitism*.

Goals of Formalism

Formalism harbored hopes that the analysis of formal systems would provide answers to many important metamathematical questions about the theories of interest. Specifically, these were the two well-known questions concerning mathematics developed in *Principia Mathematica* (see p. 25):

- The *Consistency Problem* of *PM* ≡ "Is the math developed in *PM* consistent?"
- The *Completeness Problem* of *PM* ≡ "Is the math developed in *PM* complete?"

But the ultimate goals of formalism were even more ambitious. Specifically, formalists intended to:

1. develop *all* mathematics in *one* formal axiomatic system;
2. *prove* that such mathematics is free of *all* known and unknown paradoxes.

2.3 Chapter Summary

The axiomatic method was used to develop mathematics since its beginnings. The evident axiomatic system required that basic notions and axioms be clearly confirmed by the reality. Since it was found that human experience and intuition may be misleading, the hypothetical axiomatic system was introduced. Here, axioms are only hypotheses whose fertility is more important than their link to reality. Such axiomatic systems offered more freedom in the search for interesting and useful theories. This approach was taken by Cantor when he developed his *Set Theory*. Because Cantor naively treated the existence of infinite sets, this resulted in several paradoxes in his theory.

Intuitionism, logicism, and formalism were three schools that reflected critically on the mathematical and logical notions and concepts that might be the cause of paradoxes.

Intuitionism advocated for greater rigor in the process of proving and for the non-Platonic view that the existence of mathematical objects is closely connected to the existence of their mental constructions. Intuitionism reconstructed several parts of classical mathematics that were free of all known paradoxes. But, at the same time, large parts of mathematics had to be cast off, as it seemed impossible to reconstruct them in the intuitionistic manner. Few researchers were willing to make such a sacrifice.

Logicism, the second school, developed a formal notation by which mathematics was given concise and precise expression. It also bore *Principia Mathematica*, a book that finally developed a symbolic language of mathematics and concisely formulated its rules of inference. In addition, it brought an awareness of the importance of the problems of consistency and completeness of axiomatic theories.

The third school, formalism, built on the ideas and tools developed by intuitionism and logicism, and aspired to retain all mathematics. Formalism acknowledged that the syntax and semantics of mathematical expressions should be clearly separated and dealt with in succession. It introduced the concept of the formal axiomatic system, i.e., an environment for the mechanical, syntax-oriented development of a theory. In addition, it introduced a clear distinction between a theory and its metatheory.

Chapter 3
Formalism

The form of something is its shape and structure. Something that is done in a formal way has a very ordered, organized method and style. Formalism is a style in which great attention is paid to the form rather than to the contents of things.

Abstract The great ideas and tools that intuitionism and logicism discovered in solving the crisis in mathematics were gathered by formalism in the concept of the formal axiomatic system. Later, formal axiomatic systems led to seminal discoveries about axiomatic theories and mathematics in general. Particularly important to us is the fact that formal axiomatic systems also gave rise to the need for a deeper understanding of the concepts of algorithm and computation. To appreciate this need, we devote this chapter to the understanding of formal axiomatic systems in general and describe those particular formal axiomatic systems that played a crucial role in the events to follow.

3.1 Formal Axiomatic Systems and Theories

In this section we will describe what a formal axiomatic system is and how a theory is developed in such a system. We will then show how meaning, and consequently a possible application, is given to a formally developed theory. Finally, we will describe several formal axiomatic systems and their theories that played important roles in the development of the notions of algorithm and computation.

3.1.1 What Is a Formal Axiomatic System?

A formal axiomatic system (in short *f.a.s.*) **F** is determined by three entities: a symbolic language, a set of axioms, and a set of rules of inference.

© Springer-Verlag Berlin Heidelberg 2015
B. Robič, *The Foundations of Computability Theory*,
DOI 10.1007/978-3-662-44808-3_3

Symbolic Language

The basic building blocks of the symbolic language are *symbols*. There are a count-able (potentially infinite) number of them and they constitute a set called the *alpha-bet* of the language. In the alphabet there are *individual-constant symbols* (e.g., a, b, c), *individual-variable symbols* (e.g., x, y, z), *function symbols* (e.g., f, g, h), *predi-cate symbols* (e.g., P, Q, R), *logical connectives* (e.g., $\lor$, $\land$, $\Rightarrow$, $\Leftrightarrow$, $\neg$), *quantification symbols* (e.g., $\forall$, $\exists$), and *punctuation marks* (e.g., comma ","; colon ":"; parentheses "(", ")"). Usually, but not necessarily, there is the *equality symbol* (i.e., $=$). In some cases, certain function symbols will be designated as *function-variable symbols* and certain predicate symbols will be designated as *predicate-variable symbols*. (The reasons for this naming of the symbols will become clear when we discuss their interpretation.)

From symbols one constructs larger building blocks of the language, i.e., symbols are combined into arbitrary finite sequences called *words*. Some of these are called terms and are inductively defined by the following *rule of construction*: A *term* is either an individual-constant symbol or an individual-variable symbol, or it is a word $f(t_1, t_2, \ldots, t_k)$, where t_i are terms and f is a k-ary function symbol.

A formula is defined inductively by another syntactical *rule of construction*: A *formula* is either an expression $P(t_1, t_2, \ldots, t_k)$, where t_i are terms and P is a k-ary predicate symbol, or it is one of the expressions $F \lor G$, $F \land G$, $F \Rightarrow G$, $F \Leftrightarrow G$, $\neg F$, $\forall \tau F$, $\exists \tau F$, where F and G are formulas and τ is a variable symbol.

The symbols $\forall$ and $\exists$ are called the *universal* and *existential quantification sym-bol*, respectively, while $\forall \tau$ and $\exists \tau$, where τ is a variable symbol, are called the *universal* and *existential quantifier*, respectively. If τ immediately following $\forall$ or $\exists$ can only be an individual-variable symbol, then the symbolic language is said to be of the *first order*. If, however, τ can be a function-variable symbol or a predicate-variable symbol, then the language is of the *second order*.

If $\forall \tau F$ and $\exists \tau F$ are formulas, we say that τ is *bound* in F by $\forall$ and $\exists$, respectively. An individual-, function-, or predicate-variable symbol that is not bound in a given formula is said to be *free* in that formula. A formula with at least one free variable symbol is said to be *open*. A formula with no free variable symbols is said to be *closed*; a closed formula is also called a *sentence*.

Notice that the construction of the building blocks of the language is governed ex-clusively by rules of construction that are syntactic by nature. Consequently, neither intended nor possible meanings of the building blocks interfere in their construction.

Example 3.1. (Term, Formula, Sentence) The symbols a and x are terms. If f and g are function symbols, then $f(x)$ and $g(a, f(x))$ are both terms. If P and Q are relation symbols, then $P(a, x)$ and $Q(a, x, f(x))$ are formulas; so is $P(a, x) \lor Q(a, x, f(x))$. The formula $\forall x \exists y P(x, y)$ is a sentence because its individual-variable symbols x and y are bound by $\forall$ and $\exists$, respectively. The formula $\forall x \exists y R(x, y, z)$ is open because the individual-variable symbol z is free in R. If h is a function-variable symbol, then the formula $\forall h P(a, h(a))$ belongs to a second-order language.   □

Axioms

Axioms are selected formulas. There are *logical* and *proper* (i.e., non-logical) axioms. Logical axioms are present in every formal axiomatic system, while proper axioms vary from system to system. As we will see shortly, logical axioms are intended to epitomize the principles of pure logical reflection, while proper axioms condense other special basic notions and facts.

Rules of Inference

A *rule of inference*, say $\mathscr{R}$, specifies the conditions in which, given a set $\mathcal{P}$ of formulas, called the *premises* of $\mathscr{R}$, one is allowed to *derive* another formula, say F. The formula F is called the *conclusion* drawn from premises $\mathcal{P}$ by the rule of inference $\mathscr{R}$. We also say that F *directly follows from* premises $\mathcal{P}$ by the rule of inference $\mathscr{R}$ and write

$$\mathcal{P} \overset{\mathscr{R}}{\vdash} F. \qquad \text{(Rule of inf. } \mathscr{R})$$

Two usual rules of inference are Modus Ponens and Generalization. *Modus Ponens* (*MP*) says: "If G and $G \Rightarrow F$ are premises, then the conclusion F directly follows." That is: "If G is asserted to hold, and G implies F, then also F holds." In short:

$$G, G \Rightarrow F \overset{MP}{\vdash} F. \qquad \text{(Modus Ponens)}$$

Generalization (*Gen*) says: "If $F(x)$ is a premise, then the conclusion $\forall x F(x)$ directly follows." That is: "If F holds for an unspecified x, then F holds for every x." In short:

$$F(x) \overset{Gen}{\vdash} \forall x F(x). \qquad \text{(Generalization)}$$

Example 3.2. (Inference) Greg now plays guitar. If Greg plays guitar, Becky sings. So (by *MP*), she now sings. Becky likes icecream. So (by *Gen*), she likes vanilla, lemon, ... icecream. □

Development of the Theory

When the symbolic language, axioms, and rules of construction are fixed, the *development* of the *theory* belonging to the defined formal axiomatic system **F** can start.[1] During the development, the following strict rules must be obeyed. Firstly,

---

[1] In the *Platonic view*, as soon as **F** is defined, also the theory belonging to **F** is perfectly defined: it consists of *all* those propositions that are *provable* within **F**—regardless of whether or not they have actually been proved. In this sense, the theory is "static" and defined from its very birth. Adopting the Platonic view allows us to identify f.a.s. and its theory and denote both by **F**. The development of the theory **F** is by discovering new (existing) theorems, i.e., finding their proofs. In contrast, the *constructivist view* takes the theory to consist of *all* propositions that were *proved* within **F**. At its birth, the theory only contains **F**'s axioms, and then grows by absorbing new propositions as they are proved. At any stage of its development, the theory is denoted by th(**F**).

each new notion must be defined by the basic notions or previously defined notions. Secondly, each new proposition must be *derived* (i.e., *formally proved*) before it is named a *theorem*. Here, a *derivation* (i.e., *formal proof*) of a formula F in the theory **F** is a finite sequence of formulas such that 1) the last formula of the sequence is F, and 2) each formula of the sequence is either an axiom of **F**, or it directly follows from some of the preceding formulas of the sequence by one of the rules of inference of **F**. Such a formula F is called a *theorem* of the theory **F**. That F is a theorem of **F** we denote by

$$\underset{\mathbf{F}}{\vdash} \mathbf{F}.$$

Example 3.3. (Derivation) Let us derive a simple formula $t = t$ in the formal axiomatic system **A**, called the *Formal Arithmetic*. (We will describe **A** in detail on p. 41.) In the derivation below, the left column contains enumerated formulas that appear in the derivation, and the right column explains, for each formula, how the formula was inferred. At this point, the symbols $\oplus$ and 0 should not be given any meaning although they resemble the usual symbols for addition and the number zero.

1. $t \oplus 0 = t$ Axiom 5 of **A** with x substituted by t.
2. $t \oplus 0 = t \Rightarrow (t \oplus 0 = t \Rightarrow t = t)$ Axiom 1 of **A** with x, y, z subst. by $t \oplus 0, t, t$, resp.
3. $t \oplus 0 = t \Rightarrow t = t$ *Modus Ponens* of premises 1. and 2.
4. $t = t$ *Modus Ponens* of premises 1. and 3.

Derivations are finite sequences of formulas, which are in turn finite sequences of symbols. So, formally, a derivation is a finite sequence of symbols. For example, the above derivation is the sequence

$$1.t \oplus 0 = t, 2.t \oplus 0 = t \Rightarrow (t \oplus 0 = t \Rightarrow t = t) \overset{MP}{\underset{1,2}{\vdash}} 3.t \oplus 0 = t \Rightarrow t = t \overset{MP}{\underset{1,3}{\vdash}} 4.t = t$$

$\square$

It seems that the development of a theory **F** in a formal axiomatic system is nothing more than a meaningless manipulation of symbols that runs according to a given set of strict rules. It is a kind of a game of symbols regulated by certain rules. In other words, the development of the theory **F** is strictly formal, so **F** is a rather strange-looking theory. Yet the reasons for this are rather meaningful:

It is in principle much easier to maintain and check the validity of a proof that is being constructed in a formal axiomatic system.

This is because in a formal axiomatic system deduction is more of a mechanical process, so it is less prone to human errors. Such a deduction, now called a derivation, can be checked in a purely combinatorial way by checking whether formal rules of construction and inference have been obeyed. In contrast, the situation in a non-formal axiomatic system is quite different: There, a proof is a mental process that involves the meaning of the constituents of the proof and the relations between them. As such the proof is vulnerable and prone to errors because of man's subjective capability of realization as to whether an inference is valid.

3.1.2 Interpretations and Models

As we have seen, all the aspects of meaning have been expelled from the development of a theory and postponed to a later stage. We have now come to the point where the potential applications of the developed theory can be searched for. So, now the question is: "How does a formally developed theory get connected to actuality and, at last, gain meaning?"

On one hand, the first ideas of the intended meaning *may* get involved in the very first stage of establishing a formal axiomatic system, that is, in setting up its symbolic language and choosing its proper axioms. This usually happens when a formal system is established in order to use it in an investigation of a particular, concrete field of interest. Some of the concrete fields of interest that we will discus in the next sections are concerned with logical statements, natural numbers, and sets. A chosen field of interest is called the *interpretation domain* of the formal axiomatic system (and of its theory as well). It is, therefore, reasonable to choose an alphabet and rules of construction in such a way that the resulting language is capable of a precise and comfortable description of any situation in the domain. (For example, in the previous example we used symbols $\oplus$ and $=$ with the obvious intention that these symbols will be later interpreted as the addition and equality.) In addition, together with axioms and rules of inference, the language should facilitate the analysis of the situation. Of course, when the development of the theory starts, the role of the intended meaning diminishes and syntactic issues come to the fore.

On the other hand, one may just as well define a formal axiomatic system and develop its theory irrespective of any particular field of interest. One just mechanically develops the theory through the disciplined use of formal rules of construction and inference.

In any case, eventually some meaning must be assigned to the theory if the theory is to be applied somewhere. How is this done?

Interpretation of a Theory

To assign a particular meaning to a theory **F**, one has to *interpret* **F** in a particular *mathematical structure*[2] $\mathscr{S} = (\mathcal{D}om, \cdot)$. Informally, this means that one has to choose a particular class $\mathcal{D}om$ and particular functions and relations defined on $\mathcal{D}om$, and define, for every closed formula of **F**, how the formula is to be understood as a statement about the members, functions, and relations of $\mathcal{D}om$. More formally,

---

[2] Informally, a mathematical structure is a class $\mathcal{D}om$ endowed with additional mathematical objects, such as functions and relations which are defined on $\mathcal{D}om$, and certain designated elements. For example, groups, rings, vector spaces, and partially ordered sets are structures. Formally, a mathematical structure is an ordered set $\mathscr{S} = (\mathcal{D}om, R_0, \ldots, R_k, f_0, \ldots, f_m, e_0, \ldots, e_n)$, where $\mathcal{D}om$ is a set, $R_0, \ldots, R_k$ are relations on $\mathcal{D}om$, $f_0, \ldots, f_m$ are functions from Cartesian powers of $\mathcal{D}om$ into $\mathcal{D}om$, and $e_0, \ldots, e_n$ are designated elements of $\mathcal{D}om$. For example, $(\mathbb{N}, =, +, *, 0, 1)$, the ring of natural numbers, is a structure. For brevity we used the *dot* to stand for all the particular functions, relations, and designated elements to be considered on $\mathcal{D}om$.

the *interpretation* of a theory **F** is an ordered pair $(\iota, \mathscr{S})$, where ι is a mapping $\iota : \mathbf{F} \to \mathscr{S}$ that assigns to each symbol, term, and formula of **F** its "meaning" in $\mathcal{D}om$. (See Fig. 3.1.) The "meaning" can be an element of $\mathcal{D}om$, or a function or relation defined on $\mathcal{D}om$. The class $\mathcal{D}om$ is called the *domain of the interpretation*. Usually, the mapping ι is defined inductively in accordance with **F**'s rules of construction. Further details about the definition of the interpretation are in Box 3.1.

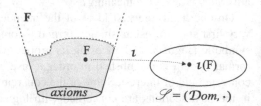

Fig. 3.1 When a theory **F** is interpreted in a structure $\mathscr{S} = (\mathcal{D}om, \cdot)$, the mapping ι assigns to each formula $F \in \mathbf{F}$ a formula $\iota(F) \in \mathscr{S}$

Box 3.1 (Interpretation).

How does the interpretation ι assign meaning to a theory **F** in a structure $\mathscr{S} = (\mathcal{D}om, \cdot)$? A symbol of **F** gets its meaning as follows:

- an individual-constant symbol c is mapped to an element of the domain: $\iota(\mathsf{c}) \in \mathcal{D}om$;
- a k-ary function symbol f is mapped to a k-ary function defined on $\mathcal{D}om$: $\iota(\mathsf{f}) : \mathcal{D}om^k \to \mathcal{D}om$;
- a k-ary predicate symbol P is mapped to a k-ary relation defined on $\mathcal{D}om$: $\iota(\mathsf{P}) \subseteq \mathcal{D}om^k$;
- logical connectives get their usual meanings ($\vee$ "or"; $\wedge$ "and"; $\Rightarrow$ "implies"; $\Leftrightarrow$ "iff"; $\neg$ "not");
- quantification symbols get their usual meanings ($\forall$ "for all"; $\exists$ "exists");
- punctuation marks get their usual meanings (comma, colon, parentheses);
- the equality symbol $=$ always gets its usual meaning (i.e., the equality relation).

The meaning of a term of **F** goes with its construction:

- a term that is an individual-constant symbol c gets the same meaning as the symbol: $\iota(\mathsf{c}) \in \mathcal{D}om$;
- a term $\mathsf{f}(\mathsf{t}_1, \mathsf{t}_2, \ldots, \mathsf{t}_k)$ is mapped to $\iota(\mathsf{f})(\iota(\mathsf{t}_1), \iota(\mathsf{t}_2), \ldots, \iota(\mathsf{t}_k))$; this is an element of $\mathcal{D}om$.

A formula of **F**, too, gets its meaning inductively:

- a formula $\mathsf{P}(\mathsf{t}_1, \mathsf{t}_2, \ldots, \mathsf{t}_k)$ is mapped to $\iota(\mathsf{P})(\iota(\mathsf{t}_1), \iota(\mathsf{t}_2), \ldots, \iota(\mathsf{t}_k))$; this is a statement that is true *iff* the elements $\iota(\mathsf{t}_i)$ of $\mathcal{D}om$ are related by the relation $\iota(\mathsf{P})$;
- a formula $F \vee G$ is mapped to the statement $\iota(F) \vee \iota(G)$. This statement is true *iff* at least one of the statements $\iota(F), \iota(G)$ is true;
- the formulas $F \wedge G$, $F \Rightarrow G$, $F \Leftrightarrow G$, and $\neg F$ are mapped to statements $\iota(F) \wedge \iota(G)$, $\iota(F) \Rightarrow \iota(G)$, $\iota(F) \Leftrightarrow \iota(G)$, and $\neg \iota(F)$, respectively. The statements are true according to the well-known rules of the *Propositional Calculus* **P** (see Appendix A, p. 297).

Finally, let $F(\mathsf{x})$ be a formula of **F** with a free individual-variable symbol x. Then:

- the formula $\forall \mathsf{x} F(\mathsf{x})$ is mapped to the statement $\iota(\forall \mathsf{x} F(\mathsf{x}))$, which is true *iff* the statement $\iota(F(\mathsf{x}))$ is true for every $\iota(\mathsf{x}) \in \mathcal{D}om$;
- the formula $\exists \mathsf{x} F(\mathsf{x})$ is mapped to the statement $\iota(\exists \mathsf{x} F(\mathsf{x}))$, which is true *iff* the statement $\iota(F(\mathsf{x}))$ is true for at least one $\iota(\mathsf{x}) \in \mathcal{D}om$.

Satisfiability and Validity

Note that *free* variable symbols have not been assigned exact meanings under the interpretation $(\iota, \mathscr{S})$. Instead, we only know that a free individual-variable symbol represents *any element* of the domain $\mathcal{D}om$. Similarly, a free function-variable symbol and a free predicate-variable symbol, when they exist, represent *any function* and *any relation* on the domain $\mathcal{D}om$, respectively.

Consequently, free variable symbols still await someone to fix their meanings. This can usually be done in many ways and, in general, fixing the meanings of free variable symbols affects the truth-value of the formula. Let us explain this in detail.

If a formula F is *closed*, i.e., it has no free variable symbols, then its interpretation $\iota(\mathrm{F})$ is a statement about the state of affairs in the domain $\mathcal{D}om$. The statement $\iota(\mathrm{F})$ is either *true* or *false*, depending on whether it is in agreement with the actual situation in $\mathcal{D}om$ or not. For example, $\mathrm{P}(\mathrm{a}, \mathrm{b})$ in Fig. 3.2 is such a formula.

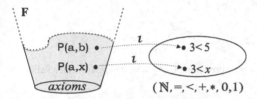

Fig. 3.2 A theory F is interpreted in the structure $(\mathbb{N}, =, <, +, *, 0, 1)$. The mapping ι assigns to the predicate symbol P the usual relation $<$ on $\mathbb{N}$, and to individual-constant symbols a and b natural numbers 3 and 5, respectively. Since the individual-variable symbol x is free in the formula $\mathrm{P}(\mathrm{a}, \mathrm{x})$, the mapping ι assigns *no* meaning (i.e., no number) to it

If, however, a formula F is *open*, then it contains free variable symbols. (For example, $\mathrm{P}(\mathrm{a}, \mathrm{x})$ in Fig. 3.2 is such a formula.) Since the interpretation $(\iota, \mathscr{S})$ did not assign meanings to these variable symbols, $\iota(\mathrm{F})$ says nothing definite about the situation in $\mathcal{D}om$. Thus $\iota(\mathrm{F})$ is neither true nor false at this point and, indeed, is not yet a statement about the state of affairs in $\mathcal{D}om$. However, as soon as all the free variable symbols *are* assigned meanings, $\iota(\mathrm{F})$ becomes either a true or a false statement about $\mathcal{D}om$. (Clearly, free individual-variable symbols are assigned particular elements of $\mathcal{D}om$, while free function-variable symbols and free predicate-variable symbols are assigned particular functions and relations on $\mathcal{D}om$, respectively.)

Later, we can reassign meanings to one or more free variable symbols of the formula. The change in the meanings of the free variable symbols generally affects the truth-value of the statement $\iota(\mathrm{F})$. Regarding this we emphasize two special cases:

- if $\iota(\mathrm{F})$ is true for *at least one* assignment of meanings to its free variable symbols, then we say that F is *satisfiable under the interpretation* $(\iota, \mathscr{S})$;
- if $\iota(\mathrm{F})$ is true for *every* assignment of meanings to its free variable symbols, then we say that F is *valid under the interpretation* $(\iota, \mathscr{S})$ and designate this by

$$\overset{(\iota, \mathscr{S})}{\models} \mathrm{F}.$$

Logical Validity

If a formal axiomatic system has been established to investigate a particular field of interest, then there is an obvious interpretation of its theory; it is called the *standard* interpretation. For example, the formal axiomatic system **A** (which will be described shortly) was defined to formalize arithmetic, so the standard interpretation of **A** uses the structure $(\mathbb{N}, =, +, *, 0, 1)$. However, given a formal axiomatic system **F**, there may be several different interpretations $(\iota, \mathscr{S})$ of its theory **F**, each of which differs in ι or $\mathscr{S}$. With regard to this, of particular importance are those formulas of **F** that are valid under *every* interpretation of **F**. Such formulas are said to be *logically valid*. (See Fig. 3.3.)

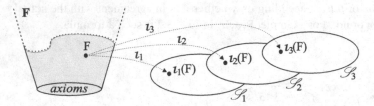

Fig. 3.3 A formula F that is valid under every interpretation $(\iota, \mathscr{S})$ is said to be logically valid

The logical validity of a formula depends only on its structure and the general properties of functions, relations, and quantifications; it is independent of any interpretation. We denote that F is a logically valid formula by

$$\models F.$$

Observe that the *logical axioms* of **F** must be logically valid. This should not be surprising because logical axioms are meant to epitomize the principles of pure logical thought, and such principles *should be* (and are) independent of the current field of man's interest, i.e., the domain of interpretation. Thus they must remain valid, irrespective of the interpretation.

Model of a Theory

What about the other kind of axiom: *proper axioms*? These are meant to abstract specific basic notions and facts typical of the current field of interest (as it is, for example, the standard interpretation of the theory). This leads us to the concept of a *model* of a theory. Given a theory **F**, it is natural to be interested only in interpretations $(\iota, \mathscr{S})$ under which *all the proper axioms* of **F** are valid. (Otherwise, **F** would be of no use under the interpretation.) Under such interpretations *all* the axioms of **F** are valid (since logical axioms are already logically valid). Each such interpretation $(\iota, \mathscr{S})$ is called the *model* of the theory **F**. Intuitively, a model of a theory is any field of interest that the theory sensibly formalizes.

Example 3.4. (Model of a Theory) The set of natural numbers with usual operations is a model of *Peano's Arithmetic* (pp. 12 and 41). A sphere is a model of elliptic geometry (Box 2.2, p. 11). So is a geoid, the shape of Earth. And our Universe is a model of Einstein's *General Relativity Theory*.   □

A theory may have several models. When a formula F of a theory **F** is valid in *each* model of **F**, the formula is said to be *valid in the theory* **F**. This is denoted by

$$\models_{\mathbf{F}} F.$$

A formula F valid in **F** represents a certain mathematical *Truth* expressible in the f.a.s. **F**. We will say that such an F represents a *Truth* in **F**. Each axiom of **F** represents a *Truth* in **F**. Clearly, we prefer theories with large powers of expression.

After we have designed a formal axiomatic system **F** and developed (some of) its theory **F**, we are interested in the existence and kinds of its models, i.e., fields that are sensibly formalized by **F**. We are confronted with questions such as "Does **F** have models? If so, how many are there? What are the differences between them? What is their applicability?" It is not very important that a model be a part of the *real, actual* world; instead, it suffices that the model behave as a *possible* part of the world. When does that happen? It turns out that the theory **F** has to be *consistent*, i.e., it must not allow a derivation of two contradictory theorems (and, hence, has no paradoxes). But the consistency of a theory is a syntactic notion, so it must be dealt with within the corresponding metatheory (metamathematics). We will return to the question of consistency soon.

3.2 Formalization of Logic, Arithmetic, and Set Theory

Some of the formally developed theories and their models that played an important role in solving the crisis in the foundations of mathematics were concerned with the following fields of interest: the structure and use of logical statements, the arithmetic of natural numbers, and the construction and use of sets. The corresponding first-order formal axiomatic systems and theories are called the *First-Order Logic* **L**, the *Formal Arithmetic* **A**, and the two *Axiomatic set theories* **NBG** and **ZFC**. In this section we will get acquainted with each of them.

Formalization of Logic

It was clear that in order to develop *any* theory in a logically unassailable way, the corresponding formal axiomatic system must offer all the necessary logical principles and tools (i.e., logical symbols, logical axioms, rules of inference). This called for serious reflection on all the principles of pure logical reasoning, which should result in a formal list of all of them. Fortunately, this was done by Boole, Frege,

Peano, Whitehead, Russell and other logicists (see Sect. 2.2.3). Formalism was able to gather all the undisputed logical principles in a formal axiomatic system called *First-Order Logic* (with equality) and denoted by **L**.

- **First-Order Logic**[3] **L.** The alphabet of the language of **L** contains a potentially infinite number of individual-variable symbols $x, y, \ldots$; the equality symbol $=$; logical connectives $\neg$, $\Rightarrow$, $\forall$; and the usual punctuation marks. Using the given logical connectives one may define additional logical connectives, such as $\wedge$, $\vee$, $\Leftrightarrow$, $\exists$, which are used as abbreviations. The symbolic language is of the first order. The terms and formulas are defined in the usual way. For instance, $\forall x \forall y (x = y \Rightarrow y = x)$ is a formula. Instead of explicit logical axioms, of which there are infinitely many, there are five *axiom schemas* that describe all of them (see Box 3.2). There are no proper axioms. Because of this, **L** is said to be a *pure logic theory*. The rules of inference are *Modus Ponens* and *Generalization*.

Box 3.2 (Logical Axioms of L).

First-Order Logic **L** has five *axiom schemas*. An axiom schema is a pattern used to build concrete logical axioms. If F, G, H are arbitrary formulas, the following formulas are logical axioms:

1) $F \Rightarrow (G \Rightarrow F)$ **2)** $(F \Rightarrow (G \Rightarrow H)) \Rightarrow ((F \Rightarrow G) \Rightarrow (F \Rightarrow H))$
3) $(\neg G \Rightarrow \neg F) \Rightarrow ((\neg G \Rightarrow F) \Rightarrow G)$ **4)** $\forall x F(x) \Rightarrow F(t)$ if t can be substituted for x in F
5) $\forall x (F \Rightarrow G) \Rightarrow (F \Rightarrow \forall x G)$ if x is not free in F

The schemas 1–3 are also axioms of the *Propositional Calculus* **P** (which is contained in **L**).

First-Order Formal Axiomatic Systems and Theories

Many other formal axiomatic systems are *extensions* of **L**. Each of them contains, besides everything that **L** has, additional *proper symbols* (i.e., constant symbols $a, b, c, \ldots$; function symbols $f, g, h, \ldots$; and predicate symbols $P, Q, R, \ldots$) and *proper axioms* (i.e., axioms that are inspired by the standard interpretation). The rules of inference are *Modus Ponens* and *Generalization*. Most often these systems use first-order language. In such cases we call them *first-order formal axiomatic systems*, and their theories we call *first-order theories*.

Especially important to us will be three first-order theories: *Formal Arithmetic* **A**, which formalizes the arithmetic of natural numbers, and the two *Axiomatic set theories* **ZFC** and **NBG**, which formalize set theory in two different ways.

Formalization of Arithmetic

It was clear that in order to develop formally any nontrivial mathematical theory, natural numbers had to be taken into account. This is because natural numbers play

---

[3] Also called *First-Order Predicate Calculus*.

a key role in the construction of other kinds of numbers (e.g., integer, rational, algebraic, real), and consequently in the development of any nontrivial mathematical theory (e.g., algebra, analysis). Fortunately, the grounding for this had already been laid; the properties of natural numbers had been described in 1889 by Peano's nine axioms (as mentioned on p. 12). The formal axiomatic system describing arithmetic is called the *Formal Arithmetic*[4] and is denoted by **A**.

- **Formal Arithmetic A.** The alphabet of the language of **A** contains all the symbols of **L** and, in addition, the following *proper symbols*: the individual-constant symbol 0, the unary function symbol $'$, and the binary function symbols $\oplus$ and $\odot$. Usually, but not necessarily, one can define symbols that are abbreviations for other symbols, e.g., $\ominus, \oslash, \ominus, \oslash$. The terms and formulas are constructed as usually. For instance, x' is a term, $x \oplus 0 = x$ is an open formula, and $\forall x \forall y \forall z (x \odot (y \oplus z) = x \odot y \oplus x \odot z)$ is a closed formula (i.e., sentence). The symbolic language is of the first order. In addition to logical axioms (actually axiom schemas), which were inherited from **L**, *Formal Arithmetic* **A** has nine *proper axioms* (see Box 3.3).[5] The proper axioms summarize the characteristic properties of natural numbers as discovered by Peano. There are no additional rules of inference besides *Modus Ponens* and *Generalization*, inherited from **L**.

  The standard model of the theory **A** is $(\iota, (\mathbb{N}, =, +, *, 0, 1))$, with the domain $\mathbb{N}$ being the set of natural numbers and the function ι, which assigns meanings to formulas in the usual way. For example, the meaning of the individual-constant symbol 0 is the natural number 0; that is, $\iota(0) = 0$. An individual-variable symbol x means any natural number, that is, $\iota(x) \in \mathbb{N}$. The meaning of the function symbol $'$ is the successor function, that is, $\iota(x') = \iota(x) + 1$. Hence, $0'$ means the natural number 1, $(0')'$ means the number 2, etc. The binary function symbols $\oplus$ and $\odot$ are interpreted, as expected, as the addition and multiplication of natural numbers. Each closed formula is mapped by ι to a proposition about natural numbers, which is either true or false. The formula $x \oplus 0 = x$ is open, because x is free in it. In the standard model, the formula means: "Adding 0 to a natural number gives the same number." Since this is true for every assignment of a natural number to x, the formula is valid under the standard interpretation of **A**.

Box 3.3 (Proper Axioms of A).

1) $\forall x \forall y \forall z (x = y \Rightarrow (x = z \Rightarrow y = z))$ 2) $\forall x \forall y (x = y \Rightarrow x' = y')$
3) $\forall x (0 \neq x')$ 4) $\forall x \forall y (x' = y' \Rightarrow x = y)$
5) $\forall x (x \oplus 0 = x)$ 6) $\forall x \forall y (x \oplus y' = (x \oplus y)')$
7) $\forall x (x \odot 0 = 0)$ 8) $\forall x \forall y (x \odot y' = (x \odot y) \oplus x)$
9) $F(0) \wedge \forall x (F(x) \Rightarrow F(x')) \Rightarrow \forall x F(x)$, for any formula F with free x (but see also Box 3.6)

Standard interpretation of axioms: **2)** Equal natural numbers have equal successors. **3)** 0 is not a successor of any natural number. **4)** If the successors of two natural numbers are equal, then the numbers are equal. **5)** Adding 0 to a natural number gives the same number. **6)** This axiom

---

[4] Also called *Peano Arithmetic* and denoted by **PA**.

[5] The proper axioms listed in Box 3.3 are not the same as those originally proposed by Peano.

describes how to add a successor of a natural number. **7)** Multiplying a natural number by 0 gives 0. **8)** This axiom describes how to multiply by the successor of a natural number. **9)** This is the *Axiom of Mathematical Induction*. It postulates the following principle. Let $F(x)$ be a relation on $\mathbb{N}$ with a free variable x. (We say that F is a *property*.) If $F(0)$ is true, and, if for any natural n, $F(n)$ implies $F(n+1)$, then $F(x)$ is true for all natural numbers x.

Formalization of Set Theory

Recall that according to Cantor's naive set theory, the set $\mathcal{S}_P = \{x \mid P(x)\}$ exists for the *arbitrary* property P. So, if we set $P \equiv$ "is a set," there is a set $\mathcal{S}_P = \mathcal{U}$ of all sets. But Russell deduced that then there exists a set which, paradoxically, at the same time is and is not a member of itself (see Sect. 2.1.3) . Hence, $\mathcal{U} = \mathcal{S}_P$, the set of all sets, is a paradoxical object as well. The obvious conclusion was that the object $\mathcal{U} = \mathcal{S}_P$ should not exist as a set. As a result, it was necessary to reconsider carefully when, for a given property P, the object $\mathcal{S}_P = \{x \mid x \text{ has the property } P\}$ has the status of a set and when it does not. Which definitions of sets are to be allowed and which are not? It was clear that if $\mathcal{U}$ had existed, it would have been a huge set. So, was it the colossal size of $\mathcal{U}$ that led to Russell's Paradox? Should we allow only those properties P that define objects $\mathcal{S}_P$ of reasonable size? But what is a "reasonable" size of a set? In light of the way mathematics should deal with large objects (sets), two views arose and led to two axiomatic set theories:

- **Axiomatic Set Theory ZFC.** The first view was advocated by Zermelo,[6] Fraenkel,[7] and Skolem.[8] Their plan was to

  find axioms that will ensure the existence of all sets needed in mathematics, and that will, at the same time, prevent the construction of too large sets.

Fig. 3.4 Ernst Zermelo   **Fig. 3.5** Abraham Fraenkel
(Courtesy: See Preface)   (Courtesy: See Preface)

---

[6] Ernst Friedrich Ferdinand Zermelo, 1871–1953, German mathematician.

[7] Abraham Halevi Fraenkel, 1891–1965, German (later Israeli) mathematician.

[8] Thoralf Albert Skolem, 1887–1963, Norwegian mathematician.

Based on this idea they gradually, during 1908–30, defined a formal axiomatic system **ZF**. In this system one can derive all the important theorems of Cantor's naive set theory, while avoiding all the *known* logical paradoxes. The developed theory is called *Zermelo-Fraenkel axiomatic set theory*. Nowadays, this is a standard set theory. When *Axiom of Choice* is added to its proper axioms, the theory is denoted by **ZFC**. (For details about the proper axioms of **ZFC**, see Box 3.4.)

- **Axiomatic Set Theory NBG.** The second view was less conservative. It was advocated by von Neumann,[9] Bernays, and Gödel.[10] Their belief was that

  paradoxes do not follow from the existence of too large sets, but from allowing every (large) set to be a member of some other set.

Fig. 3.6 John von Neumann   **Fig. 3.7** Paul Bernays   **Fig. 3.8** Kurt Gödel
(Courtesy: See Preface)   (Courtesy: See Preface)   (Courtesy: See Preface)

During 1925–40 they gradually defined a formal axiomatic system **NBG**. In addition to the two usual basic notions (i.e., the set and the relation $\in$), **NBG** introduced one more basic notion, *class*, which is a generalization of the notion of set. The advantage of this formal axiomatic system is that there are only a finite number of axioms (because there are no axiom schemas). The theory developed in this system is called *von Neumann-Bernays-Gödel's set theory*. (For further details see Box 3.5.)

What is the relationship between **ZF** and **NBG**? It was found that whatever can be proved in **ZF** can also be proved in **NBG**. The opposite holds only for the formulas of **NBG** that are also formulas of **ZF**. (This is because the notion of the class is unknown to **ZF**.) Because of this, **NBG** is said to be a *conservative extension of* the theory **ZF**. Theorem 3.1 condenses all of this.

Theorem 3.1. *For any formula F of* **ZF** *it holds that* $\underset{\text{NBG}}{\vdash} F$ *iff* $\underset{\text{ZF}}{\vdash} F$.

---

[9] John von Neumann, 1903–1957, Hungarian–American mathematician.

[10] Kurt Gödel, 1906–1978, Austrian–American logician, mathematician, and philosopher.

Let us note that the notion of a class is often used in **ZF** too, but it is not formally defined. For example, while an object $\{w \mid w \in z \wedge P(w)\}$ is surely a set when z is a set (by *Separation Schema*), an object $\{w \mid P(w)\}$ is called a class for safety's sake (as it may not exist as a set). Thus, one can talk about "the class of all sets" knowing that "the set of all sets" does not exist.

Box 3.4 (ZFC, Zermelo-Fraenkel Axiomatic Set Theory).

Formal Axiomatic System **ZFC**. The alphabet has symbols from **L** and two proper symbols: an individual-constant symbol $\emptyset$ and a binary relation symbol $\in$. The terms and formulas are as usual. The language is of the first order. The rules of inference are *Modus Ponens* and *Generalization*. In addition to the logical axioms of **L**, there are nine proper axioms:

1) $\forall x \forall y (\forall w(w \in x \Leftrightarrow w \in y) \Rightarrow x = y)$ *(Axiom of Extensionality)*
2) $\forall y \exists z \forall w(w \in z \Leftrightarrow w \in y \wedge P(w))$, for any P with no free z.   *(Separation Schema)*
3) $\forall x \forall y \exists z \forall w(w \in z \Leftrightarrow w = x \vee w = y)$ *(Axiom of Pair)*
4) $\forall y \exists z \forall w(w \in z \Leftrightarrow \exists x(w \in x \wedge x \in y))$ *(Axiom of Union)*
5) $\forall u \forall v \forall w(F(u, v) \wedge F(u, w) \Rightarrow v = w) \Longrightarrow \forall y \exists z \forall w(w \in z \Leftrightarrow \exists x(x \in y \wedge F(x, w)))$,
    for any F with no free z.   *(Substitution Schema)*
6) $\exists z(\emptyset \in z \wedge \forall x \in z : x \cup \{x\} \in z)$ *(Axiom of Infinity)*
7) $\forall y \exists z \forall w(w \in z \Leftrightarrow w \subseteq x)$ *(Axiom of Power Set)*
8) $\forall y \neq \emptyset \; \exists x \in y : x \cap y = \emptyset$ *(Axiom of Regularity)*
9) $\forall y (\forall x \in y : x \neq \emptyset \Rightarrow \exists f \in (\cup y)^y \forall x \in y : f(x) \in x)$ *(Axiom of Choice)*

Standard Interpretation. The domain of the standard interpretation was described by von Neumann, so we denote it by $\mathcal{V}$. Von Neumann insisted that $\mathcal{V}$ contains *exactly all the sets*. Thus, if x is in $\mathcal{V}$, then x is a set. Now, if an element of x had not itself been a set, it would not have been in $\mathcal{V}$, and this would have led to trouble. To avoid this, von Neumann required that each element of a set be itself a set. Such sets are called *hereditary*. Thus, $\mathcal{V}$ contains exactly all the hereditary sets. But now it seems that there are useful sets, such as $\{0, 1, 2\}$, which are not hereditary. We will see shortly that this is not so.

Remarks. In the following we describe, for each proper axiom, the motivation for adding it to the axioms of **ZF**, its meaning, and its consequences. When interpreting a proper axiom, bear in mind that the individual-variable symbols mean hereditary sets.

1)   *Axiom of Extensionality: A set is completely determined by its elements.*
     Consequences: Two sets with the same elements are equal. $\mathcal{V}$ contains exactly hereditary sets. There is at most one empty set, $\emptyset$. Motivation for axiom 2: Does $\emptyset$ exist?
2)   *Separation Schema: The set $\{w \mid w \in z \wedge P(w)\}$ exists.*
     Comment: z is any set and P is any property defined by a formula of **L**. Consequences: $\emptyset$ is a set. (Proof: $P(w) \equiv \neg(w = w)$). For any sets $\mathcal{A}$ and $\mathcal{B}$, also $\mathcal{A} \cap \mathcal{B} \stackrel{\text{def}}{=} \{w \mid w \in \mathcal{A} \wedge w \in \mathcal{B}\}$ and $\mathcal{A} \setminus \mathcal{B} \stackrel{\text{def}}{=} \{w \mid w \in \mathcal{A} \wedge w \notin \mathcal{B}\}$ are sets. Motivation for axiom 3: We need more sets.
3)   *Axiom of Pair: For any x and y there is a set containing exactly x and y.*
     Consequences: Ordered pair $(x, y) \stackrel{\text{def}}{=} \{\{x\}, \{x, y\}\}$ is a set. $\{\emptyset\}$ is a set (as $\{\emptyset\} = \{\emptyset, \emptyset\}$); so is $\{\{\emptyset\}\}$ (as $\{\{\emptyset\}\} = \{\{\emptyset\}, \{\emptyset\}\}$); and so on. Defining $0 \stackrel{\text{def}}{=} \emptyset$, $1 \stackrel{\text{def}}{=} \{\emptyset\}$, and $2 \stackrel{\text{def}}{=} \{\emptyset, \{\emptyset\}\}$, we obtain the numbers $0, 1, 2$ and count to two. Motivation for axiom 4: We cannot define larger numbers in this way, because we cannot construct sets with more than two members. So we need more sets.

4) *Axiom of Union: For any family y of sets x there is a set z that is the union of the sets x.*
Consequences: Now we can construct sets with more than two elements and define any natural number, e.g., $3 \stackrel{\text{def}}{=} 2 \cup \{2\} = \{\emptyset, \{\emptyset\}, \{\emptyset, \{\emptyset\}\}\}$, using the definition $n + 1 \stackrel{\text{def}}{=} n \cup \{n\}$. The definition is applicable on every $n \in \mathbb{N}$, where $\mathbb{N}$ denotes the collection of all natural numbers. Motivation: Is $\mathbb{N}$ a set? We will postulate this in the *Axiom of Infinity* (see below). However, it turns out that $\mathbb{N}$ alone does not allow for the development of a full theory of ordinals and for the use of transfinite induction. This is why we first introduce axiom 5.

5) *Substitution Schema: If the domain of a function is a set, then its range is also a set.*
Comment: In the schema, $F(x, y)$ denotes a function $x \mapsto y$. Consequences: There exist certain well-ordered sets, i.e., ordinal numbers.

6) *Axiom of Infinity: There is an inductive infinite set y.*
Comment: A set y is defined to be *inductive* if $\emptyset \in y \wedge \forall x (x \in y \Rightarrow x \cup \{x\} \in y)$. A set y is defined to be *infinite* if it equipolent to a proper subset of y. Consequence: $\mathbb{N}$ is a set. Motivation for axiom 7: Some sets cannot be constructed (e.g., the set of all subsets of a set).

7) *Axiom of Power Set: For any set x there is a set y containing all the subsets of x as members.*
Motivation: The axioms of pair, union, and power set allow for the construction of larger sets from smaller ones. Thus, a set exists if it can be constructed only from $\emptyset$ and $\mathbb{N}$, which are the only sets whose existence was postulated by axioms. What about "irregular" sets, such as Russell's $\mathcal{R}$? At this point we can still construct a set x, where $x \in y \in x$ for some set y. We should confine constructions so that only "regular" (i.e., reasonable) sets will exist. Axiom 8 takes care of this.

8) *Axiom of Regularity: Any nonempty set x contains an element y so that x and y have no common elements.*
Consequence: There can be no set x such that $x \in y \in x$ for some set y (else, we would have $x \in \{x, y\} \cap y$ and $y \in \{x, y\} \cap x$, implying that $\{x, y\}$ would not contain an element sharing no elements with $\{x, y\}$, in contradiction with axiom 8). This prevents Russell's Paradox.

9) *Axiom of Choice: For any family z of nonempty sets there is a function that assigns to each member w of z a member of w.*

Box 3.5 (NBG, von Neumann-Bernays-Gödel's Axiomatic Set Theory).

Basic Ideas. In addition to the two usual basic notions of a set and a membership relation $\in$, there is also the notion of a *class*. Each set is also a class, but some classes are not sets. Classes that are not sets are called *proper classes*. A characteristic of the proper class is that it is not a member of any class (and hence, of any set). The intention of such a definition of a class is now clear: proper classes should represent collections that are too large to be sets, and non-proper classes (that is, sets) should represent all the reasonably large sets that are used in mathematics.
   Drawing a distinction between sets and proper classes enables us to prevent paradoxes. Let us see how this works on Russell's Paradox. Define the class $\mathcal{R} = \{\mathcal{S} \mid \mathcal{S} \text{ is a set} \wedge \mathcal{S} \notin \mathcal{S}\}$. Like every class, $\mathcal{R}$ either is or is not a member of itself. Let us see whether $\mathcal{R}$, even as a class, gives rise to Russell's Paradox $\mathcal{R} \in \mathcal{R} \Leftrightarrow \mathcal{R} \notin \mathcal{R}$. If $\mathcal{R} \in \mathcal{R}$, then $\mathcal{R}$ is a set that is a member of itself, and hence $\mathcal{R} \notin \mathcal{R}$. This is a contradiction. Assume now that $\mathcal{R} \notin \mathcal{R}$. This means that $\mathcal{R}$ is not a set or $\mathcal{R} \in \mathcal{R}$. The latter alternative is impossible because of the assumption, which leaves us the first alternative: $\mathcal{R}$ is not a set. Therefore, $\mathcal{R}$ is a *proper class*. We have seen that this deduction, which in Cantor's naive set theory led to Russell's Paradox, now luckily ends up with the conclusion that $\mathcal{R}$ is a proper class. In a similar way Burali-Forti's and Cantor's paradoxes are eliminated. So how was this system formally defined?

Formal Axiomatic System **NBG**. There are many similarities with **ZF**. The alphabet has all the symbols of **L** and two proper constant symbols $\emptyset$ and $\in$. The terms and formulas are built as usually. The symbolic language is of the first order. The rules of inference are *Modus Ponens* and *Generalization*. In addition to the logical axioms inherited from **L**, there are proper axioms. How were these selected?

Recall, that Cantor's *Axiom of Abstraction* postulated that "Every property P defines a set S_P." As we have seen, the authors of **ZF** limited the properties P so that S_P are reasonable and not too large. In contrast, the authors of **NBG** argued as follows:

> If we demanded that S_P be a *class*, then S_P might not be a set (but be a proper class). Hence, the fear of too large *sets* might become superfluous. But then, could P again be an *arbitrary* property? Could we declare the following generalization of the *Axiom of Abstraction*: "Every property P defines a *class* S_P"?

It turned out that such an axiom is bad, for it would allow $\mathcal{R}' = \{ S \mid S \text{ is a class } \wedge\ S \notin S \}$ to be a class, and this class would again lead to Russell's Paradox $\mathcal{R}' \in \mathcal{R}' \Leftrightarrow \mathcal{R}' \notin \mathcal{R}'$. Thus, more caution was needed in order to generalize the *Axiom of Abstraction*. The result of the search is the following **Axiom of Class Existence**: *Every property P of sets defines a class.* Hence, a class cannot be determined by a property of proper classes, but only by a property of sets. This finally leads to an informal definition of a class:

> *A class is a collection of sets that have in common a property P:* $\{ S \mid S \text{ is a set } \wedge\ P(S) \}$.

Of course, the sets S must exist in the first place. This is ensured by other axioms (as in **ZF**). For this **NBG** has three groups of proper axioms.

The first group initially consisted only of the *Axiom of Class Existence* to establish the notion of a class. This axiom is actually an axiom schema, because it represents an infinite number of axioms, one for each property P of sets. It turned out that the schema can be replaced by only eight axioms! These axioms now constitute the first group. In the second group is the following **Axiom of Extensionality**: *Two classes are equal if they have the same elements.* The third group consists of axioms that, similarly to **ZF**, postulate the existence of *sets* obtained either ex nihilo (such as $\emptyset$ and $\mathbb{N}$) or by a construction from existing ones.

Second-Order Formal Axiomatic Systems and Theories

It turned out that certain properties of mathematical objects cannot be defined in a first-order symbolic language. So, the basic notions and axioms referring to such properties cannot be stated in these languages. Consequently, there are no first-order theories about such objects. In such cases, it often turns out that the quantifiers $\forall$ and $\exists$ should be applicable to function-variable symbols and/or relation-variable symbols, something that is not allowed in first-order symbolic languages. For instance, first-order languages do not enable us to define the *completeness* of the set $\mathbb{R}$ of real numbers, and the concepts of *torsion group* and *mathematical induction* (see Box 3.6 for further details). If, however, the action of quantifiers is expanded to function-variable or relation-variable symbols, we obtain a second-order symbolic language, a second-order formal axiomatic system, and a second-order theory.

Unfortunately, second-order theories are not as useful as it seems. This is because they lack some important properties that are characteristic of first-order theories. For example, for any theory it is important to know whether the theory has models, how many there are, what their properties are and what the relations between them are. Such questions are dealt with in *model theory*. Two of its important theorems are the *Compactness Theorem*[11] and the "downward" *Löwenheim-Skolem Theorem*[12]. But, in general, the theorems do not hold in second-order theories. Thus, second-order languages and theories may be more powerful in their expression, but they are less amenable to a metamathematical treatment.

As we will see, the deficiencies of second-order theories had no opportunity to manifest themselves and influence formalism, because it was not long before a disappointment about first-order theories came as a result of Gödel's *Incompleteness Theorems*.

Box 3.6 (Expression of First-Order Languages).

We give examples of where a second-order language is needed.

- *Mathematical Induction.* The *Axiom of Mathematical Induction* is usually written as the first-order defining formula $F(0) \wedge \forall i(F(i) \Rightarrow F(i+1)) \Longrightarrow \forall n F(n)$ to which an *explanation* is added stating that $F(x)$ can be *any* formula with x as a free variable; see Box 3.3. (The formula $F(x)$ describes a property of x.) But observe that the defining formula is actually an axiom schema; only after F has been substituted with an actual formula is a particular axiom (i.e., one of infinitely many) obtained. In order to write in a symbolic language that the principle of mathematical induction holds *for any formula F*, we have to add $\forall F$ to the defining formula. This gives us a *second-order* formula $\forall F(F(0) \wedge \forall i(F(i) \Rightarrow F(i+1)) \Longrightarrow \forall n F(n))$. First-order symbolic language is too weak to describe completely the principle of mathematical induction.

- *Completeness of* $\mathbb{R}$. A fundamental property of the set $\mathbb{R}$ of real numbers is *completeness*: every nonempty subset of $\mathbb{R}$ that is bounded above has a least upper bound. How can we express this in a symbolic language? Let us start generally. Let $\mathbb{R}$ be the domain of an interpretation, and B an arbitrary property of the sets $\mathcal{S} \subseteq \mathbb{R}$. We can describe the fact that B holds for every set $\mathcal{S} \subseteq \mathbb{R}$ by the formula $\forall \mathcal{S} : B(\mathcal{S})$. But this formula does not belong to any first- or second-order language, because $\forall$ refers to *sets* of elements of the domain. By viewing subsets of $\mathbb{R}$ as sets $\mathcal{S}_P = \{x \in \mathbb{R} \mid P(x)\}$, where P are predicates, the formula transforms into the *second-order* formula $\forall P : B(\mathcal{S}_P)$, because $\forall$ now binds the predicate-variable P. If we fix the property B to $B(\mathcal{S}_P) =$ "if $\mathcal{S}_P$ is bounded above then it has an l.u.b.," we obtain the second-order formula $\forall P(\exists b \forall x(P(x) \Rightarrow x \leqslant b) \Longrightarrow \exists \ell \forall u(\forall x(P(x) \Rightarrow x \leqslant u) \Leftrightarrow \ell \leqslant u))$ stating that $\mathbb{R}$ is complete. That is: For every P, if $\mathcal{S}_P$ is bounded above by b, then $\mathcal{S}_P$ has an l.u.b. (an ℓ which is $\leqslant$ than any upper bound u of $\mathcal{S}_P$).

- *Torsion Groups.* Let $(G, \cdot)$ be a group with unity e. We say that G is a *torsion* group if for every $a \in G$ there is an $n \geqslant 1$, such that $a^n = e$. How can we define the property $P(G) \equiv$ "G is a torsion group"? Let the domain of interpretation be $(G, \cdot)$. Let us try the seemingly obvious: $P(G) \equiv \forall a \in G \exists n \geqslant 1 \, a^n = e$. Notice that the interpretation of this formula is a proposition that also considers natural numbers—but these are *not* in the domain of interpretation. To fix

---

[11] *Compactness Theorem:* A first-order theory has a model if every finite part of the theory does.

[12] *Löwenheim-Skolem Theorem:* If a theory has a model, then it has a countable model. The generalization is called the "upward" *Löwenheim-Skolem Theorem* and states: If a first-order theory has an infinite model, then for every infinite cardinal κ it has a model of size κ. Since such a theory is unable to pin down the cardinality of its infinite models, it cannot have exactly one model (up to isomorphism).

that we might define $P(G) \equiv \forall a \in G \ (a = e \ \lor \ a \cdot a = e \ \lor \ a \cdot a \cdot a = e \ \lor \ ...)$ and thus avoid mentioning natural numbers. However, this formula is no longer finite and, hence, not in a first-order symbolic language. In any case, it turns out that there is no finite set of first-order formulas that would define the torsion groups (i.e., that would model the torsion groups).

3.3 Chapter Summary

A formal axiomatic system is determined by a symbolic language, a set of axioms, and a set of rules of inference. The axioms are logical or proper. All of this is the initial theory associated with the formal axiomatic system. The theory is then systematically extended into a larger theory. The development is formal: new notions must be defined by existing ones, and propositions must be formally proved, i.e., derived from axioms or previously proved formulas. Each proved formula is a theorem of the theory. Such a syntax-oriented and rigorous development of the theory is a mechanical process and because of this better protected from man's fallibility.

At any stage of the development, the theory can be interpreted in a chosen field of interest, called the domain of interpretation. The interpretation defines how a formula must be understood as a statement about the elements of the domain. Each interpretation of a theory under which all the axioms of the theory are valid is called a model of the theory. A theory may have several different models. A model may not be a part of the real world.

Formal axiomatic systems both protected the development of theories from man's fallibility and preserved the freedom given by the hypothetical axiomatic. The three particular fields of mathematics whose formal axiomatic systems and theories played a crucial role in the events that followed are logic, arithmetic, and Cantor's set theory. The corresponding formal axiomatic systems are *First-Order Logic* **L**, *Formal Arithmetic* **A**, and the two *Axiomatic Set Theories* **ZF** and **NBG**.

Chapter 4
Hilbert's Attempt at Recovery

If something is consistent, no part of it contradicts or conflicts with any other part. If something is complete, it contains all the parts that it should contain. If something is decidable, we can establish the fact of the matter after considering the facts.

Abstract *Hilbert's Program* was a promising formalistic attempt to recover mathematics. It would use formal axiomatic systems to put mathematics on a sound footing and eliminate all the paradoxes. Unfortunately, the program was severely shaken by Gödel's astonishing and far-reaching discoveries about the general properties of formal axiomatic systems and their theories. Thus Hilbert's attempt fell short of formalists' expectations. Nevertheless, although shattered, the program left open an important question about the existence of a certain algorithm—a question that was to lead to the birth of *Computability Theory*.

4.1 Hilbert's Program

In this section we will describe *Hilbert's Program*. In order to understand the goals of the program, we will first define the fundamental metamathematical problems of the formal axiomatic systems and their theories. Then we will describe the goals of the program and Hilbert's intentions that influenced the program.

4.1.1 Fundamental Problems of the Foundations of Mathematics

The rigor and syntactic orientation of formal axiomatic systems not only protected them from man's fallibility, but also enabled a precise definition and investigation of various metamathematical problems, i.e., questions about their theories. Naturally, these questions were closely linked to the burning question of protecting mathematics from paradoxes. They are called the *problems of the foundations of mathematics*.

Of special importance to the history of the notion of algorithm and *Computability Theory* will be the following four problems of the foundations of mathematics:

© Springer-Verlag Berlin Heidelberg 2015   49
B. Robič, *The Foundations of Computability Theory*,
DOI 10.1007/978-3-662-44808-3_4

1. **Consistency Problem.** Let **F** be a first-order theory. (So we can use the logical connective ¬.) Suppose that there is a closed formula F in **F** such that both F and ¬F are derivable in **F**. Then the contradictory formula F ∧ ¬F immediately follows! We say that such a theory is *inconsistent*. But we can readily show that in an inconsistent theory *any* formula of the theory can be derived! (See Box 4.1.)

Box 4.1 (Derivation in an Inconsistent Theory).

Suppose that formulas F and ¬F are derivable in **F** and let A be an *arbitrary* formula in **F**. Then we have the following derivation of A:

1. F	Supposition.
2. ¬F	Supposition.
3. F ⇒ (¬A ⇒ F)	Ax. 1 of **L** (Box 3.2) with ¬A instead of G (i.e.,¬A/G)
4. ¬A ⇒ F	From 1. and 3. by *MP*.
5. ¬F ⇒ (¬A ⇒ ¬F)	Ax. 1 of **L** (Box 3.2) with ¬F/F and ¬A/G.
6. ¬A ⇒ ¬F	From 2. and 5. by *MP*.
7. (¬A ⇒ ¬F) ⇒ ((¬A ⇒ F) ⇒ A)	Ax. 3 of **L** (Box 3.2) with A/G.
8. (¬A ⇒ F) ⇒ A	From 6. and 7. by *MP*.
9. A	From 4. and 8. by *MP*.

An inconsistent theory has no cognitive value. This is why we seek consistent theories (Fig. 4.1). So the following metamathematical question is important:

Consistency Problem: "Is a theory **F** *consistent?"*

For example, in 1921, Post proved that the *Propositional Calculus* **P** is consistent.

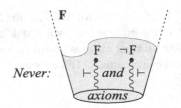

Fig. 4.1 In a consistent theory **F** for no formula F both F and ¬F are derivable in **F**

2. **Syntactic Completeness Problem.** Let **F** be a consistent first-order theory and F an arbitrary closed formula of **F**. Since **F** is consistent, F and ¬F are not both derivable in **F**. But, what if neither F nor ¬F is derivable in **F**? In such a case we say that F is *independent* of **F** (as it is neither provable nor refutable in **F**). This situation is undesirable. We prefer that *at least* one of F and ¬F be derivable in **F**. When this is the case for every closed formula of **F**, we say that **F** is *syntactically complete* (see Fig. 4.2). Thus, in a consistent and syntactically complete theory every closed formula is *either* provable *or* refutable. Informally, there are no "holes" in such theory, i.e., no closed formulas independent of the theory. So, the next metamathematical question is important:

*Syntactic Completeness Problem: "Is a theory **F** syntactically complete?"*

The answer tells us whether **F** guarantees that, for no formula of **F**, the search for either a proof or a refutation of the formula is *a priori* doomed to fail. For instance, we know that the *Propositional Calculus* **P** and *First-Order Logic* **L** are *not* syntactically complete.

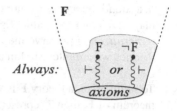

Fig. 4.2 In a syntactically complete theory **F** it holds, for every formula F, that F or ¬F is derivable in **F**

3. **Decidability Problem.** Let **F** be a consistent and syntactically complete first-order theory, and F an arbitrary formula of **F**. The derivation of F or ¬F may be highly intricate. Consequently, the search for a formal proof or refutation of F is inevitably dependent on our ingenuity. As long as F is neither proved nor refuted, we can be sure that this is because of the lack of our ingenuity (because **F** is syntactically complete). Now suppose that there existed an *algorithm*—call it *decision procedure*—capable of answering—in *finite* time, and for *any* formula F of **F**—the question "Is F derivable in **F**?" Such a decision procedure would be considered *effective* because it could *decide*, for any formula, whether or not it is a theorem of **F**. When such a decision procedure exists, we say that the theory **F** is *decidable*. (See Fig. 4.3.) So, the metamathematical question, called the

*Decidability Problem: "Is a theory **F** decidable?"*

is important because the answer tells us whether **F** allows for a systematic (i.e., mechanical, algorithmic) and effective search of formal proofs. In a decidable theory we can, at least *in principle*, develop the theory without investing our ingenuity and creativity. For instance, the *Propositional Calculus* **P** is known to be a decidable theory; the corresponding decision procedure uses the well-known *truth-tables* and was discovered in 1921 by Post.[1]

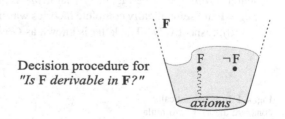

Fig. 4.3 In a decidable theory **F** there is a decision procedure (algorithm) that tells, for arbitrary formula F, whether or not F is derivable in **F**

Decision procedure for
*"Is F derivable in **F**?"*

---

[1] Emil Leon Post, 1897–1954, American mathematician, born in Poland.

4. **Semantic Completeness Problem.** Let **F** be a consistent first-order theory. When interpreting **F**, we are particularly interested in the formulas that are *valid in the theory* **F**, i.e., formulas that are valid in *every* model of **F**. Such formulas represent *Truths* in **F** (see p. 39). Now, we know that all the axioms of **F**, both logical and proper, represent *Truths* in **F**. If, in addition, the rules of inference of **F** preserve the property "to represent a *Truth* in **F**," then also every theorem of **F** represents a *Truth* in **F**. When this is the case, we say that **F** is *sound*. Informally, in a sound theory we cannot deduce something that is not a *Truth*, so the theory may have cognitive value. Specifically, it can be proved that *Modus Ponens* and *Generalization* preserve the *Truth*-ness of formulas. So we can assume that the theories we are interested in are sound.

To summarize, a theory **F** is *sound* when the following holds: If a formula F is a theorem of **F**, then F represents a *Truth* in **F**; in short

$$\text{If } \vdash_{\mathbf{F}} F \text{ then } \models_{\mathbf{F}} F. \qquad (\mathbf{F} \text{ is sound})$$

However, the opposite may not hold: A sound theory **F** may contain a formula that represents a *Truth* in **F**, yet the formula is *not* derivable in **F**. This situation can arise when **F** lacks some axiom(s).

Of course, we would prefer a sound theory whose axioms suffice for deriving *every Truth*-representing formula in the theory. When this is the case, the theory is said to be *semantically* complete. Thus, a theory **F** is *semantically complete* when the following holds: A formula F is a theorem of **F** *if and only if* F represents a *Truth* in **F**; in short

$$\vdash_{\mathbf{F}} F \text{ if and only if } \models_{\mathbf{F}} F. \qquad (\mathbf{F} \text{ is semantically complete})$$

The metamathematical question

> *Semantic Completeness Problem: "Is a theory* **F** *semantically complete?"*

is of the greatest importance because the answer tells us whether the syntactic property "to be a theorem of **F**" coincides with the semantic property "to represent a *Truth* in **F**" (see Fig. 4.4.) That *Propositional Calculus* **P** and *First-Order Logic* **L** are semantically complete theories was proved by Post (1921) and Gödel (1930), respectively. (The latter is known as *Gödel's Completeness Theorem*.)

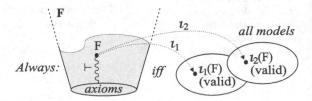

Fig. 4.4 In a semantically complete theory **F** a formula F is derivable if and only if F is valid in **F** (i.e., is valid in every model of **F**)

4.1.2 Hilbert's Program

Let us now return to the foundational crisis of mathematics at the beginning of twentieth century. During 1920–28, Hilbert gradually formed a list of goals—called *Hilbert's program*—that should be attained in order to base the whole of mathematics on new foundations that would prevent paradoxes.

Fig. 4.5 David Hilbert
(Courtesy: See Preface)

Hilbert's Program consisted of the following goals:

A. find an f.a.s. **M** capable of deriving all the theorems of mathematics;
B. prove that the theory **M** is semantically complete;
C. prove that the theory **M** is consistent;
D. construct an algorithm that is a decision procedure for the theory **M**.

Note that the goal D asks for a *constructive* proof that **M** is decidable, i.e., a proof by exhibiting a decision procedure for **M**. Let us denote this procedure by D_{Entsch} since Hilbert called the *Decidability Problem* for **M** the *Entscheidungsproblem*.

Intention

Hilbert's intention was that, having attained the goals A, B, C, D, every mathematical statement would be mechanically verifiable. How could they be verified? We should first write the statement as a sentence, i.e., a closed formula F of **M** (hence goal A). How would we find out whether the statement represented by F is a mathematical *Truth*? If **M** were a semantically complete theory (hence goal B), we would be sure that F represents a *Truth* in **M** *iff* F is a theorem of **M**. Therefore, we could focus on syntactic issues only. If **M** were a consistent theory (hence goal C), the formulas F and ¬F could not both be theorems. Finally, we would apply the decision procedure D_{Entsch} (hence goal D) to find out which of F and ¬F is a theorem of **M**. Notice that Hilbert expected that **M** would be syntactically complete.

What is more, by using the decision procedure D_{Entsch}, mathematical statements could be algorithmically classified into *Truths* and non-*Truths*. There would be no need for human ingenuity in mathematical research; one would just systematically generate mathematical statements (i.e., sentences), check them by D_{Entsch}, and collect only those that are *Truths*.[2]

Finitism

Hilbert expected that the consistency and semantic completeness of **M** could be proved only by analyzing the *syntactic* properties of **M** and its formulas. To avoid deceptive intuition, he defined the kind of reasoning, called *finitism*, one should preferably use in such an analysis. Here, Hilbert approached the intuitionist view of infinity. In particular, proofs of the goals B and C should be *finitist* in the sense that they should use finite objects and methods that are constructive, at least in principle. For instance, the analysis should avoid actual infinite sets, the use of the *Law of Excluded Middle* in certain existence proofs, and the use of *transfinite induction*. (We will informally describe what transfinite induction is in Box 4.7 on p. 62.)

4.2 The Fate of Hilbert's Program

After Hilbert proposed his program, researchers started investigating how to attain the goals A, B, C, and D. While the research of the formalization of mathematics (goal A) and the decidability of mathematics (goal D) seemed to be promising, it took only a few years before Gödel discovered astonishing and far-reaching facts about the semantic completeness (goal B) and consistency (goal C) of formally developed mathematics. In this section we will give a detailed explanation of how this happened.

4.2.1 Formalization of Mathematics: Formal Axiomatic System **M**

So, what should the sought-for formal axiomatic system **M** look like? Preferably it would be a first-order or, if necessary, second-order formal axiomatic system. Probably it would contain one of the formal axiomatic systems **ZFC** or **NBG** in order to introduce sets. Perhaps it would additionally contain some other formal axiomatic systems that formalize other fields of mathematics (analysis, for example). Despite these open questions, it was widely believed that **M** should inevitably contain the following two formal axiomatic systems (see Fig. 4.6):

---

[2] Today, one would use a computer to perform these tasks. Of course, when Hilbert proposed his program, there were no such devices, so everything was a burden on the human processor.

1. *First-Order Logic* **L**. This would bring to **M** all the tools needed for the logically unassailable development of the theory **M**, that is, all mathematics. The trust in **L** was complete after the consistency of **L** had been proved with finitist methods, and after Gödel and Herbrand had proved the semantic completeness of **L** —Herbrand even with finitist methods.

2. *Formal Arithmetic* **A**. This would bring natural numbers to **M**. Since natural numbers play a key role in the construction of other kinds of numbers (e.g., rational, irrational, real, complex), they are indispensable in **M**.

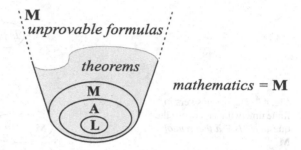

mathematics = **M**

Fig. 4.6 Mathematics as a theory **M** belonging to the formal axiomatic system **M**

4.2.2 Decidability of **M**: Entscheidungsproblem

Recall that the goal of the *Entscheidungsproblem* was: Construct an algorithm D_{Entsch} that will, for any formula F of **M**, decide whether F is derivable in **M**; in short, whether $\vdash_{\mathbf{M}} F$. (See Fig. 4.7).

Hopes that there was such a D_{Entsch} were raised by the syntactic orientation of formal axiomatic systems and their theories, and, specifically, by their view of a derivation (formal proof) as a finite sequence of language constructs built according to a finite number of syntactic rules. At first sight, the search for a derivation of a formula F could proceed, at least in principle, as follows:

> *systematically generate finite sequences of symbols of* **M**, *and*
> *for each newly generated sequence*
> *check whether the sequence is a proof of F in* **M**;
> *if so, then answer* YES *and halt.*

If, in truth, F were derivable in **M**, and if a few reasonable assumptions held (see Box 4.2), then the procedure would find a formal proof of F. However, if in truth F were not derivable in **M**, the procedure would never halt, because it would keep generating and checking the candidate sequences. But notice that if a newly generated sequence is not a proof of F, it may still be a proof of ¬F. So we check this possibility too. We obtain the following improved procedure:

> *systematically generate finite sequences of symbols of* **M**, *and*
> *for each newly generated sequence*
>   *check whether the sequence is a proof of* F *in* **M**;
>   *if so, then answer* YES *and halt*
>   *else check whether the sequence is a proof of* ¬F *in* **M**;
>   *if so, then answer* NO *and halt.*

Assuming that either F or ¬F is provable in **M**, the procedure always halts. In Hilbert's time there was wide belief that **M** would be syntactically complete.

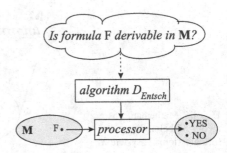

Fig. 4.7 D_{Entsch} answers in finite time with YES or NO the question *"Is F a theorem of* **M***?"*

Box 4.2 (Recognition of Derivations).

Is a sequence of symbols a derivation of F in **M**? Since the sequence is finite, it can only be composed of *finitely* many formulas of **M**. There are also *finitely* many rules of inference in **M** that can connect these formulas in a syntactically correct derivation of F. Assuming that we can find out in finite time whether a formula (contained in the sequence) directly follows from a finite set of premises (contained in the sequence) by a rule of inference of **M**, we can decide in finite time whether the sequence is a derivation of F in **M**. To do this, we must systematically check a finite number of possible triplets (formula, set of premises, rule).

Here, an assumption is needed. Since a premise can also be an axiom of **M**, we must assume that there is a procedure capable of deciding in finite time whether a formula is an axiom of **M**. (Today, we say that such a theory is *computably axiomatizable*.)

If the theory **M** were *consistent*, then at most one of F and ¬F would be derivable in **M**. If, in addition, **M** were *syntactically complete*, then at least one of F and ¬F would be derivable in **M**. Consequently, for an arbitrary F of such an **M**, the procedure would halt in a finite time and answer either YES (i.e., F is a theorem of **M**) or NO (i.e., ¬F is a theorem of **M**). So, **M** would be *decidable* and the above procedure would be the decision procedure D_{Entsch}. Thus we discovered the following relationship:

> *if*   **M** is consistent *and* **M** is syntactically complete
> *then* there is a decision procedure for **M**, i.e., **M** is decidable.

NB *These questions caused the birth of Computability Theory, which took over the research connected with the goal D of* Hilbert's Program, *and finally provided answers to these and many other questions.*

We will return to the questions about the decision procedure in the next chapter. Later we will describe how the new theory solved the *Entscheidungsproblem* (see Theorem 9.2 on p. 203). For the present, we continue describing what happened to the other two all-important goals (B and C) of *Hilbert's Program*.

4.2.3 Completeness of M: Gödel's First Incompleteness Theorem

So, how successful was proving the semantic completeness of **M** (goal B)? In 1931, hopes of finding such a proof were dashed by 25-year-old Gödel. He proved the following metamathematical theorem.

Fig. 4.8 Kurt Gödel
(Courtesy: See Preface)

Theorem 4.1. (First Incompleteness Theorem) *If the Formal Arithmetic* **A** *is consistent, then it is semantically incomplete.*

Informally, the *First Incompleteness Theorem* tells us that if **A** is a consistent theory, then it is not capable of proving *all Truths* about natural numbers; there are statements about natural numbers that are true, but are unprovable within **A**.

Gödel proved this theorem by constructing an independent formula G in **A** (that is, a formula G such that neither G nor ¬G is derivable in **A**). In addition, he proved that G represents a *Truth* about natural numbers (that is, that the interpretation of G is true in the standard model $(\mathbb{N}, =, +, *, 0, 1)$ of **A**). What is more, he proved that even if G were added to proper axioms of **A**, the theory **A**′ belonging to the extended formal axiomatic system would still be semantically incomplete—now because of some other formula G′ independent of **A**′ yet true in the standard model. Finally, he proved the following generalization: *Any* consistent extension of the set of axioms of **A** gives a semantically incomplete theory. (For a more detailed explanation of Gödel's proof, see Box 4.4 on p. 59.)

Informally, the generalization tells us that *no* consistent theory that includes **A** is capable of proving *all 𝒯ruths* about the natural numbers; there will *always* be statements about the natural numbers that are true, yet unprovable within that theory.

4.2.4 Consequences of the First Incompleteness Theorem

Gödel's discovery revealed unexpected limitations of the axiomatic method and seriously undermined *Hilbert's Program*. Since **M** was supposed to be a consistent extension of **A**, it would inevitably be semantically incomplete! This means that the mathematics developed as a formal theory **M** would be like a "Swiss cheese full of holes" (Fig. 4.9), with some of the mathematical *𝒯ruths* dwelling in the holes, inaccessible to usual mathematical reasoning (i.e., logical deduction in **M**).

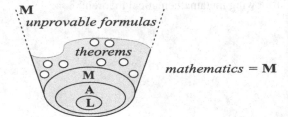

Fig. 4.9 Mathematics developed in the formal axiomatic system **M** would not be semantically complete

mathematics = **M**

The independent formulas G, G′, ... of the proof of the *First Incompleteness Theorem* are constructed in such a shrewd way that they express their own undecidability and, at the same time, represent *𝒯ruths* in **M** (see Box 4.4).

But in other holes of **M** sit independent formulas that tell us nothing about their own validity in **M**. When such a formula is brought to light and its independence of the theory is uncovered, we may declare that either the formula or its negation represents a *𝒯ruth* in **M** (mathematics). In either case our choice does not affect the consistency of the theory **M**. However, the choice may not be easy, because either choice may have a model that behaves as a possible part of mathematics. An example of this situation is the formula that represents (i.e., describes) the *Continuum Hypothesis*, which we encountered on p. 16. (See more about this in Box 4.3.)

Remark. Nevertheless, all of this still does not mean that such *𝒯ruths* will *never* be recognized. Note that the *First Incompleteness Theorem* only says that for recognizing such *𝒯ruths* the *axiomatic method* is too weak. So there still remains a possibility that such *𝒯ruths* will be proven (recognized) with some *other* methods surpassing the axiomatic method in its proving capability. Such methods might use non-finitist tools or any other tools yet to be discovered.

Box 4.3 (Undecidability of the *Continuum Hypothesis*).

Intuitively, the *Continuum Hypothesis* (CH) conjectures that there is no set with more elements than natural numbers and fewer than real numbers. In short, there is no cardinal between $\aleph_0$ and c.

In 1940, Gödel proved the following metamathematical theorem: *If* ZFC *is consistent, then* CH *cannot be refuted in* ZFC. Then, in 1963, Cohen[3] proved: *If* ZFC *is consistent, then* CH *cannot be proved in* ZFC. Thus, CH is independent of ZFC. Gödel's and Cohen's proofs show that neither CH nor ¬CH is a *Truth* in ZFC. Discussions about whether or not to add CH to ZFC (and hence mathematics) still continue.

There is a similar situation with the generalization of CH. Let α be any ordinal and $\aleph_\alpha$ and $\aleph_{\alpha+1}$ cardinalities of a set and its power set, respectively. The *Generalized Continuum Hypothesis* (GCH) conjectures: There is no cardinal between $\aleph_\alpha$ and $\aleph_{\alpha+1}$, i.e., $2^{\aleph_\alpha} = \aleph_{\alpha+1}$.

Box 4.4 (Proof of the *First Incompleteness Theorem*).

The theorem states: *Every consistent theory that contains* A *is essentially incomplete*. That is: Every consistent and sufficiently strong theory has countably infinitely many formulas that are true statements about natural numbers but are not derivable in the theory.

How did Gödel prove this? His first breakthrough was the idea to *transform metamathematical statements about the theory* A *into formulas of the very theory* A. In this way, each statement *about* A would become a formula *in* A, and therefore accessible to a formal treatment *within* A. In particular, the metamathematical statement $\mathscr{G}$ saying that "a given formula of A is not provable in A" would be transformed into a formula of A. Gödel's second breakthrough was *the construction of this formula and the use of it in proving its own undecidability*. The main steps of the proof are:

1. *Arithmetization of* A. A *syntactic object* of A is a symbol, a term, a formula, or any finite sequence of formulas (e.g., formal proof).

    First, Gödel showed that with every syntactic object X one can associate a precisely defined natural number $\gamma(X)$ —today called the *Gödel number* of X. Different syntactic objects have different Gödel numbers, but there are natural numbers that are not Gödel numbers. The computation of $\gamma(X)$ is straightforward. Also straightforward is testing to see whether a number is Gödel's and, if so, constructing the syntactic object from it. (See Problems on p. 64.)

    Secondly, Gödel showed that with every *syntactic relation* (defined on the set of all syntactic objects) there is associated a precisely defined *numerical relation* (defined on $\mathbb{N}$). In particular, with the syntactic relation $D(X,Y) \equiv$ "X is a derivation of formula Y" is associated a numerical relation $D \subseteq \mathbb{N}^2$, such that $D(X,Y)$ *iff* $D(\gamma(X),\gamma(Y))$. All this enabled Gödel to describe A only with natural numbers and numerical relations. We say that he *arithmetized*[4] A.

2. *Arithmetization of Metatheory* $\overline{A}$. Gödel then arithmetized the metatheory $\overline{A}$, i.e., the theory about A. A metatheoretical proposition in $\overline{A}$ is a statement $\mathscr{F}$ which (in natural language and using special symbols like $\vdash$ and $\models$) states something about the syntactic objects and syntactic relations of A. Since Gödel was already able to substitute these with natural numbers and numerical relations, he could translate $\mathscr{F}$ into a statement referring only to natural numbers and numerical relations. But notice that such a statement belongs to the theory A and is, therefore, representable by a formula of A! We see that Gödel was now able to transform metatheoretical statements of $\overline{A}$ into formulas of A.

---

[3] Paul Cohen, 1934–2007, American mathematician.

[4] Recall (p. 6) that Leibniz had a similar idea, though that he aimed to arithmetize man's reflection.

3. *Gödel's Formula*. Now Gödel could do the following: 1) he could transform any metamathematical statement $\mathscr{F}$ about formulas of **A** into a formula F about natural numbers; 2) since natural numbers can represent syntactic objects, he could interpret F as a statement about syntactic objects; and 3) since formulas themselves are syntactic objects, he could make it so that *some formula is a statement about itself*. How did he do that?

Let $\mathscr{G}(\text{H})$ be a metamathematical statement defined by $\mathscr{G}(\text{H}) \equiv$ "Formula H is not provable in **A**." To $\mathscr{G}(\text{H})$ corresponds a formula in **A**; let us denote it by $\text{G}(h)$, where $h = \gamma(\text{H})$. The formula $\text{G}(h)$ states the same as $\mathscr{G}(\text{H})$, but in number-theoretic vocabulary. Specifically, $\text{G}(h)$ states: "The formula with Gödel number h is not provable in **A**."

Now, also $\text{G}(h)$ is a formula, so it has a Gödel number, say g. What happens if we take $h := g$ in $\text{G}(h)$? The result is $\text{G}(g)$, *a formula that asserts about itself that it is not provable in* **A**. Clever, huh? To improve readability, we will from now on write G instead of $\text{G}(g)$.

4. *Incompleteness of* **A**. Then, Gödel proved: G *is provable in* **A** *iff* $\neg$G *is provable in* **A**. (In the proof of this he used so-called ω-consistency; later, Rosser showed that usual consistency suffices.) Now suppose that **A** is consistent. So, G and $\neg$G are not both provable. Then it remains that *neither* is provable. Hence, **A** *is syntactically incomplete*. Because there is no proof of G, we see that what G asserts is in fact *true*. Thus, G is true in the standard model $(\mathbb{N}, =, +, *, 0, 1)$ of **A**, yet it is not provable in **A**. In other words, **A** is *semantically incomplete*.

5. *Incompleteness of Axiomatic Extensions of* **A**. The situation is even worse. Because G represents a $\mathscr{T}$*ruth* about natural numbers, it seems reasonable to admit it to the set of axioms of **A**, hoping that the extended formal axiomatic system will result in a better theory $\mathbf{A}^{(1)}$. Indeed, $\mathbf{A}^{(1)}$ *is* consistent (assuming **A** is). But again, there is a formula $\text{G}^{(1)}$ of $\mathbf{A}^{(1)}$ (not equivalent to G) which is independent of $\mathbf{A}^{(1)}$. So, $\mathbf{A}^{(1)}$ is syntactically incomplete. What is more, $\text{G}^{(1)}$ is true in the standard model $(\mathbb{N}, =, +, *, 0, 1)$. Hence, $\mathbf{A}^{(1)}$ is semantically incomplete.

If we insist and add $\text{G}^{(1)}$ to the axioms of $\mathbf{A}^{(1)}$, we get a consistent yet semantically incomplete theory $\mathbf{A}^{(2)}$ containing an independent formula $\text{G}^{(2)}$ (not equivalent to any of $\text{G}, \text{G}^{(1)}$) which is true in $(\mathbb{N}, =, +, *, 0, 1)$. Gödel proved that we can continue in this way indefinitely, but each extension $\mathbf{A}^{(i)}$ will yield a consistent and semantically incomplete theory (because of some formula $\text{G}^{(i)}$, which is not equivalent to any of the formulas $\text{G}, \text{G}^{(1)}, \ldots, \text{G}^{(i-1)}$). $\qquad\square$

4.2.5 Consistency of M: Gödel's Second Incompleteness Theorem

What about the consistency of the would-be theory **M** (goal C)? Hilbert believed that it would suffice to prove the consistency of *Formal Arithmetic* **A** only. Then, the consistency of other formalized fields of mathematics (due to their construction from **A**) and, finally, the consistency of all formalized mathematics **M** would follow. Thus, the proof of the consistency of **M** would be *relative* to **A**. But this also means that, eventually, the consistency of **A** should be proved with *its own means* and *within* **A** *alone*. In other words, the proof should be constructed without the use of other fields of **M**—except for the *First-Order Logic* **L**—because, at that time, their consistency (being relative) would not be established beyond any doubt. *Formal Arithmetic* **A** should demonstrate its own consistency! We say that the proof of the consistency of **A** should be *absolute*. A method that tried to prove the consistency of **A** is described in Box 4.5.

Box 4.5 (Absolute Proof of Consistency).

We have seen in Box 4.1 (p. 50) that for any first-order theory **F** the following holds: *If there is a formula* F *such that both* F *and* ¬F *are provable in* **F***, then* arbitrary *formula* A *of* **F** *is provable in* **F**. In an inconsistent system everything is provable. Now we see: the consistency of **F** would be proved if we found a formula B of **F** that is *not* provable in **F**.

The question now is, how do we find such a formula B? We can use the following method. Let *P* be any property of the formulas of **F** such that 1) *P* is shared by all the axioms of **F**, and 2) *P* is preserved by the rules of inference of **F** (i.e., if *P* holds for the premises of a rule, it does so for the conclusion). Obviously, theorems of **F** have the property *P*. Now, if we find in **F** a formula B that *does not* have the property *P*, then B is not a theorem of **F** and, consequently, **F** is consistent. Obviously, we must construct such a property *P* that will facilitate the search for B.

Using this method the consistency of *Propositional Calculus* **P** was proved as well as the consistency of *Presburger Arithmetic* (i.e., arithmetic where addition is the only operation).

But in 1931 Gödel also buried hopes that an absolute proof of the consistency of **A** would be found. He proved:

Theorem 4.2. (Second Incompleteness Theorem) *If the Formal Arithmetic* **A** *is consistent, then this cannot be proved in* **A**.

The proof of the theorem is described in Box 4.6.

In other words, **A** cannot demonstrate its own consistency.

Box 4.6 (Proof of the *Second Incompleteness Theorem*).

The theorem says: *If* **A** *is consistent, then we cannot prove this using only the means of* **A**.

In proving this theorem Gödel used parts of the proof of his first theorem. Let 𝒞 be the following metamathematical statement: 𝒞 ≡ "**A** is consistent." This statement too is associated with a formula of **A**—denote it by C—which says the same thing as 𝒞, but by using number-theoretic vocabulary. Gödel then proved that the formula C ⇒ G is provable in **A**. Now, if C were provable in **A**, then (by *Modus Ponens*) also G would be provable in **A**. But in his *First Incompleteness Theorem* Gödel proved that G is *not* provable in **A** (assuming **A** is consistent). Hence, also C is not provable in **A** (if **A** is consistent).   □

4.2.6 Consequences of the Second Incompleteness Theorem

Gödel's discovery revealed that proving the consistency of *Formal Arithmetic* **A** would require means that are more complex—and therefore *less transparent*—than those available in **A**. Of course, a less transparent object or tool may also be more disputable and more controversial, at least in view of Hilbert's finitist recommendations.

In any case, in 1936, Gentzen[5] proved the consistency of **A** by using *transfinite induction* in addition to usual finitist tools. (See the description of transfinite induction in Box 4.7.) Following Gentzen, several other non-finitist consistency proofs of **A** were found. Finally, the belief was accepted that arithmetic **A** is in fact consistent, and that Hilbert's finitist methods may sometimes be too strict.

Did these non-finitist consistency proofs of **A** enable researchers to prove (relative to **A**, as expected by Hilbert) the consistency of other formalized fields of mathematics and, ultimately, of all mathematics **M**? Unfortunately, no. Namely, there is a generalization of the *Second Incompleteness Theorem* stating: *If a consistent theory* **F** *contains* **A**, *then the consistency of* **F** *cannot be proved within* **F**. Of course, this also holds when **F** := **M**, the would-be f.a.s. for all mathematics. This was the second heavy blow to *Hilbert's Program*.

To prove the consistency of all mathematics, one is forced to use external means (non-finitist, metamathematical, or others yet to be discovered), which may be disputable in view of the finitist philosophy of mathematics. But fortunately, the *Second Incompleteness Theorem* does not imply that the formally developed mathematics **M** would be *in*consistent. It only tells us that the chances of proving the consistency of such a mathematics in *Hilbert's way* are null.[6]

Box 4.7 (Transfinite Induction).

This is a method of proving introduced by Cantor. Let us first recall mathematical induction. If $a_0, a_1, \ldots$ is a sequence of objects (e.g., real numbers) and P is a property sensible of objects a_i, then to prove that *every* element of the sequence has this property, we use mathematical induction as follows: 1) we must prove that $P(a_0)$ holds, and 2) we must prove that $P(a_n) \Rightarrow P(a_{n+1})$ holds for an arbitrary natural number n. In other words, if P holds for an element a_n, then it holds for its *immediate successor* a_{n+1}.

To make the description of transfinite induction more intuitive, let us take a sequence $a_0, a_1, \ldots$, where a_i are real numbers and there is no index after which all the elements are equal. Suppose that the sequence converges and a^* is the limit. (If we take, for example, $a_n = \frac{n}{n+1}$, we have $a^* = 1$.) The limit a^* is not a member of the sequence, because $a^* \neq a_n$ for every *natural* number n. But we can consider a^* to be an *infinite (in order)* element of the sequence, that is, the element that comes

---

[5] Gerhard Karl Erich Gentzen, 1909–1945, German mathematician and logician.

[6] Even if the mathematics is inconsistent, there are attempts to overcome this. A recent approach to accommodate inconsistency of a theory in a sensible manner is *paraconsistent logic*. The approach challenges the classical result from Box 4.1 (p. 50) that from contradictory premises *anything* can be inferred. "Mathematics is not the same as its foundations," advocate the researchers of paraconsistency, "so contradictions may not necessarily affect all the 'practical' mathematics." They have shown that in certain theories, called *paraconsistent*, contradictions may be allowed to arise, but they need not infect the whole theory. In particular, a paraconsistent axiomatic set theory has been developed that includes cardinals and ordinals and is capable of supporting the core of mathematics. Further developments of different fields of mathematics, including arithmetic, in paraconsistent logics are well underway.

after every element a_n, where $n \in \mathbb{N}$. Since ω is the smallest ordinal that is larger than any natural number (see p. 17), we can write $a^* = a_\omega$. The sequence can now be extended by a_ω and denoted

$$a_0, a_1, \ldots ; a_\omega \ .$$

Now, what if we wanted to prove that the property P holds for every element of this *extended* sequence? It is obvious that mathematical induction cannot possibly succeed, because it does not allow us to infer $P(a_\omega)$. The reason is that a_ω is not an immediate successor of any a_n, $n \in \mathbb{N}$, so we cannot prove $P(a_n) \Rightarrow P(a_\omega)$ for any natural n.

An ordinal that is neither 0 nor the immediate successor of another ordinal is called the *limit ordinal*. There are infinitely many limit ordinals, with ω being the smallest of them. Mathematical induction fails at each limit ordinal. The *transfinite induction* remedies that.

Principle of Transfinite Induction: Let $(\mathcal{S}, \preccurlyeq)$ be a well-ordered set and P a property sensible for its elements. Then P holds for *every* element of $\mathcal{S}$ if the following condition is met:

• P holds for $y \in \mathcal{S}$ if P holds for *every* $x \in \mathcal{S}$ such that $x \prec y$.

Transfinite induction is a generalization of mathematical induction. It can be used to prove that a given property P holds for all ordinals (or all elements of a well-ordered set; see Appendix A). Normally it is used as follows:

1. *Suppose* that P does not hold for all ordinals.
2. Therefore, there is the smallest ordinal, say α, for which we have $\neg P(\alpha)$.
3. Then we try to deduce a contradiction.
4. If we succeed, we conclude: P holds for every ordinal.

4.3 Legacy of Hilbert's Program

The ideas of Whitehead and Russell put forward in their *Principia Mathematica* (Sect. 2.2.3) proved to be unrealistic. Mathematics cannot be founded on logic only.

Also, *Hilbert's Program* (Sect. 4.1.2) failed. The mechanical, syntax-directed development of mathematics within the framework of formal axiomatic systems may be safe from paradoxes, yet this safety does not come for free. The mathematics developed in this way suffers from semantic incompleteness and the lack of a possibility of proving its consistency. All this makes Hilbert's ultimate idea of the mechanical development of mathematics questionable.

Aspiration and Inspiration

Consequently, it seems that the research in mathematics cannot avoid human inspiration, ingenuity and intuition (albeit deceptive). See Fig. 4.10. A fortiori, Leibniz's idea (see p. 6) of replacing human reflection by mechanical and mechanized arithmetic is just an illusion. Mathematics and other axiomatic sciences selfishly guard their *Truths*; they admit to these *Truths* only humans who, in addition to demonstrating a strong aspiration for knowledge, demonstrate sufficient inspiration and ingenuity.

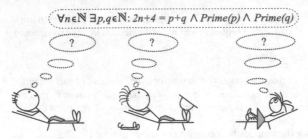

$$\forall n \in \mathbb{N} \ \exists p, q \in \mathbb{N}: \ 2n+4 = p+q \ \wedge \ Prime(p) \ \wedge \ Prime(q)$$

Fig. 4.10 Research cannot avoid inspiration, ingenuity and intuition

4.4 Chapter Summary

Hilbert proposed a promising recovery program to put mathematics on a sound footing and eliminate all the paradoxes. To do this, the program would use formal axiomatic systems and their theories. More specifically, Hilbert aimed to define a formal axiomatic system **M** such that the theory **M** developed in it would contain the whole of mathematics. The theory **M** would also comply with several fundamental requirements: it would be consistent, semantically complete, and decidable. In addition, Hilbert required that a decision procedure for **M** should be devised, i.e., an *algorithm* should be constructed capable of deciding, for any formula of **M**, whether the formula represents a mathematical *Truth*.

It was soon realized that **M** must contain at least *First-Order Logic* **L** and *Formal Arithmetic* **A**. This, however, enabled Gödel to discover that there can be no such **M**! Although Gödel's discovery shattered *Hilbert's Program*, *the problem of finding an algorithm that is a decision procedure for a given theory remained topical.*

Problems

Definition 4.1. (Gödel numbering) The arithmetization of the *Formal Arithmetic* **A** associates each syntactic object X of **A** with the corresponding **Gödel number** $\gamma(\text{X}) \in \mathbb{N}$. The function γ can be defined in the following way.

First, we associate with Gödel numbers the following symbols: logical connectives, "¬",1; "∨", 2; "⇒",3; the quantification symbol, "∀",4; the equality symbol, "=",5; the individual-constant symbol, "0",6; the unary function symbol for the successor function, "′",7; punctuation marks, "(",8; ")",9; ",",10.

Second, individual-variable symbols are associated with prime numbers greater than 10 in order of increasing magnitude: $x,11$; $y,13$; $z,17$; and so on.

Third, predicate symbols are associated with squares of prime numbers greater than 10: $\text{P},11^2$; $\text{Q},13^2$; $\text{R},17^2$; and so on. ($\text{P},\text{Q},\text{R}$ etc. represent relations, e.g., $<, \leqslant, >$, etc., respectively.)

Fourth, a formula F is a sequence of symbols, $\text{F} = \text{s}_1\text{s}_2\ldots\text{s}_k$, and its Gödel number $\gamma(\text{F})$ is the product $p_1^{\gamma(\text{s}_1)} p_2^{\gamma(\text{s}_2)} \ldots p_k^{\gamma(\text{s}_k)}$, where p_i is the ith prime number. For example, the axiom $\forall x \forall y (x = y \Rightarrow x' = y')$ has Gödel number $2^4 3^{11} 5^4 7^{13} 11^8 13^{11} 17^5 19^{13} 23^3 29^{11} 31^7 37^5 41^{13} 43^7 47^9$.

Finally, a proof is a sequence of formulas $\text{F}_1, \text{F}_2, \ldots, \text{F}_n$; its Gödel number is $p_1^{\gamma(\text{F}_1)} p_2^{\gamma(\text{F}_2)} \ldots p_n^{\gamma(\text{F}_n)}$.

Remark. Gödel's original arithmetization was more succinct: "0",1; "′",3; "¬",5; "∨",7; "∀",9; "(",11; ")",13. The symbols such as ∧, ⇒, =, ∃ are only abbreviations and can be represented by the previous ones. Individual-variable symbols were associated with prime numbers greater than 13, and predicate symbols with squares of these prime numbers. Gödel also showed that, for $k \geqslant 2$, k-ary function and predicate symbols can be represented by the previous symbols.

4.1. Prove: The function $\gamma : \Sigma^* \to \mathbb{N}$ is injective. (*Remark.* Σ denotes the set of symbols of **A**.)
[*Hint.* Use the *Fundamental Theorem of Arithmetic*.]

4.2. Let $n \in \mathbb{N}$. Describe how we can decide whether or not there exists an $F \in \mathbf{A}$ such that $n = \gamma(F)$.

4.3. Let $n = \gamma(F)$ for some $F \in \mathbf{A}$. Describe how we can reconstruct $F = \gamma^{-1}(n)$.

4.4. Informally describe how we can decide, for any sequence $F_1, F_2, \ldots, F_n$ of formulas of **A**, whether or not the sequence is a derivation of F_n in **A**.
[*Hint.* See Box 4.2 on p. 56.]

Bibliographic Notes

The following excellent monographs were consulted and are fine sources for the subjects covered in Part I of this book:

- A general treatment of the *axiomatic method* is Tarski [176]. Here, after an introduction to mathematical logic, the use of the axiomatic method in the development of theories is described. Examples of such theories are given (e.g., elementary number theory), and several metatheoretical questions are defined. The axiomatization of geometry is given in Hilbert [66].
- Introductions to *mathematical logic* are Robbin [138], Mendelson [105], and Stoll [174], and recently Rautenberg [134] and Smullyan [162]. In-depth and more demanding are Kleene [83], Shoenfield [150], Manin [99], and recently Ebbinghaus et al. [41]. A different perspective of the *First-Order Logic* is presented in Smullyan [161]. *Second-Order* languages and theories are discussed in Ebbinghaus et al. [41], Boolos and Jeffrey [16], and Boolos et al. [15]. In Slovenian, an in-depth treatise of mathematical logic is Prijatelj [132]. We combined this with Prijatelj [131].
- The relation between mathematics and *metamathematics* is described in depth in Kleene [82].
- Burali-Forti described his paradox in [19]. Russell described his paradox to Frege in a letter [143]. English translations of both can be found in van Heijenoort [63]; here, also other English translations of *seminal publications* between 1879 and 1931 can be found. Explanations of a range of paradoxes can be found in Sainsbury [147].
- Initial *intuitionistic ideas* about the development of mathematics can be found in Brouwer [18]. See also Kleene [82]. A more demanding and in-depth description of intuitionism is Heyting [64]. But see also a historical account of the development of intuitionistic reasoning in Moschovakis [107].
- The main ideas of *logicism* were published in Boole [14], Frege [47], Peano [117], and Whitehead and Russell [187]. Russell's achievements and impact on modern science and philosophy are described in Irvine [73].
- *Formalism* and Hilbert's finitist program were finalized and stated in Hilbert and Ackermann [67]. Here, first-order formal axiomatic systems were introduced. The need for an absolute proof of the consistency of *Formal Arithmetic* was explained in Hilbert [65]. A description of the basic ideas of formalism is in Kleene [82]. For a historical account of the development of Hilbert's view of mathematical proof, his achievements, and their impact, see Sieg [154, 155].
- An in-depth discussion of *second-order* formal axiomatic systems and their theories can be found in Ebbinghaus et al. [41]. Logicism, intuitionism, and formalism are extensively described in George and Velleman [50].

- Basics of *model theory* can be found in all of the above monographs on mathematical logic. An introductory but concise account of model theory is Tent and Ziegler [177]. For an in-depth treatment of models of first-order theories, see Chang and Keisler [22].
- The truth-table method and the proofs that *Propositional Calculus* **P** is a consistent, decidable, and semantically complete theory appeared in Post [120]. See also Tarski [176]. The semantic completeness of *First-Order Logic* **L** was proved by Gödel [52].
- Introductions to *axiomatic set theory* ZFC are Halmos [61], Suppes [175], and Goldrei [56]. For an advanced-level treatment of ZFC, see Levy [95], Winfried and Weese [188], and recently Kunen [88] and Jech [76]. About the axiomatic set theory NBG, see Mendelson [105]. An in-depth historical account of alternative set theories such as NBG is Holmes et al. [70]. The rich history of set theory from Cantor to Cohen is described in Kanamori [78]. That *Continuum Hypothesis* is independent of ZFC is shown in Cohen [26], and that it is independent of NBG is shown in Smullyan and Fitting [163]. A historical account of the problems of continuum is Steprāns [171]. For an in-depth explanation of the methods used to construct independence proofs in ZFC, see Kunen [87]. The *Axiom of Choice* is extensively analyzed and discussed in Jech [75] and Moore [106].
- The roots of Hilbert's foundational work are systematically traced to the radical transformation of mathematics in Sieg [155].
- Gödel's *Incompleteness Theorems* appeared in [53]. For in-depth yet user-friendly explanations of both theorems, see Nagel and Newman [112], Smullyan [160], Gödel [55] and Smith [159]. An in-depth discussion of Gödel's *Incompleteness Theorems* as well as his *Completeness Theorem* from the historical perspective is van Atten and Kennedy [182]. Gödel's impact on modern mathematics and logic is presented in Baaz et al. [9].
- Kleene [82] gives a brief account of the use of *transfinite induction* in Gentzen's proof of the consistency of *Formal Arithmetic*.
- The ideas of *paraconsistent logic* are described in Priest [130].
- Many concepts of modern mathematics that we use throughout this monograph are nicely discussed in Stewart [172]. Mathematical logic until 1977 is extensively described in Barwise [11]. For an extensive and up-to-date treatment of set theory, see Jech [76]. The *philosophical aspects* of mathematical logic, set theory, and mathematics in general can be found in Benacerraf and Putnam [12], Machover [97], George and Velleman [50], and Potter [128]. Two expositions of the interplay between mathematical logic and computability are Epstein and Cornielli [43] and Chaitin [21].

Part II
CLASSICAL COMPUTABILITY THEORY

Our intuitive understanding of the concept of the algorithm, which perfectly sufficed for millennia, proved to be insufficient as soon as the non-existence of a certain algorithm had to be proven. This triggered the search for a model of computation, a formal characterization of the concept of the algorithm. In this part we will describe different competing models of computation. The models are equivalent, so we will adopt the Turing machine as the most appropriate one. We will then describe the Turing machine in greater detail. The existence of the universal Turing machine will be proven and its impact on the creation and development of the general-purpose computer will be explained. Then, several basic yet crucial theorems of *Computability Theory* will be deduced. Finally, the existence of incomputable problems will be proven, a list of such problems from practice will be given, and several methods for proving the incomputability of problems will be explained.

Chapter 5
The Quest for a Formalization

A model of a system or process is a theoretical description that can help you understand how the system or process works.

Abstract The difficulties that arose at the beginning of the twentieth century shook the foundations of mathematics and led to several fundamental questions: "What is an algorithm? What is computation? What does it mean when we say that a function or problem is computable?" Because of *Hilbert's Program*, intuitive answers to these questions no longer sufficed. As a result, a search for appropriate definitions of these fundamental concepts followed. In the 1930s it was discovered—miraculously, as Gödel put it—that all these notions can be formalized, i.e, mathematically defined; indeed, they were formalized in several completely different yet equivalent ways. After this, they finally became amenable to mathematical analysis and could be rigorously treated and used. This opened the door to the seminal results of the 1930s that marked the beginning of *Computability Theory*.

5.1 What Is an Algorithm and What Do We Mean by Computation?

We have seen (Sect. 4.2.2) that *if* **M** is consistent and syntactically complete *then* there exists a decision procedure for **M**. But the hopes for a complete and unquestionably consistent **M** were shattered by Gödel's Theorems. Does this necessarily mean that there cannot exist a decision procedure for **M**? Did this put an end to the research on the *Entscheidungsproblem*? In this case, no. Namely, the proofs of Gödel's Theorems only used logical notions and methods; they did not involve loose notions of algorithm and computation. Specifically, the *First Incompleteness Theorem* revealed that there are mathematical 𝒯*ruths* that cannot be *derived* in **M**. But it was not obvious that there could be no other way of *recognizing* every mathematical 𝒯*ruth*. (After all, Gödel himself was able to determine that his formulas $G, G^{(1)}, \ldots$ are true although undecidable.) Because of this, the *Decidability Problem* for **M** and, in particular, the *Entscheidungsproblem* with its quest for a decision procedure (algorithm) D_{Entsch}, kept researchers' interest. Yet, the *Entscheidungsproblem* problem proved to be much harder than expected (see p. 55). What is an algorithm, anyway?

© Springer-Verlag Berlin Heidelberg 2015
B. Robič, *The Foundations of Computability Theory*,
DOI 10.1007/978-3-662-44808-3_5

It became clear that the problem could not be solved unless the intuitive, loose definition of the concept of the algorithm was replaced by a formal, rigorous definition.

5.1.1 Intuition and Dilemmas

So, let us return to the intuitive definition of the algorithm (Definition 1.1 on p. 4), which was at Hilbert's disposal: An algorithm for solving a given problem is a recipe consisting of a finite number of instructions which, if strictly followed, leads to the solution of the problem. When Hilbert set the goal *"Find an algorithm which is a decision procedure for* **M**," he, in essence, asked us to *conceive* an appropriate recipe (functioning, at least in principle, as a decision procedure for **M**) by using only common sense, logical inference, knowledge, experience, and intuition (subject only to finitist restrictions). In addition, the recipe was to come with an idea of how it would be executed (at least in principle). We see that the notions of the algorithm and its execution were entirely intuitive.

But there were many questions about such an understanding of the algorithm and its execution. What would be the *kind* of basic instructions used to compose algorithms? In particular, would they execute in a discrete or a continuous way? Would their execution and results be predictable (i.e., deterministic) or probabilistic (i.e., dependent on random events)? These questions were relevant in view of the discoveries being made in physics at the time.[1]

Which instructions should be *basic*? Should there be only *finitely* many of them? Would they suffice for composing *any* algorithm of interest? If there were *infinitely* many basic instructions, would that not demand a processor of unlimited capability? Would that be realistic? But if the processor were of limited capability, could it be *universal*, i.e., capable of executing *any* algorithm of interest?[2]

Then there were more down-to-earth questions. How should the *processor* be constructed in order to execute the algorithms? Where would the processor keep the algorithms and where would the input data be? Should data be of arbitrary or limited size? Should storage be limited or unlimited? Where and how would ba-

---

[1] *Is nature discrete or continuous?* At the beginning of the twentieth century it was discovered that energy exchange in nature seems to be *continuous* at the macroscopic level, but is *discrete* at the microscopic level. Specifically, energy exchange between matter and waves of frequency v is only possible in discrete portions (called quanta) of sizes nhv, $n = 1, 2, 3 \ldots$, where h is Planck's constant. Notice that some energy must be consumed during the instruction execution.

Is nature predictable or random? Nature at the macroscopic level (i.e., nature dealt with by classical physics) is predictable. That is, each event has a cause, and when an event seems to be random, it is only because we lack knowledge of its causes. Such randomness of nature is said to be *subjective*. In contrast, there is *objective* randomness in microscopic nature (i.e., nature dealt with by quantum physics). Here, an event may be entirely unpredictable, having no cause until it happens. Only a probability of occurrence can be associated with the event.

So, how do all these quantum phenomena impact instruction execution? (Only recently have quantum algorithms appeared; they strive to use these phenomena in problem solving.)

[2] Recall that such a universality of a processor was Babbage's goal nearly a century ago (p. 6).

sic instructions execute? Where would the processor keep the intermediate and final results? In addition, there were questions about the ways of dealing with the *algorithm–processor* pairs. For example, should it be possible to encode these pairs with natural numbers in order to enable their rigorous, or even metatheoretical, treatment?[3] Should the execution time (e.g., number of steps) of the algorithm be easily derivable from the description of the algorithm, the input data, and the processor?

5.1.2 The Need for Formalization

The "big" problem was this: How does one answer the question "Is there an algorithm that solves a given problem?" when it is not clear *what* the algorithm is?

Fig. 5.1 To decide whether something exists we must first understand what it should be

To prove that there *is* an algorithm that solves the problem, it sufficed to construct *some* candidate recipe and show that the recipe meets all the conditions (i.e., the recipe has a finite number of instructions, which are reasonably difficult, that can be mechanically followed and executed by any processor, be it human or machine, leading it, in finite time, to the solution of the problem). The loose, intuitive understanding of the concept of the algorithm was no obstacle for such a constructive existence proof.

In contrast, proving that there is *no* algorithm for the problem was a much bigger challenge. A non-existence proof should reject *every* possible recipe by showing that it does not meet all the conditions necessary for an algorithm to solve the problem. However, to accomplish such a proof, a *characterization* of the concept of the algorithm was needed. In other words, a property had to be found such that *all algorithms* and *algorithms only* have this property. Such a property would then be characteristic of algorithms. In addition, a precise and rigorous definition of the processor, i.e., the environment capable of executing algorithms, had to be found. Only then would the necessary condition for proving the non-existence of an algorithm be fulfilled. Namely, having the concept of the algorithm characterized, one would be in a position to systematically (i.e., with mathematical methods) eliminate all the

---

[3] This idea was inspired by *Gödel numbers*, introduced in his *First Incompleteness Theorem*.

infinitely many possible recipes by showing that none of them could possibly fulfill the conditions necessary for an algorithm to solve the problem.

A definition that formally describes and characterizes the basic notions of algorithmic computation (i.e., the algorithm and its environment) is called the **model of computation**.

Remark. From now on and until stated otherwise, we will use quotation marks to refer to the key notions as they were understood and used in the research of the time, i.e., intuitively. Thus, "algorithm," "computation," and "computable." For example: In 1928, Hilbert asked to construct an "algorithm" D_{Entsch} that would answer the question $\underset{M}{\vdash}?F$ for the arbitrary formula $F \in \mathbf{M}$.

5.2 Models of Computation

In this section we will describe the search for an appropriate model of computation. The search started in 1930. The goal was to find a model of computation that would characterize the notions of "algorithm" and "computation." Different ideas arose from the following question:

What could a model of computation take as an example?

On the one hand, it was obvious that man is capable of complex "algorithmic computation," yet there was scarcely any idea how he does this. On the other hand, while the operation of mechanical machines of the time was well understood, it was far from complex, human-like "algorithmic computation."

As a result, three attempts were made: modelling the computation after *functions*, after *humans*, and after *languages*. Each direction proposed important models of computation. In this section we will describe them in detail.

5.2.1 Modelling After Functions

The first direction focused on the question

What does it mean when we say that we "compute" the value of a function $f : A \to B$, or when we say that the function is "computable"?

To get to an answer, it was useful to confine the discussion to functions that were as simple as possible. It seemed that such functions were the *total numerical functions* $f : \mathbb{N}^k \to \mathbb{N}$, where $k \geqslant 1$. If f is such a function, then for *any* k-tuple $(x_1, \ldots, x_k)$ of natural numbers, there is a unique natural number called the value of f at $(x_1, \ldots, x_k)$ and denoted by $f(x_1, \ldots, x_k)$. (Later, we will see that the requirement for the totality of functions had to be omitted.)

After this restriction, the search for a definition of "computable" total numerical functions began. It was obvious that any such definition should fulfill two requirements:

1. *Completeness Requirement:* the definition should include *all* the "computable" total numerical functions, *and nothing else.*
2. *Effectiveness Requirement:* the definition should make evident, for each such function f, an *effective procedure* for computing the value $f(x_1,\ldots,x_k)$. Here, an **effective procedure** is defined to be any finite set of instructions written in any language that

   a. completes in a *finite* number of steps;
   b. returns *some* answer, that is, some natural number;
   c. returns the *right* answer, that is, the value $f(x_1,\ldots,x_k)$; and
   d. it does so for *all* instances of the problem, that is, for any $(x_1,\ldots,x_k) \in \mathbb{N}^k$.

The first requirement asked for a characterization of the "computable" total numerical functions. Only if the second requirement was fulfilled would the defined functions be considered algorithmically computable. Notice that the notion of the effective procedure is a refinement of the intuitive notion of the algorithm (Definition 1.1 on p. 4) but is still an intuitive, informal notion. Of course, an algorithm for f would be an effective procedure disclosed by f's definition.

Important definitions in this direction were proposed by Gödel and Kleene (recursive functions), Herbrand and Gödel (general recursive functions), and Church (λ-calculus).

Recursive Functions

In the proof of his Second Theorem, Gödel introduced numerical functions, the construction of which resembled the derivations of theorems in formal axiomatic systems and their theories. More precisely, Gödel fixed three simple *initial functions*, $\zeta : \mathbb{N} \to \mathbb{N}$, $\sigma : \mathbb{N} \to \mathbb{N}$, and $\pi_i^k : \mathbb{N}^k \to \mathbb{N}$ (called *zero*, *successor*, and *projection* function, respectively), and two *rules of construction* (called *composition* and *primitive recursion*) for constructing new functions from the initial and previously constructed ones.[4] (There are more details in Box 5.1.)

The functions constructed from ζ, σ, and π by finitely many applications of composition and primitive recursion are said to be *primitive recursive*. Although Gödel's primary intention was to use them in proving his *Second Incompleteness Theorem*, they displayed a property much desired at that time. Namely, the construction of a primitive recursive function is also an effective procedure for computing its values. So, a construction of such a function seemed to be the formal definition of the "algorithm," and Gödel's definition of primitive recursive functions seemed to be the wished-for definition of the "computable" total numerical functions.

However, Ackermann and others found the total numerical functions (called the Ackermann functions; see Problem 5.5 on p. 99) which were "computable" beyond any doubt, and yet they were not primitive recursive. Thus, Gödel's definition did not meet the *Completeness Requirement.*

---

[4] The resemblance between function construction and theorem derivation is obvious: initial functions correspond to axioms, and rules of construction correspond to rules of inference.

Fig. 5.2 Kurt Gödel
(Courtesy: See Preface)

Fig. 5.3 Stephen Kleene
(Courtesy: See Preface)

This deficiency of Gödel's definition was eliminated in 1936 by Kleene.[5] He added to Gödel's definition a third rule of construction, called the μ-*operation*. (See Box 5.1.) Kleene assumed that the μ-operation would be applied to construct exclusively *total* functions, although it could also return *partial* functions (i.e., $f : \mathbb{N}^k \to \mathbb{N}$ that may be undefined for some k-tuples in $\mathbb{N}^k$). The total numerical functions that can be constructed from ζ, σ, and π by finitely many applications of composition, primitive recursion, *and* μ-operation are said to be *recursive*.

The class of recursive functions proved to contain any conceivable total numerical function. So, Gödel-Kleene's definition became a *plausible* formal definition (i.e., formalization) of a "computable" total numerical function. Consequently, a construction of a recursive function became a *plausible* formalization of the notion of the "algorithm." All this was gathered in the following model of computation.

Model of Computation (Gödel-Kleene's Characterization):

- *An "algorithm"* is a construction of a recursive function.
- *A "computation"* is a calculation of a value of a recursive function that proceeds according to the construction of the function.
- *A "computable" function* is a recursive function.

Box 5.1 (Recursive Functions).

Informally, a function is said to be *recursive* if it is either an initial function, or it has been constructed from initial or previously constructed functions by a finite application of three rules of construction. *Remark*: For brevity, we will write in this box $\vec{n}$ instead of $n_1, \ldots, n_k$.
The *initial functions* are:

a. $\zeta(n) = 0$, for every natural n (*Zero* function)
b. $\sigma(n) = n + 1$, for every natural n (*Successor* function)
c. $\pi_i^k(\vec{n}) = n_i$, for arbitrary $\vec{n}$ and $1 \leqslant i \leqslant k$ (*Projection* function)

---

[5] Stephen Cole Kleene, 1909–1994, American mathematician.

The *rules of construction* are:

1. *Composition.* Let the given functions be $g : \mathbb{N}^m \to \mathbb{N}$ and $h_i : \mathbb{N}^k \to \mathbb{N}$, for $i = 1, \ldots, m$. Then the function $f : \mathbb{N}^k \to \mathbb{N}$ defined by

$$f(\overrightarrow{n}) \stackrel{\text{def}}{=} g(h_1(\overrightarrow{n}), \ldots, h_m(\overrightarrow{n}))$$

is said to be constructed by *composition* of the functions g and h_i, $i = 1, \ldots, m$.

2. *Primitive Recursion.* Let the given functions be $g : \mathbb{N}^k \to \mathbb{N}$ and $h : \mathbb{N}^{k+2} \to \mathbb{N}$. Then the function $f : \mathbb{N}^{k+1} \to \mathbb{N}$ defined by

$$f(\overrightarrow{n}, 0) \stackrel{\text{def}}{=} g(\overrightarrow{n})$$
$$f(\overrightarrow{n}, m+1) \stackrel{\text{def}}{=} h(\overrightarrow{n}, m, f(\overrightarrow{n}, m)), \text{ for } m \geqslant 0$$

is said to be constructed by *primitive recursion* from the functions g and h.

3. μ-*Operation.* Let the given function be $g : \mathbb{N}^{k+1} \to \mathbb{N}$. Then the function $f : \mathbb{N}^k \to \mathbb{N}$ defined by

$$f(\overrightarrow{n}) \stackrel{\text{def}}{=} \mu x g(\overrightarrow{n}, x)$$

is said to be constructed by the μ-*operation* from the function g. Here, the μ-operation is defined as follows: $\mu x g(\overrightarrow{n}, x) \stackrel{\text{def}}{=}$ least $x \in \mathbb{N}$, such that $g(\overrightarrow{n}, x) = 0 \wedge g(\overrightarrow{n}, z)$ is defined for $z = 0, \ldots, x$. (Note that $\mu x g(\overrightarrow{n}, x)$ may be undefined, e.g., when g is such that $g(\overrightarrow{n}, x) \neq 0$ for every $x \in \mathbb{N}$.)

The *construction* of a recursive function f is a finite sequence $f_1, f_2, \ldots, f_\ell$, where $f_\ell = f$ and each f_i is either one of the initial functions ζ, σ, π, or is constructed by one of the rules 1,2,3 from its predecessors in the sequence. Taken formally, the construction is a finite sequence of symbols. But there is also a practical side of construction: After fixing the values of the input data, we can mechanically and effectively compute the value of f simply by following its construction and calculating the values of the intermediate functions. A construction is an algorithm.

Example 5.1. (Addition) Let us construct the function $\text{sum}(n_1, n_2) \stackrel{\text{def}}{=} n_1 + n_2$. We apply the following idea: to compute $\text{sum}(n_1, n_2)$ we first compute $\text{sum}(n_1, n_2 - 1)$ and then its successor. The computation involves primitive recursion, which terminates when $\text{sum}(n_1, 0)$ should be computed. In the latter case, the sum is just the first summand, n_1. In the list below, the left-hand column contains initial and constructed functions needed to implement the idea, and the right-hand column explains, for each function, why it is there or how it was constructed. The function sum is f_5.

1. $\pi_1^1(x)$ to extract its argument and make it available for use
2. $\pi_3^3(x, y, z)$ to introduce the third variable, which will eventually be the result $x + y$
3. $\sigma(x)$ to increment its argument
4. $f_4(x, y, z)$ to increment the third argument; constructed by composition of 3. and 2.
5. $f_5(x, y)$ to compute $x + y$; constructed by primitive recursion from 1. and 4.

Now, the construction of sum is a sequence of functions and information about the rules applied: $f_1 = \pi_1^1(n_1); f_2 = \pi_3^3(n_1, n_2, n_3); f_3 = \sigma(n_1); f_4(n_1, n_2, n_3) [\text{rule } 1, f_3, f_2]; f_5(n_1, n_2) [\text{rule } 2, f_1, f_4].$ The function f_5 is constructed by primitive recursion from functions 1 and 4, so we have: $f_5(n_1, n_2) \stackrel{5.}{=} f_4(n_1, n_2 - 1, f_5(n_1, n_2 - 1)) \stackrel{4.}{=} f_5(n_1, n_2 - 1) + 1 \stackrel{5.}{=} \ldots \stackrel{4.}{=} f_5(n_1, 0) + n_2 \stackrel{1.}{=} \pi_1^1(n_1) + n_2 = n_1 + n_2.$ Given n_1 and n_2, say $n_1 = 2$ and $n_2 = 3$, we can compute $f(2, 3)$ by following the construction and computing the values of functions: $f_1 = \pi_1^1(n_1) = 2; f_2 = \pi_3^3(n_1, n_2, n_3) = n_3;$ $f_3 = \sigma(n_1) = 3; f_4(n_1, n_2, n_3) = \sigma(\pi_3^3(n_1, n_2, n_3)) = \sigma(n_3) = n_3 + 1; f_5(n_1, n_2) = f_5(2, 3) = f_4(2, 2, f_5(2, 2)) = f_5(2, 2) + 1 = f_4(2, 1, f_5(2, 1)) + 1 = f_5(2, 1) + 2 = f_4(2, 0, f_5(2, 0)) + 2 = f_5(2, 0) + 3 = 2 + 3 = 5.$ □

General Recursive Functions

In 1931, the then 23-year-old Herbrand[6] investigated how to define the total numerical functions $f : \mathbb{N}^k \to \mathbb{N}$ using systems of equations. Before suffering a fatal accident while mountain climbing in the French Alps, he explained his ideas in a letter to Gödel. We cite Herbrand (with the function names changed):

> If f denotes an unknown function and $g_1, \ldots, g_k$ are known functions, and if the g's and f are substituted in one another in the most general fashions and certain pairs of the resulting expressions are equated, then, if the resulting set of functional equations has one and only one solution for f, f is a recursive function.

Fig. 5.4 Jacques Herbrand
(Courtesy: See Preface)

Gödel noticed that Herbrand did not make clear what the rules for computing the values of such an f would be. He also noted that such rules would be the same for all the functions f defined in Herbrand's way. Thus, Gödel improved on Herbrand's idea in two steps. First, he added two conditions to Herbrand's idea:

• A system of equations must be in *standard* form, where f is only allowed to be on the left-hand side of the equations, and it must appear as

$$f(g_i(\ldots), \ldots, g_j(\ldots)) = \ldots$$

• A system of equations must guarantee that f is a *well-defined* function (i.e., has precisely defined values) which is *total* on $\mathbb{N}^k$.

Let us denote by $\mathscr{E}(f)$ a system of equations that fulfills the two conditions and defines a function $f : \mathbb{N}^k \to \mathbb{N}$. Secondly, Gödel started to search for the rules by which $\mathscr{E}(f)$ is used to compute the values of f. In 1934, he realized that there are only two such rules:

1. In an equation, all occurrences of a variable can be *substituted* by the same number (i.e., the value of the variable).
2. In an equation, an occurrence of a function can be *replaced* by its value.

A function $f : \mathbb{N}^k \to \mathbb{N}$ for which there is a system $\mathscr{E}(f)$ is said to be *general recursive*.

---

[6] Jacques Herbrand, 1908–1931, French logician and mathematician.

It seemed that any conceivable "computable" total numerical function could be defined and effectively computed in a mechanical fashion by some system of equations of this kind. The *Completeness Requirement* and *Effectiveness Requirement* of such a definition of "computable" functions seemed to be satisfied. As a result, Herbrand and Godel's ideas were merged in the following model of computation.

Model of Computation (Herbrand-Gödel's Characterization):

- *An "algorithm"* is a system of equations $\mathscr{E}(f)$ for some f.
- A *"computation"* is a calculation of a value of a general recursive function f that proceeds according to $\mathscr{E}(f)$ and the rules 1,2.
- A *"computable" function* is a general recursive function.

λ-calculus

We start with two examples that give us the necessary motivation.

Example 5.2. (Motivation) What is the value of the term $(5-3)*(6+4)$? To get an answer, we first rewrite the term in the prefix form, $*(-(5,3),+(6,4))$, and getting rid of parentheses we obtain $*-5\ 3+6\ 4$. This sequence of symbols *implicitly represents the result* (number 20) *by describing a recipe for calculating it*. Let us call this sequence the *initial term*. Now we can compute the result by a series of *reductions* (elementary transformations) of the initial term: $*-5\ 3+6\ 4 \rightarrow *2+6\ 4 \rightarrow *2\ 10 \rightarrow 20$. For example, in the first reduction we *applied* $-$ to 5 and 3 and then replaced the calculated *subterm* $-5\ 3$ by its value 2. Note that there is a different series of reductions which ends with the same result: $*-5\ 3+6\ 4 \rightarrow *-5\ 3\ 10 \rightarrow *2\ 10 \rightarrow 20$. $\qquad\square$

Example 5.3. (Motivation) Usually, a function $f : \mathbb{N} \to \mathbb{N}$ is defined by the equation $f(x) = [...x...]$, where the right-hand side is an expression containing x. Alternatively, we might define f by the expression $f = \lambda x.[...x...]$, where λx would by convention indicate that x is a variable in $[...x...]$. Then, instead of writing $f(x), x = a$, we could write $\lambda x.[...x...]a$, knowing that a should be substituted for each occurrence of x in $[...x...]$. For example, instead of writing $f(x) = x^y$ and $g(y) = x^y$, which are two different functions, we would write $\lambda x.x^y$ and $\lambda y.x^y$, respectively. Instead of writing $f(x), x=3$ and $g(y), y=5$ we would write $(\lambda x.x^y)3$ and $(\lambda y.x^y)5$, which would result in 3^y and x^5, respectively. Looking now at x^y as a function of two variables, we would indicate this by $\lambda xy.x^y$. Its value at $x=3, y=5$ would then be $((\lambda xy.x^y)3)5 = (\lambda y.3^y)5 = 3^5$. $\qquad\square$

During 1931–1933, Church[7] proposed, based on similar ideas, a model of computation called the λ-*calculus*. We briefly describe it. (See Box 5.2 for the details.)

Let f be a function and $a_1, \ldots, a_n$ its arguments. Each a_i can be a number or another function with its own arguments. Thus, functions can nest within other functions. Church proposed a way of describing f and $a_1, \ldots, a_n$ as a finite sequence of symbols that *implicitly represents the value* $f(a_1, \ldots, a_n)$ *by describing a recipe for calculating it*. He called this sequence the *initial λ-term*. The result $f(a_1, \ldots, a_n)$ is

---

[7] Alonzo Church, 1903–1995, American mathematician and logician.

computed by a systematic transformation of the initial λ-term into a *final* λ-term that *explicitly* represents the value $f(a_1,\ldots,a_n)$. The transformation is a series of elementary transformations, called *reductions*. Informally, a reduction of a λ-term applies one of its functions, say g, to g's arguments, say $b_1,\ldots,b_m$, and replaces the terms representing g and $b_1,\ldots,b_m$ with the term representing $g(b_1,\ldots,b_m)$.

Fig. 5.5 Alonzo Church
(Courtesy: See Preface)

Generally, there are several different transformations of a λ-term. However, Church and Rosser[8] proved that the final λ-term is practically independent of the order in which the reductions are made; specifically, the final λ-term, when it exists, is defined up to the renaming of its variables.

Church called functions that can be defined as λ-terms, *λ-definable*. It seemed that any conceivable "computable" total numerical function was λ-definable and could effectively be computed in a mechanical manner. So this definition of "computable" functions seemed to fulfill the *Completeness* and *Effectiveness Requirements*. For these reasons, Church proposed the following model of computation.

Model of Computation (Church's Characterization):

- *An "algorithm"* is a λ-term.
- *A "computation"* is a transformation of an initial λ-term into the final one.
- *A "computable" function* is a λ-definable function.

Box 5.2 (λ-calculus).

Let $f,g,x,y,z,\ldots$ be variables. Well-formed expressions will be called λ-terms.

A *λ-term* is a well-formed expression defined inductively as follows:

a.   a variable is a λ-term (called *atom*);
b.   if M is a λ-term and x a variable, then $(\lambda x.M)$ is a λ-term (built from M by *abstraction*);
c.   if M and N are λ-terms, then (MN) is a λ-term (called the *application of M to N*).

---

[8] John Barkley Rosser, 1907–1989, American mathematician and logician.

Remarks. Informally, abstraction exposes a variable, say x, that occurs (or even doesn't occur) in a λ-term M and "elevates" M to a *function* of x. This function is denoted by $\lambda x. M$ and we say that x is now *bound* in M by λx. A variable that is not bound in M is said to be *free* in M. A λ-term can be interpreted as a function or an argument of another function. Thus, in general, the λ-term (MN) indicates that the λ-term M (which is interpreted as a function) can be *applied* to the λ-term N (which is interpreted as an argument of M). (We also say that M can *act* on N.) By convention, the application is a left-associative operation, so (MNP) means $((MN)P)$. We can overrun, however, the convention by using parentheses. The outer parentheses are often omitted, so (MN) means MN. Any λ-term of the form $(\lambda x. M)N$ is called a β-*redex* (for reasons to become known shortly). A λ-term may contain zero or more β-redexes.

λ-terms can be transformed into other λ-terms. A transformation is a series of one-step transformations called β-*reductions*. There are two rules to do a β-reduction:

1. α-*conversion* (denoted $\mapsto_\alpha$) renames a bound variable in a λ-term;
2. β-*contraction* (denoted $\to_\beta$) transforms a β-redex $(\lambda x. M)N$ into a λ-term obtained from M by substituting N for every bound occurrence of x in M. Stated formally: $(\lambda x. M)N \equiv M[x := N]$.

Remarks. Intuitively, a β-contraction is an actual *application* (i.e., *acting*) of a function to its arguments. However, before a β-contraction is started, we must apply all the necessary α-conversions to M to avoid unintended effects of the β-contraction, such as unintended binding of N's free variables in M. When a λ-term contains no β-redexes, it cannot further be β-reduced. In this case the term is said to be β-*normal form* (β-*nf*). Intuitively, such a λ-term contains no function to apply.

A *computation* is a transformation of an initial λ-term t_0 with a sequence of β-reductions, that is

$$t_0 \to_\beta t_1 \to_\beta t_2 \to_\beta \cdots$$

If t_i is a member of this sequence, we say that t_0 is β-reducible to t_i and denote this by $t_0 \twoheadrightarrow_\beta t_i$. The computation terminates if and when some β-nf t_n is reached. This λ-term is said to be *final*.

A non-final λ-term may have several β-redexes. Each of them is a candidate for the next β-contraction. Since the selection of this usually affects the subsequent computation, we see that, in general, there exist different computations starting in t_0. Hence the questions: "Which of the possible computations is the 'right' one? Which of the possibly different final λ-terms is the 'right' result?" Fortunately, Church and Rosser proved that the order of β-reductions does not matter that much. Specifically, the final λ-term—when it exists—is defined up to the α-conversion (i.e., up to the renaming of its bound variables). In other words: if different computations terminate, they return practically equal results. This is the essence of the following theorem.

Theorem. (Church-Rosser) If $t_0 \twoheadrightarrow_\beta U$ and $t_0 \twoheadrightarrow_\beta V$, then there is W such that $U \twoheadrightarrow_\beta W$ and $V \twoheadrightarrow_\beta W$.

Natural numbers are represented by (rather unintuitive) β-nfs c_i, called *Church's numerals*:

$$
\begin{aligned}
c_0 &\equiv \lambda fx. f^0 x &&\equiv \lambda f.(\lambda x. x) \\
c_1 &\equiv \lambda fx. f^1 x &&\equiv \lambda f.(\lambda x. fx) \\
c_2 &\equiv \lambda fx. f^2 x &&\equiv \lambda f.(\lambda x. f(fx)) \\
&\;\;\vdots \\
c_n &\equiv \lambda fx. f^n x &&\equiv \lambda f.(\lambda x. f(\ldots f(fx)\ldots))
\end{aligned}
$$

A function $f : \mathbb{N}^k \to \mathbb{N}$ is λ-*definable* if there is a λ-term F such that

$$
\text{if } f(n_1,\ldots,n_k) =
\begin{cases}
m & \text{then } Fc_{n_1}\ldots c_{n_k} \twoheadrightarrow_\beta c_{n_m}; \\
\text{undefined} & \text{then } Fc_{n_1}\ldots c_{n_k} \text{ has no } \beta\text{-nf.}
\end{cases}
$$

Example 5.4. (Addition) The function $\text{sum}(n_1, n_2) = n_1 + n_2$ can be defined in λ-calculus by a λ-term $S \equiv \lambda abfx.af(bfx)$. To prove this, we must check that $Sc_{n_1}c_{n_2} \twoheadrightarrow_\beta c_{n_1+n_2}$ for any n_1, n_2. So we compute: $Sc_{n_1}c_{n_2} \equiv (Sc_{n_1})c_{n_2} \equiv ((\lambda abfx.af(bfx))c_{n_1})c_{n_2} \to_\beta \underline{(\lambda bfx.c_{n_1}f(bfx))c_{n_2}} \to_\beta$
$\to_\beta \lambda fx.c_{n_1}f(c_{n_2}fx) \equiv \lambda fx.\underline{(\lambda fx.f^{n_1}x)}f((\lambda fx.f^{n_2}x)fx) \to_\beta \lambda fx.(\lambda x.f^{n_1}x)((\lambda fx.f^{n_2}x)fx) \to_\beta$
$\to_\beta \lambda fx.\underline{(\lambda x.f^{n_1}x)}((\lambda x.f^{n_2}x)x) \to_\beta \lambda fx.\underline{(\lambda xf^{n_1}x)}(f^{n_2}x) \to_\beta \lambda fx.f^{n_1}f^{n_2}x \equiv \lambda fx.\underline{f^{n_1+n_2}x} \equiv c_{n_1+n_2}$ □

5.2.2 Modelling After Humans

The second direction in the search for an appropriate model of computation was an attempt to model computation after humans. The idea was to abstract man's activity when he mechanically solves computational problems. A seminal proposal of this kind was made by Turing,[9] whose model of computation was inspired both by human "computation" and a real mechanical device. In what follows we give an informal description of the model and postpone rigorous treatment of it to later sections (see Sect. 6.1).

Turing Machine

Turing took the quest for a definition of mechanical computation at face value. It has been suggested that he was inspired by his mother's typewriter. In 1936, the then 24-year-old conceived his own model of computation, which he called the *a-machine* (automatic machine). Today, the model is called the *Turing machine* (or *TM* in short).

Fig. 5.6 Alan Turing
(Courtesy: See Preface)

The Turing machine consists of several components (see Fig. 5.7):

1. a *control unit* (corresponding to the human brain);
2. a potentially infinite *tape* divided into equally sized *cells* (corresponding to the paper used during human "computation");
3. a *window* that can move over any cell and makes it accessible to the control unit (corresponding to the human eye *and* the hand with a pen).

---

[9] Alan Mathison Turing, 1912–1954, British mathematician, logician, and computer scientist.

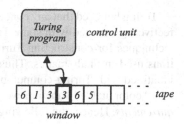

Fig. 5.7 Turing machine
(*a*-machine)

The control unit is always in some *state* from a finite set of states. Two of the states
are the *initial* and the *final* state. There is also a *program* in the control unit (corresponding to the "algorithm" that a human uses when "computing" the solution of
a given problem). We call it the *Turing program*. Different Turing machines have
different Turing programs.

Before the machine is started, the following must be done:

1. an *input word* (i.e., the input data written in an alphabet Σ) is written on the tape;
2. the window is shifted to the beginning of the input word;
3. the control unit is set to the initial state.

From then on, the machine operates independently, in a purely mechanical fashion, step by step, as directed by its Turing program. At each step, the machine reads
the symbol from the cell under the window into the control unit and, based on this
symbol and the current state of the control unit, does the following:

1. writes a symbol to the cell under the window (while deleting the old symbol);
2. moves the window to one of the neighboring cells or leaves the window as it is;
3. changes the state of the control unit.

The machine *halts*, if the control unit enters the final state or if its Turing program
has no instruction for the next step.

How can Turing machines compute the values of functions, say numerical functions? Turing had this idea. Take any Turing machine T and any $k \in \mathbb{N}$. Then, *define*
a function $f_T^{(k)}$ as follows:

*If the input word to T represents numbers $a_1, \ldots, a_k \in \mathbb{N}$, and T halts, and after
halting the tape contents make up a word over Σ that represents a natural number,
say b, then let b be the value of the function $f_T^{(k)}$ at $a_1, \ldots, a_k$; that is,*

$$f_T^{(k)}(a_1, \ldots, a_k) \stackrel{\text{def}}{=} b.$$

We can view $f_T^{(k)}$ as a function that maps k words over Σ into a word over Σ. The
values of $f_T^{(k)}$ can be mechanically computed simply by executing the Turing program of T. Today, we say that the function $f_T^{(k)}$ is *Turing-computable*. Actually, *any
function f for which there is a Turing machine T such that $f = f_T^{(k)}$, is said to be
Turing-computable.*

Turing believed that any conceivable "computable" function $f : \mathcal{A} \to \mathcal{B}$ can be effectively computed with some Turing machine. To show this, he developed several techniques for constructing Turing machines and found machines for many functions used in mathematics. Thus, it seemed that "computable" functions could be identified with Turing-computable functions. In sum, it seemed that this definition of a "computable" function would fulfill the *Completeness* and *Effectiveness Requirements*. Hence, the following model of computation.

Model of Computation (Turing's Characterization):

- *An "algorithm"* is a Turing program.
- *A "computation"* is an execution of a Turing program on a Turing machine.
- *A "computable" function* is a Turing-computable function.

Box 5.3 (Memorizing During Computation).

It seems that there is an important difference between a human and a Turing machine. A human can see only a finite portion of the scribbles under his or her nose, yet is able to remember some of the previously read ones and use them in the "computation." It seems that the Turing machine's control unit—lacking explicit storage—does not allow for this.

However, this is not true. We will see later that the control unit can *simulate* finite storage. The basic idea is that the control unit memorizes the symbol that has been read from the tape by changing to a state that corresponds to the symbol. This also allows the Turing machine to take into account the memorized symbol during its operation. In order to implement this idea, the states and instructions of the Turing machine have to be appropriately defined and interpreted during the execution. (Further details will be given in Sects. 6.1.2 and 6.1.3.)

Example 5.5. (Addition) That sum(n_1, n_2) is Turing-computable shows Example 6.1 (p. 105). □

5.2.3 Modelling After Languages

The third direction in the search for an appropriate model of computation focused on modelling after languages. The idea was to view human mathematical activity as the transformation of a sequence of words (description of a problem) into another sequence of words (solution of the problem), where the transformation proceeds according to certain rules. Thus, the "computation" was viewed as a sequence of steps that gradually transform a given input expression into an output expression. The rules that govern the transformation are called *productions*. Each production describes how a current expression should be partitioned into fixed and variable sub-expressions and, if the partition is possible, how the sub-expressions must be changed and reordered to get the next expression. Productions are of the form

$$\alpha_0 x_1 \alpha_1 x_2 \ldots \alpha_{n-1} x_n \alpha_n \to \beta_0 x_{i_1} \beta_1 x_{i_2} \ldots \beta_{m-1} x_{i_m} \beta_m$$

where α_i, β_j are fixed sub-expressions and x_k are variables whose values are arbitrary sub-expressions of the current expression. Based on these ideas, Post and Markov proposed two models of computation, called the *Post machine* and the *Markov algorithms*.

The Post Machine

In the 1920s, Post investigated the decidability of logic theories. He viewed the derivations in theories as man's language activity and, for this reason, developed his *canonical systems*. (We will return to these in Sect. 6.3.2.) The research on canonical systems forced Post to invent an abstract machine, now called the *Post machine*.

Fig. 5.8 Emil Post
(Courtesy: See Preface)

The Post machine is similar to the Turing machine:[10] it has a control unit with a program, a tape for input words, and a movable window over the tape (see Fig. 5.9). But there are differences as well: 1) the Post machine can only *read* a tape symbol; and 2) it uses a *queue* of symbols. In each step, the control unit reads the symbol under the window and consumes the symbol from the head of the queue, and then—based on the two symbols and the current state of the control unit—does the following:

1. adds a symbol to the end of the queue;
2. moves the window to a neighboring cell;
3. changes the state of the control unit.

A program of the Post machine can be viewed as a directed graph, where vertices contain instructions and arcs are used to pass the data between the instructions.[11] We call it the *Post program*.

The computation starts at the input vertex, whose instruction reads the input word from the tape and feeds it to the graph. In general, when a vertex has a word on one of its incoming arcs, it consumes the word, processes it by the associated instruction, and puts the result on one of the output arcs. Hence, during the computation the input word travels from one vertex to another and changes at each vertex. The basic

---

[10] Turing and Post proposed their models independently of each other.

[11] In the late twentieth century, a similar model of computation, the *data-flow graph*, was used.

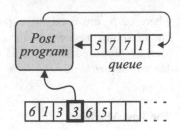

Fig. 5.9 Post machine

instructions are *start* (i.e., read the input word from the tape), *assign s* (i.e., append the symbol s to the word), *test* (i.e., output the tail of the word to the arc corresponding to the head of the word), and *accept/reject* (i.e., accept or reject the word). The machine *halts* if the word reaches a vertex with the *accept* or *reject* instruction.

Now we proceed to the model of computation in a similar way as we did with the Turing machine. Let P be an arbitrary Post machine and $k \in \mathbb{N}$. Define a k-ary function $f_P^{(k)}$ as follows: If the input word to P represents natural numbers $a_1, \ldots, a_k$, and P halts, and the word after halting represents a natural number b, then define $f_P^{(k)}(a_1, \ldots, a_k) \overset{\text{def}}{=} b$. A function f is said to be *Post-computable* if there is a Post machine P such that $f = f_P^{(k)}$ (i.e., P computes the values of f). Of course, $f_P^{(k)}$ can also be viewed as a function mapping k words over Σ into a word over Σ. Again, the computation of such a function is mechanical and seems to be effective.

This brings us to the following model of computation.

Model of Computation (Post's Characterization):

- *An "algorithm"* is a Post program.
- *A "computation"* is an execution of a Post program on a Post machine.
- *A "computable" function* is a Post-computable function.

Markov Algorithms

Later, in the Soviet Union, similar reasoning was applied by Markov.[12] In 1951, he described a model of computation that is now called the *Markov algorithm*.

A Markov algorithm is a finite sequence M of *productions*

$$
\begin{aligned}
\alpha_1 &\to \beta_1 \\
\alpha_2 &\to \beta_2 \\
&\vdots \\
\alpha_n &\to \beta_n.
\end{aligned}
$$

---

[12] Andrey Andreyevič Markov, Jr., 1903–1979, Russian mathematician.

where α_i, β_i are words over a given alphabet Σ. The sequence M is also called the *grammar*. A production $\alpha_i \rightarrow \beta_i$ is said to be *applicable* to a word w if α_i is a subword of w. If such a production is actually applied to w, it transforms w so that it replaces the leftmost occurrence of α_i in w with β_i.

An execution of a Markov algorithm is a sequence of steps that gradually transform a given *input word* via a sequence of *intermediate words* into some *output word*. At each step, the last intermediate word is transformed by the first applicable production of M. Some productions are said to be *final*. If the last applied production was final, or if there was no production to apply, then the execution halts and the last intermediate word is the *output word*.

Fig. 5.10 Andrey Markov
(Courtesy: See Preface)

Markov algorithms can be used to compute functions. To see this, let M be an arbitrary Markov algorithm and $k \in \mathbb{N}$. Define a k-ary function $f_M^{(k)}$ as follows: If the input word to M contains the data $a_1, \dots, a_k$, and M halts, and the output word is denoted by b, then define $f_M^{(k)}(a_1, \dots, a_k) \stackrel{\text{def}}{=} b$. (The data encoded in the input and output words can be natural numbers, so Markov algorithms can compute numerical functions as well.) Now we can define: A function f is said to be *Markov-computable* if there is a Markov algorithm M such that $f = f_M^{(k)}$ (that is, M computes the values of f).

Many functions were shown to be Markov-computable. The values of each can be mechanically and effectively calculated by the corresponding Markov algorithm. This suggested the following model of computation.

Model of Computation (Markov's Characterization):

- *An "algorithm"* is a Markov algorithm (grammar).
- *A "computation"* is an execution of a Markov algorithm.
- *A "computable" function* is a Markov-computable function.

5.2.4 Reasonable Models of Computation

Although the described models of computation are completely different, they share an important property: they are *reasonable*. Clearly, we cannot prove this because reasonableness is not a formal notion. But we can give some facts in support.

First we introduce new notions. An *instance* of a model of computation is called the *abstract computing machine*. Examples of such a machine M are:

- a particular construction of a recursive function (with the rules for using the construction to calculate the function values);
- a particular system of equations $\mathscr{E}(f)$ (with substitution and replacement rules);
- a particular λ-term (with α-conversion and β-reduction rules);
- a particular Turing program (in a Turing machine);
- a particular Post program (in a Post machine); and
- a particular Markov grammar (with the rules of production application).

As the "computation" on an abstract computing machine M goes on, the status (e.g., contents, location, value, state) of each M's component may be changing. At any step of the "computation," the statuses of the relevant components of M make up the so-called *internal configuration* of M at that step. Informally, this is a snapshot of M at a particular point of its "computation."

Now we return to the reasonableness of the models. For any abstract computing machine M belonging to any of the proposed models of computation, it holds that

1. M is of *limited capability*. This is because:

    a. it has finitely many different basic instructions;

    b. each basic instruction takes at least one step to be executed;

    c. each basic instruction has a finite effect, i.e., the instruction causes a limited change in the current internal configuration of M.

2. There is a *finite-size description of M*, called the *code of M* and denoted by $\langle M \rangle$.
3. The code $\langle M \rangle$ is *effective* in the sense that, given an internal configuration of M, the code $\langle M \rangle$ enables us to "algorithmically" compute the internal configuration of M after the execution of an instruction of M.
4. M is *unrestricted*, i.e., it has *potentially infinite resources* (e.g., time and space).

The proposed models of computation are *reasonable* from the standpoint of *Computability Theory*, which is concerned with the questions "What is an algorithm? What is computation? What can be computed?" Firstly, this is because these models do not offer unreasonable computational power (see item 1 above). Secondly, the answers to the questions are not influenced by any limitation of the computing resources except that a finite computation must use only a finite amount of each of the resources (see item 4). Hence, any problem that cannot be computed even with unlimited resources, will remain such if the available resources become limited.[13]

---

[13] In *Computational Complexity Theory*, a large special part of *Computability Theory*, the requirements (1c) and (4) are stiffer: In (1c) the effect of each instruction must be *reasonably large* (not just finite); and in (4) *the resources are limited*, so their consumption is highly important.

5.3 Computability (Church-Turing) Thesis

The diversity of the proposed models of computation posed the obvious question: "Which model (if any) is the *right* one?" Since each of the models fulfilled the *Effectiveness Requirement*, the issue of effectiveness was replaced by the question "Which model is the most natural and appealing?" Of course, the opinions were subjective and sometimes different. As with the *Completeness Requirement*, it was not obvious whether the requirement was *truly* fulfilled by *any* of the models. In this section we will describe how the *Computability Thesis*, which in effect states that *each* of the proposed models fulfills this requirement, was born.

5.3.1 History of the Thesis

Church. By 1934, Kleene managed to prove that every conceivable "computable" numerical function was λ-definable. For this reason, in the spring of 1934, Church conjectured that "computable" functions are exactly λ-definable functions. In other words, Church suspected that the intuitive concepts of algorithm and computation are appropriately formalized by his model of computation. He stated this in the following thesis:

Church Thesis. *"algorithm"* $\longleftrightarrow$ λ-term

Church presented the thesis to Gödel, the authority in mathematical logic of the time, but he rejected it as unsatisfactory. Why? At that time, Gödel was reflecting on the relation between "computable" functions and general recursive functions. He suspected that the latter might be a formalization of the former, yet he was well aware of the fact that such an equivalence could not possibly be proved, because it would equate two concepts of which one is formal (i.e., precisely defined) and the other is informal (i.e., intuitive). In his opinion, researchers needed to continue analyzing the intuitive concepts of algorithm and computation. Only after their intrinsic components and properties were better understood would it make sense to propose a thesis of this kind.

Shortly after, Kleene, Church, and Rosser proved the equivalence between λ-calculus and the general recursive functions. Thus, every λ-definable function is general recursive, and vice versa. Since λ-calculus (being somewhat unnatural) was not well accepted in the research community, Church restated his thesis in the terminology of the equivalent model (general recursive functions). In 1936, Church finally published his thesis.

This, however, did not convince Gödel (and Post). In their opinion, the equivalence of the two models still did not indicate that they fulfilled the *Completeness Requirement*, i.e., that they fully captured the intuitive concepts of algorithm and computation.

Turing. Independently, Turing pursued his research in England. He found that every conceivable "computable" numerical function was Turing-computable. He also proved that the class of Turing-computable functions remains the same under many generalizations of the Turing machine. Consequently, he suspected that "computable" functions are exactly Turing-computable functions. In other words, he suspected that the intuitive concepts of algorithm and computation are appropriately formalized by his model of computation. In 1936, he published his thesis.

> **Turing Thesis.** *"algorithm"* $\longleftrightarrow$ *Turing program*

Gödel accepted the Turing machine as the model of computation that convincingly formalizes the concepts of "algorithm" and "computation." He was convinced[14] by the simplicity and generality of the Turing machine, its mechanical working, its resemblance to human activity when solving computational problems, Turing's reasoning and analysis of "computable" functions, and his argumentation that "computable" functions are exactly Turing-computable functions.

In 1937, Turing also proved that Church's and his model are equivalent in the sense that what can be computed by one can also be computed by the other. That is,

$$\lambda\text{-}definable \Longleftrightarrow Turing\text{-}computable$$

He proved this by showing that each of the two models can simulate the basic instructions of the other.

Because of all of this, the remaining key researchers accepted the Turing machine as the most appropriate model of computation. (The reader is advised to compare Examples 5.1, 5.4, and 6.1.) Soon, the Turing machine met general approbation.

5.3.2 The Thesis

As the two theses were equivalent, they were merged into the *Church-Turing Thesis*. More recently, this has also been given the neutral name *Computability Thesis* (CT).

> **Computability Thesis.** *"algorithm"* $\longleftrightarrow$ *Turing program (or equivalent model)*

Gradually, it was also proved that other models of computation are equivalent to the Turing machine or some other equivalent model. The diversity of the equivalent models strengthened the belief in the *Computability Thesis*.[15]

---

[14] If the reader has doubts, he or she may compare Examples 5.1 (p.75), 5.4 (p.80), and 6.1 (p.105).

[15] Interestingly, a similar situation arose in physics ten years before. To explain the consequences of the quantization of energy and the unpredictability of events in microscopic nature, two theories were independently developed: matrix mechanics (Werner Heisenberg, 1925) and wave mechanics (Erwin Schrödinger, 1926). Though the two theories were completely different, Schrödinger proved that they are equivalent in the sense that physically they mean the same. This and their capability of accurately explaining and predicting physical phenomena strengthened the belief in the quantum explanation of microscopic nature.

If the *Computability Thesis* is in truth correct, then, in principle, this cannot be proved. (One should give rigorous arguments that an informal notion is equivalent to another, formal one—but this was the very goal of the formalization described above.) In contrast, if the thesis is wrong, then there may be a proof of this. (One should conceive a "computable" function and prove that the function is not Turing-computable.) However, until now no one has found such a proof. Consequently, most researchers believe that the thesis holds.[16]

The *Computability Thesis* proclaimed the following formalization of intuitive basic concepts of computing:

Formalization. *Basic intuitive concepts of computing are formalized as follows:*

   "algorithm" ⟷ *Turing program*

   "computation" ⟷ *execution of a Turing program on a Turing machine*

"computable function" ⟷ *Turing-computable function*

NB *The Computability Thesis established a bridge between the intuitive concepts of "algorithm," "computation," and "computability" on the one hand, and their formal counterparts defined by models of computation on the other. In this way it finally opened the door to a mathematical treatment of these intuitive concepts.*

The situation after the acceptance of the *Computability Thesis* is shown in Fig. 5.11.

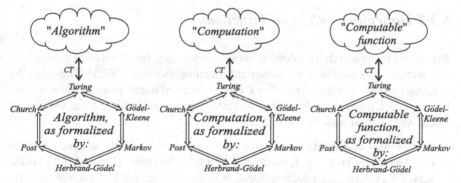

Fig. 5.11 Different formalizations of intuitive concepts are equivalent. By the *Computability Thesis* (CT), they also adequately capture the intuitive concepts

---

[16] To this day, several new models of computation have been proposed and proved to be equivalent to the Turing machine. Some of the notable ones are the *register machine*, which, in various variants such as RAM and RASP, epitomizes modern computers (we will describe RAM in Sect. 6.2.7); the *cellular automaton* (Conway 1970), which is inspired by the development of artificial organisms; and *DNA-calculus* (Adleman 1993), which uses DNA molecules to compute.

At last we can refine Fig. 1.2 (p. 4) to Fig. 5.12, and Definition 1.1 (p. 4) to Definition 5.1.

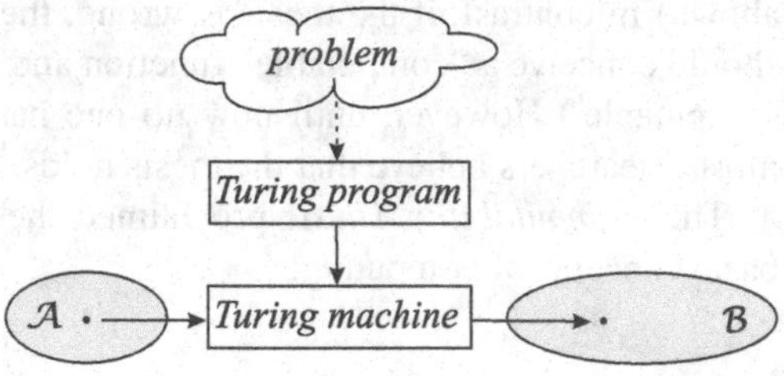

Fig. 5.12 A Turing program directs a Turing machine to compute the solution to the problem

Definition 5.1. (Algorithm Formally) The **algorithm** for solving a problem is a Turing program that leads a Turing machine from the input data of the problem to the corresponding solution.

Remark. Since the concepts of "algorithm" and "computation" are now formalized, we no longer need to use quotation marks to distinguish between their intuitive and formal meanings. In contrast, with the concept of "computable" function we will be able to do this only after we have clarified *which functions we must talk about.* This is the subject of the next two subsections.

5.3.3 Difficulties with Total Functions

Recall that in the search for a definition of "computable" functions some researchers pragmatically focused on *total* numerical functions (see Sect. 5.2.1). These are the functions $f : \mathbb{N}^k \to \mathbb{N}$ that map *every* k-tuple of natural numbers into a natural number. However, it became evident that the restriction to total functions was too severe. There were several reasons for this.

1. **The Notion of Totality.** Focusing on total functions tacitly assumed that we can decide, for arbitrary f, whether f is a total function. This would be necessary, for example, after each application of the μ-operation in recursive function construction (see Box 5.1, p. 74). But finding out whether f is total may not be a finite process; in the extreme case, we must check individually, for each k-tuple in $\mathbb{N}^k$, whether f is defined. (Only later, in Sect. 9.4, will we be able to prove this. For starters, see Box 5.4 for a function which is potentially of this kind.) Unfortunately, this meant that the concept of a recursive function—an important model of computation—had a serious weak point: it was founded on a notion (totality) that is disputable in view of the finitist philosophy of mathematics.

Box 5.4 (Goldbach's Conjecture).

In 1742, Goldbach proposed the following conjecture G:

$G \equiv$ *"Every even integer greater than 2 is the sum of two primes."*

For example, $4 = 2+2$; $6 = 3+3$; $8 = 3+5$; $10 = 3+7$. But Goldbach did not prove G. What is more, in spite of many attempts of a number of prominent mathematicians, to this date, the conjecture remains an open problem. In particular, no pattern was found such that, for every natural n, $4+2n = p(n)+q(n)$ and $p(n)$, $q(n)$ are primes. (See discussion in Box 2.5, p. 21.)

Let us now define the Goldbach function $g : \mathbb{N} \to \mathbb{N}$ as follows:

$$g(n) \stackrel{\text{def}}{=} \begin{cases} 1, & \text{if } 4+2n \text{ is the sum of two primes;} \\ \text{undefined,} & \text{otherwise.} \end{cases}$$

Is g total? This question is equivalent to the question of whether G holds. Yet, all attempts to answer either of the questions have failed. What is worse, there is a possibility that G is one of the undecidable *Truths* of arithmetic, whose existence was proved by Gödel (see Sect. 4.2.3).

Does this mean that, lacking any pattern $4+2n = p(n)+q(n)$, the only way of finding out whether g is total, is by checking, for each natural n individually, whether $4+2n$ is the sum of two primes? Indeed, in 2012, G was verified by computers for n up to $4 \cdot 10^{18}$. However, if g is in truth total, this is not a finite process, and will never return the answer "g is total." In sum: for certain functions we might not be able to decide whether or not they are total.

2. **Diagonalization.** An even more serious consequence of the restriction to total functions was discovered by a method called *diagonalization*.

By *Computability Thesis* every "computable" function is also recursive (see p. 74). But is it really so? It was immediately noticed that there are only *countably* infinitely many recursive functions (i.e., as many as different constructions), whereas it was known that there are *uncountably* many numerical functions (see Appendix A, p. 302). So there are many numerical functions that are not recursive. Hence, the question was whether there is a numerical function which is by our standards "computable," and yet not recursive. If there is, how do we find it when all "computable" functions that researchers have managed to conceive have turned out to be recursive?

The success came with a method called *diagonalization*. We will describe this method in depth later, in Sect. 9.1, but we can still briefly describe the idea of the construction. Using diagonalization, a certain function g was defined in a somewhat unusual way (typical of diagonalization), but it was still perfectly evident how one could compute its values. So, g was "computable." But, the definition of g was such that it implied a contradiction *if g was supposed to be total*. Consequently, the assumption that g was recursive had to be dropped. (Recall that recursive functions are by definition total.) Thus, g was the "computable"-and-not-recursive function they were looking for.

Details are:

a. Constructions of recursive functions are finite sequences of symbols over a finite alphabet. The set of all constructions can be well-ordered in *shortlex order*.[17] Consequently, given arbitrary $n \in \mathbb{N}$, the nth construction is precisely defined (in shortlex order). Let f_n denote the recursive function whose construction is nth in this order.

b. Define a function $g : \mathbb{N}^{k+1} \to \mathbb{N}$ as follows:

$$g(n, a_1, \ldots, a_k) \stackrel{\text{def}}{=} f_n(a_1, \ldots, a_k) + 1.$$

c. *The function g is "computable."* The algorithm for calculating its values for arbitrary $(a_1, \ldots, a_k) \in \mathbb{N}^k$ is straightforward:
   • find the nth construction; then
   • use the construction to compute $f_n(a_1, \ldots, a_k)$; and then
   • add 1 to the result.

d. Is g recursive? *Suppose it is.* Then, there exists a construction of g. So, g is one of the functions $f_1, f_2, \ldots$, i.e., $f_m = g$ for some $m \in \mathbb{N}$.

e. Let us now focus on the value $g(m, m, \ldots, m)$. The definition of g gives $g(m, m, \ldots, m) = f_m(m, \ldots, m) + 1$. On the other hand, we have (by d above) $f_m(m, \ldots, m) = g(m, m, \ldots, m)$. From these two equations we obtain

$$g(m, m, \ldots, m) = g(m, m, \ldots, m) + 1$$

and, consequently, $0 = 1$, which is a contradiction. Our supposition that g is recursive implies a contradiction. Thus, g *is not recursive!*

The conclusion is inescapable: *There are "computable" functions which are not recursive!* At first, researchers thought that the definition of recursive functions (see Box 5.1, p. 74) should be extended by additional initial functions and/or additional rules of construction. It soon became clear that this would not prevent the contradiction. Namely, *any definition* of recursive functions which defines only *total* functions leads in the same way as above to the function g and the contradiction $g(m, m, \ldots, m) = g(m, m, \ldots, m) + 1$.

Did this refute the *Computability Thesis*? Fortunately not so; the thesis only had to be slightly adapted. Namely, it was noticed that no contradiction would have occurred if the construction of *partial* functions had also been allowed. Recall that the values of a partial function can be undefined for arbitrary arguments. If partial functions were allowed, the value $g(m, m, \ldots, m)$ could be undefined, so the equation $g(m, m, \ldots, m) = g(m, m, \ldots, m) + 1$ would not be contradictory, because undefined plus one is still undefined.

---

[17] Sort the constructions by increasing length, and those of the same length lexicographically.

3. **Experience.** Certain well-known numerical functions were not total. For instance, rem : $\mathbb{N}^2 \to \mathbb{N}$, defined by rem$(m,n)$ = "remainder from dividing m by n." This function is not total because it is not defined for pairs $(m,n) \in \mathbb{N} \times \{0\}$. Nevertheless, rem could be treated as "computable" because of the following algorithm: if $n > 0$, then compute rem(m,n) and halt; otherwise, return a warning (e.g., $-1 \notin \mathbb{N}$). Notice that the algorithm returns the value whenever rem is defined; otherwise, it warns that rem is not defined.

4. **Computable Total Extensions.** We can view the algorithm in the previous paragraph such that it computes the values of some other, but total function rem*, defined by

$$\text{rem}^*(m,n) \stackrel{\text{def}}{=} \begin{cases} \text{rem}(m,n) & \text{if } (m,n) \in \mathbb{N} \times (\mathbb{N} - \{0\}); \\ -1 & \text{otherwise.} \end{cases}$$

The function rem* has the same values as rem, whenever rem is defined. We say that rem* is a *total extension* of rem to $\mathbb{N} \times \mathbb{N}$. Now, if every "computable" partial function had a "computable" total extension, then there would be no need to consider partial functions in order to study "computability." However, we'll see that this is not the case: there are "computable" partial functions that have no "computable" total extensions. This is yet another reason for the introduction of *partial* functions in the study of "computable" functions.

5.3.4 Generalization to Partial Functions

The difficulties with "computable" total functions indicated that the definition of a "computable" function should be founded on *partial* functions instead of only total functions. So, let us recall the basic facts about partial functions and introduce some useful notation. We will be using Greek letters to denote partial functions, e.g., φ.

Definition 5.2. (Partial Function) We say that $\varphi : \mathcal{A} \to \mathcal{B}$ is a **partial function** if φ may be undefined for some elements of $\mathcal{A}$. (See Fig. 5.13.) If φ is defined for $a \in \mathcal{A}$, we write

$$\varphi(a)\!\downarrow;$$

otherwise we write $\varphi(a)\!\uparrow$. The set of all the elements of $\mathcal{A}$ for which φ is defined is the **domain** of φ, denoted by

$$\text{dom}(\varphi).$$

Hence dom$(\varphi) = \{a \in \mathcal{A} \mid \varphi(a)\!\downarrow\}$. Thus, for partial φ we have dom$(\varphi) \subseteq \mathcal{A}$. In the special case when dom$(\varphi) = \mathcal{A}$, we omit the adjective partial and say that φ is a **total** function (or just a function). When it is clear that a function is total, we denote it by a Latin letter, e.g., f, g. The expression

$$\varphi(a)\!\downarrow= b$$

says that φ is defined for $a \in \mathcal{A}$ and its value is b. The set of all elements of $\mathcal{B}$, which are φ-images of elements of $\mathcal{A}$, is the **range** of φ, denoted by

$$\text{rng}(\varphi).$$

Hence $\text{rng}(\varphi) = \{b \in \mathcal{B} \mid \exists a \in \mathcal{A} : \varphi(a)\!\downarrow= b\}$. The function φ is **surjective** if $\text{rng}(\varphi) = \mathcal{B}$, and it is **injective** if different elements of $\text{dom}(\varphi)$ are mapped into different elements of $\text{rng}(\varphi)$. Partial functions $\varphi : \mathcal{A} \to \mathcal{B}$ and $\psi : \mathcal{A} \to \mathcal{B}$ are said to be **equal**, and denoted by

$$\varphi \simeq \psi$$

if they have the same domains and the same values; that is, for every $x \in \mathcal{A}$ it holds that $\varphi(x)\!\downarrow \Longleftrightarrow \psi(x)\!\downarrow$ and $\varphi(x)\!\downarrow \Longrightarrow \varphi(x) = \psi(x)$.

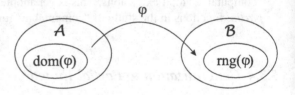

Fig. 5.13 Partial function $\varphi : \mathcal{A} \to \mathcal{B}$ is defined on the set $\text{dom}(\varphi) \subseteq \mathcal{A}$

The improved definition of a "computable" function (the definition that would consider partial functions in addition to total ones) was expected to retain the intuitive appeal of the previous definition (see p. 89). Fortunately, this was not difficult to achieve. One had only to take into account all the possible outcomes of the computing function's values for arguments where the function is not defined.

So let $\varphi : \mathcal{A} \to \mathcal{B}$ be a partial function, $a \in \mathcal{A}$ and suppose that $\varphi(a)\!\uparrow$. There are two possible outcomes when an attempt is made to compute $\varphi(a)$:

1. the computation halts and returns a nonsensical result (not belonging to $\text{rng}(\varphi)$);
2. the computation does not halt.

In the first outcome, the nonsensical result is also the signal (i.e., warning) that φ is not defined for a. The second outcome is the trying one: neither do we receive the result nor do we receive the warning that $\varphi(a)\!\uparrow$. As long as the computation goes on, we can only wait and hope that it will soon come—not knowing that all is in vain (Fig. 5.14). The situation is even worse: we will prove later, in Sect. 8.2, that there is no general way to find out whether all is in vain. In other words, we can never find out that we are victims of the second outcome.

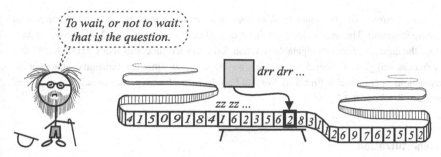

Fig. 5.14 What does one do if a computation on a Turing machine is still going on?

Since deciding which of the two outcomes will take place is generally impossible, we will not try to distinguish between them. We will not concern ourselves with what is going on when an attempt is made to compute an undefined function value. We will be satisfied with a new definition which, in essence, says that

A *partial* function is "computable" if there is an algorithm
which can compute its value *whenever the function is defined.*

When it is known that such a function is *total*, we will drop the adjective partial. We can now give a formalization of the concept of "computable" partial function.

Formalization. (cont'd from p. 89) *The intuitive concept of "computable" partial function* $\varphi : \mathcal{A} \to \mathcal{B}$ *is formalized as follows:*

φ is **"computable"** *if there exists a TM that can compute the value*
$$\varphi(x) \text{ for any } x \in \operatorname{dom}(\varphi)$$
$$\text{and } \operatorname{dom}(\varphi) = \mathcal{A};$$

φ is **partial "computable"** *if there exists a TM that can compute the value*
$$\varphi(x) \text{ for any } x \in \operatorname{dom}(\varphi);$$

φ is **"incomputable"** *if there is no TM that can compute the value*
$$\varphi(x) \text{ for any } x \in \operatorname{dom}(\varphi).$$

Note that Fig. 5.11 (p. 89) is valid for the new definition of "computable" functions.

Remarks. 1) Since the concept of a "computable" function is now formalized, we will no longer use quotation marks to distinguish between its intuitive and formal meanings. 2) From now on, when we will say that a function is *computable*, it will be tacitly understood that it is total (by definition). And when we say that a function is *partial computable* (or *p.c.* in short), we will be aware of the fact that the function may or may not be total. 3) In the past, the naming of computable and partial computable functions was not uniform. In 1996, a suggestion was made by Soare to modernize and unify the terminology. According to this, we use the terms *computable* function (instead of the older term *recursive* function) and *partial computable* function (instead of the older term *partial recursive* function). The reasons for the unification will be described on p. 142. 4) Above, we have defined the concept of an incomputable function. But, do such functions really exist? A clue that they do exist was given in Sect. 5.3.3, where it was noticed that numerical functions outnumber the constructions of computable functions. We will be able to construct a particular such function later (see Sects. 8.2.3 and 8.3.1).

A Generalization

Later, a slight generalization of the above formalization will ease our expression. Let $\varphi : A \to B$ be a partial function and $S \subseteq A$ an arbitrary set. Observe that even if φ is incomputable, there may still exist a Turing machine that is capable of computing the value $\varphi(x)$ for arbitrary $x \in S \cap \text{dom}(\varphi)$. (See Fig. 5.15.) In this case, we will say that φ is *partial computable on the set* S.[18] If, in addition, $S \subseteq \text{dom}(\varphi)$, then we will say that φ is *computable on the set* S.[19] Here is the official definition.

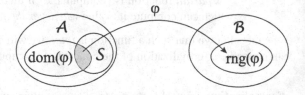

Fig. 5.15 $\varphi : A \to B$ is partial computable (p.c.) on $S \subseteq A$ if it can be computed everywhere on $S \cap \text{dom}(\varphi)$

Definition 5.3. (Computability on a Set) Let $\varphi : A \to B$ be a partial function and $S \subseteq A$. We say that:

φ is **computable on** S *if* there exists a TM that can compute the value
$$\varphi(x) \text{ for any } x \in S \cap \text{dom}(\varphi)$$
$$\text{and } S \subseteq \text{dom}(\varphi);$$

φ is **partial computable on** S *if* there exists a TM that can compute the value
$$\varphi(x) \text{ for any } x \in S \cap \text{dom}(\varphi);$$

φ is **incomputable on** S *if* there is *no* TM that can compute the value
$$\varphi(x) \text{ for any } x \in S \cap \text{dom}(\varphi).$$

If we take $S = A$, the definition transforms into the above formalization.

---

[18] Equivalently, $\varphi : A \to B$ is *p.c. on the set* $S \subseteq A$ if the restriction $\varphi|_S$ is a p.c. function.

[19] Equivalently, $\varphi : A \to B$ is *computable on the set* $S \subseteq \text{dom}(\varphi)$ if $\varphi|_S$ is a computable function.

5.3.5 Applications of the Thesis

The *Computability Thesis* (CT) is useful in proving the existence or non-existence of certain "algorithms." Such proofs are called *proofs by CT*. There are two cases:

1. Suppose we want to prove that a given function φ is p.c. We might try to construct a TM and prove that it is capable of computing the values of φ. However, this approach is cumbersome and prone to mistakes. Instead, we can do the following:

    a. informally describe an "algorithm" that "computes" the values of φ;
    b. refer to the CT (saying: by CT the "algorithm" *can* be replaced by *some* Turing program; hence φ is p.c.).

2. Suppose we want to prove that a function φ is "incomputable." To do this, we

    a. prove that φ is not Turing-computable (i.e., there exists no TM for computing the values of φ);
    b. refer to the CT (saying: then, by CT, φ is "incomputable").

5.4 Chapter Summary

Hilbert's Program left open the *Entsheidungsproblem*, the problem that was calling for an algorithm that would, for any mathematical formula, decide whether the formula can be derived.

Soon it became clear that the problem could not be solved unless the intuitive, loose definition of the concept of the algorithm was replaced by a rigorous, formal definition. Such a definition, called the model of computation, should characterize the notions of "algorithm," "computation" and "computable."

In 1930, the search for an appropriate model of computation started. Different ideas arose and resulted in several totally different models of computation: the recursive functions, general recursive functions, λ-calculus, the Turing machine, the Post machine and Markov algorithms. Although diverse, the models shared two important properties, namely that they were *reasonable* and they fulfilled the *Effectiveness Requirement*.

The Turing machine was accepted by many as the most appropriate model of computation. Surprisingly, it turned out that all the models are equivalent in the sense that what can be computed by one can also be computed by the other.

Finally, the *Computability Thesis* equated the informally defined concepts of "algorithm," "computation" and "computable" with the counterparts that were formally defined by the models of computation. In effect, the thesis declared that all these models of computation also fulfill the *Completeness Requirement*. It was also found that the definition of a "computable" function must be founded on partial functions instead of only on total functions.

All in all, the *Computability Thesis* made it possible to mathematically treat the intuitive concepts of computation.

Problems

5.1. Prove that these functions are primitive recursive:

(a) $\text{const}_j^k(n_1,\ldots,n_k) \overset{\text{def}}{=} j$, for $j \geqslant 0$ and $k \geqslant 1$;

(b) $\text{add}(m,n) \overset{\text{def}}{=} m+n$;

(c) $\text{mult}(m,n) \overset{\text{def}}{=} mn$;

(d) $\text{power}(m,n) \overset{\text{def}}{=} m^n$;

(e) $\text{fact}(n) \overset{\text{def}}{=} n!$

(f) $\text{tower}(m,n) \overset{\text{def}}{=} \left. m^{m^{m^{\cdot^{\cdot^{\cdot^{m}}}}}} \right\} n \text{ levels}$

(g) $\text{minus}(m,n) \overset{\text{def}}{=} m \dot{-} n = \begin{cases} m-n & \text{if } m \geqslant n; \\ 0 & \text{otherwise.} \end{cases}$

(h) $\text{div}(m,n) \overset{\text{def}}{=} m \div n = \lfloor \frac{m}{n} \rfloor$;

(i) $\text{floorlog}(n) \overset{\text{def}}{=} \lfloor \log_2 n \rfloor$

(j) $\log^*(n) \overset{\text{def}}{=}$ the smallest k such that $\overbrace{\log(\log(\cdots(\log(n))\cdots))}^{k \text{ times}} \leqslant n$;

(k) $\gcd(m,n) \overset{\text{def}}{=}$ greatest common divisor of m and n;

(l) $\text{lcm}(m,n) \overset{\text{def}}{=}$ least common multiplier of m and n;

(m) $\text{prime}(n) \overset{\text{def}}{=}$ the nth prime number;

(n) $\pi(x) \overset{\text{def}}{=}$ the number of primes not exceeding x;

(o) $\phi(n) \overset{\text{def}}{=}$ the number of positive integers not exceeding n relatively prime to n (Euler function);

(p) $\max^k(n_1,\ldots,n_k) \overset{\text{def}}{=} \max\{n_1,\ldots,n_k\}$, for $k \geqslant 1$;

(q) $\min^k(n_1,\ldots,n_k) \overset{\text{def}}{=} \min\{n_1,\ldots,n_k\}$, for $k \geqslant 1$;

(r) $\text{neg}(x) \overset{\text{def}}{=} \begin{cases} 0 & \text{if } x \geqslant 1; \\ 1 & \text{if } x = 0. \end{cases}$

(s) $\text{and}(x,y) \overset{\text{def}}{=} \begin{cases} 1 & \text{if } x \geqslant 1 \wedge y \geqslant 1; \\ 0 & \text{otherwise.} \end{cases}$

(t) $\text{or}(x,y) \overset{\text{def}}{=} \begin{cases} 1 & \text{if } x \geqslant 1 \vee y \geqslant 1; \\ 0 & \text{otherwise.} \end{cases}$

(u) $\text{if-then-else}(x,y,z) \overset{\text{def}}{=} \begin{cases} y & \text{if } x \geqslant 1; \\ z & \text{otherwise.} \end{cases}$

(v) $\text{eq}(x,y) \overset{\text{def}}{=} \begin{cases} 1 & \text{if } x = y; \\ 0 & \text{otherwise.} \end{cases}$

(w) $\text{gr}(x,y) \overset{\text{def}}{=} \begin{cases} 1 & \text{if } x > y; \\ 0 & \text{if } x \leqslant y. \end{cases}$

(x) $\text{geq}(x,y) \overset{\text{def}}{=} \begin{cases} 1 & \text{if } x \geqslant y; \\ 0 & \text{if } x < y. \end{cases}$

(y) $\text{ls}(x,y) \overset{\text{def}}{=} \begin{cases} 1 & \text{if } x < y; \\ 0 & \text{if } x \geqslant y. \end{cases}$

(z) $\text{leq}(x,y) \overset{\text{def}}{=} \begin{cases} 1 & \text{if } x \leqslant y; \\ 0 & \text{if } x > y. \end{cases}$

5.2. Prove: If $f : \mathbb{N} \to \mathbb{N}$ is primitive recursive, then so is $g(n,m) \stackrel{\text{def}}{=} f^{(n)}(m) = \overbrace{f(f(\cdots(f(m))\cdots))}^{n \text{ times}}$.

5.3. Prove: Every primitive recursive function is total.

5.4. Prove: Every partial recursive function can be obtained from the initial functions ζ, σ, π_i^k by a finite number of compositions and primitive recursions and *at most one* μ-operation.

5.5. A version of the **Ackermann function** $A : \mathbb{N}^2 \to \mathbb{N}$ is defined as follows:

$$A(m,0) = m+1$$
$$A(0,n+1) = A(1,n)$$
$$A(m+1,n+1) = A(A(m,n+1),n).$$

(a) Prove: A is a recursive function.

(b) Try to compute $A(k,k)$ for $k = 0,1,2,3$.

(c) The function A grows faster than *any* primitive recursive function in the following sense: For every primitive recursive $f(n)$, there exists an $n_0 \in \mathbb{N}$ such that $f(n) < A(n,n)$ for all $n > n_0$. Can you prove that? (*Remark.* This is why A is *not primitive* recursive.)

Bibliographic Notes

- For a treatise of *physical aspects* of information and computation, see Feynman [46]. The use of quantum phenomena in computation is described in Aaronson [1].
- *Primitive* recursive functions were first described in Gödel [53], and *general* recursive functions in Gödel [54]. Based on these, *recursive functions* were defined in Kleene [79].
- Ackermann functions were defined in Ackermann [2] and were extensively studied in the 1930s by Péter [118]. See also Kleene [82, §55]
- The λ-*calculus* was introduced in Church [23]. Hindley and Seldin [68] is a nice introduction to the λ-*calculus* and the related Haskell Curry's model of computation called *combinatory logic* (CL). For an extensive treatise of the two models, see Barendregt [10].
- Turing introduced his *a-machine* in [178]. The equivalence of the λ-calculus and the Turing machine was proved in Turing [179].
- The *Post machine* was introduced in Post [121]. Post suspected (but did not prove) that it is equivalent to the Turing machine and the λ-calculus.
- *Markov algorithms* were described in Markov [100, 101].
- A lot of information about the reasons for searching for the models of computation is given in Kleene [82]. A comprehensive historical view of the development of models of computation, their properties, and applications can be found in Odifreddi [116] and Adams [3]. A concise introduction to models of computation, from the standard Turing machine and recursive functions to the modern models inspired by quantum physics, is Fernández [45]. For the proofs that all these models of computation are equivalent, see, for example, Machtey and Young [98] (for the Turing machine, partial recursive functions, RAM, and Markov algorithms) and Enderton [42] (for the register machine).
- Church published his thesis in [25] and Turing proposed his thesis in [178]. The genesis of the *Computability Thesis*, Gödel's reservations about Church's Thesis, and flaws that were recently (1988) discovered in Church's development of his thesis, are detailed in Gandy [49], Sieg [153], and Soare [165, 167].
- Based on his paper [120], Post later introduced *canonical* and *normal systems* in [122] and [123, 124]. Here he introduced his thesis about set generation.
- Several cited papers are in Cooper and van Leeuwen [30], van Heijenoort [63], and Davis [35].

- Many historical notes about different models of computation are scattered in a number of excellent general monographs on *Computability Theory*. For such monographs, see the Bibliographic Notes to Chapter 6 on p. 141.

Chapter 6
The Turing Machine

A machine is a mechanically, electrically, or electronically
operated device for performing a task. A program is a sequence
of coded instructions that can be inserted into a machine to
control its operation.

Abstract The Turing machine convincingly formalized the concepts of "algorithm,"
"computation," and "computable." It convinced researchers by its simplicity, generality, mechanical operation, and resemblance to human activity when solving computational problems, and by Turing's reasoning and analysis of "computable" functions and his argumentation that partial "computable" functions are exactly Turing-
computable functions. Turing considered several variants that are generalizations of
the basic model of his machine. But he also proved that they add nothing to the
computational power of the basic model. This strengthened the belief in the Turing
machine as an appropriate model of computation. Turing machines can be encoded
and consequently enumerated. This enabled the construction of the universal Turing
machine that is capable of computing anything that can be computed by any other
Turing machine. This seminal discovery laid the theoretical grounds for several all-
important practical consequences, the general-purpose computer and the operating
system being the most notable. The Turing machine is a versatile model of computation: it can be used to compute values of a function, or to generate elements of a set,
or to decide about the membership of an object in a set. The last led to the notions
of decidable and semi-decidable sets that would later prove to be very important in
solving general computational problems.

6.1 Turing Machine

Most researchers accepted the Turing machine (TM) as the most appropriate model
of computation. We described the Turing machine briefly in Sect. 5.2.2. In this section we will go into detail. First, we will describe the basic model of the TM. We
will then introduce several other variants that are generalizations of the basic model.
Finally, we will prove that, from the viewpoint of general computability, they add
nothing to the computational power of the basic model. This will strengthen our
belief in the basic Turing machine as a simple yet highly powerful model of computation.

© Springer-Verlag Berlin Heidelberg 2015
101
B. Robič, *The Foundations of Computability Theory*,
DOI 10.1007/978-3-662-44808-3_6

Fig. 6.1 Alan Turing
(Courtesy: See Preface)

6.1.1 Basic Model

Definition 6.1. (Turing Machine) The basic variant of the **Turing machine**
has the following components (Fig. 6.2): a *control unit* containing a *Turing
program*; a *tape* consisting of *cells*; and a movable *window* over the tape,
which is connected to the control unit. (See also the details below.)

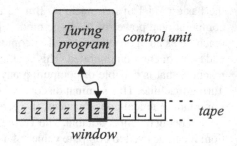

Fig. 6.2 Turing machine
(basic model)

The details are:

1. The *tape* is used for writing and reading the input data, intermediate data, and
   output data (results). It is divided into equally sized cells, and is potentially infinite in one direction (i.e., whenever needed, it can be extended in that direction
   with a finite number of cells).

   In each cell there is a *tape symbol* belonging to a finite *tape alphabet* $\Gamma =
   \{z_1, \ldots, z_t\}, t \geqslant 3$. The symbol z_t is special, for it indicates that a cell is empty; for
   this reason it is denoted by $\sqcup$ and called the *empty space*. In addition to $\sqcup$ there
   are at least two[1] additional symbols: 0 and 1. We will take $z_1 = 0$ and $z_2 = 1$.

---

[1] The reasons for at least two additional symbols are mostly practical (leaving out of consideration
the non-polynomial relation between the lengths of unary and binary representation of data, which
is important in *Computational Complexity Theory*). Only once, in Sect. 8.3.1, we will come across
Turing machines that need just *one* additional tape symbol (which will be the symbol 1). There,
we will simply ignore the other additional symbol.

The input data is contained in the *input word*. This is a word over some finite *input alphabet* Σ, such that $\{0,1\} \subseteq \Sigma \subseteq \Gamma - \{\sqcup\}$. Initially, all the cells are empty (i.e., each contains $\sqcup$) except for the leftmost cells, which contain the input word.

The *control unit* is always in some *state* belonging to a finite *set of states* $Q = \{q_1, \ldots, q_s\}$, where $s \geqslant 1$. We call q_1 the *initial state*. Some states are said to be *final*; they are gathered in the set $F \subseteq Q$. All the other states are non-final. When the index of a state will be of no importance, we will use q_{yes} and q_{no} to refer to *any* final and non-final state, respectively.

There is a program called the *Turing program* (*TP*) in the control unit. The program directs the components of the machine. It is characteristic of the particular Turing machine, that is, different TMs have different TPs. Formally, a Turing program is a partial function $\delta : Q \times \Gamma \to Q \times \Gamma \times \{\text{Left}, \text{Right}, \text{Stay}\}$. It is also called the *transition function*. We can view δ as a table $\Delta = Q \times \Gamma$, where the component $\Delta[q_i, z_r] = (q_j, z_w, D)$ if $\delta(q_i, z_r) = (q_j, z_w, D)$ is an instruction of δ, and $\Delta[q_i, z_r] = 0$ if $\delta(q_i, z_r)\uparrow$ (see Fig. 6.3). Without loss of generality we assume that $\delta(q_{no}, z)\downarrow$ for some $z \in \Gamma$, and $\delta(q_{yes}, z)\uparrow$ for all $z \in \Gamma$. That is, there is always a transition from a non-final state, and none from a final state.

There is a *window* that can move over any single cell, thus making the cell accessible to the control unit. The control unit can then *read a symbol* from the cell under the window, and *write a symbol* to the cell, substituting the previous symbol. In one step, the window can only move to the neighboring cell.

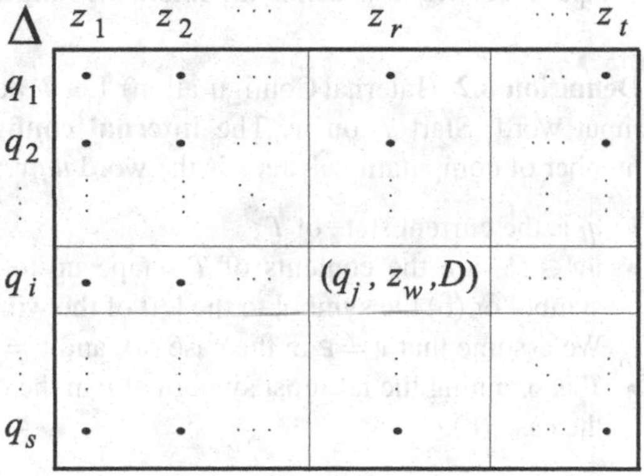

Δ	z_1	z_2	$\cdots$	z_r	$\cdots$	z_t
q_1	$\cdot$	$\cdot$	$\cdots$	$\cdot$		
q_2	$\cdot$	$\cdot$	$\cdots$	$\cdot$		$\cdot$
$\vdots$						
q_i	$\cdot$	$\cdot$	$\cdots$	(q_j, z_w, D)	$\cdots$	
$\vdots$						
q_s	$\cdot$	$\cdot$	$\cdots$	$\cdot$	$\cdots$	$\cdot$

Fig. 6.3 Turing program δ represented as a table Δ. Instruction $\delta(q_i, z_r) = (q_j, z_w, D)$ is described by the component $\Delta[q_i, z_r]$

2. Before the Turing machine is started, the following must take place:

   a. an input word is written to the beginning of the tape;
   b. the window is shifted to the beginning of the tape;
   c. the control unit is set to the initial state.

3. From now on the Turing machine operates independently, in a mechanical stepwise fashion as instructed by its Turing program δ. Specifically, if the TM is in a state $q_i \in Q$ and it reads a symbol $z_r \in \Gamma$, then:

- *if* q_i is a final state, *then* the TM halts;
- *else, if* $\delta(q_i, z_r)\uparrow$ (i.e., TP has no next instruction), *then* the TM halts;
- *else, if* $\delta(q_i, z_r)\downarrow = (q_j, z_w, D)$, *then* the TM does the following:
  a. changes the state to q_j;
  b. writes z_w through the window;
  c. moves the window to the next cell in direction $D \in \{\text{Left}, \text{Right}\}$, or leaves the window where it is ($D = \text{Stay}$).

Formally, a Turing machine is a seven-tuple $T = (Q, \Sigma, \Gamma, \delta, q_1, \sqcup, F)$. To fix a particular Turing machine, we must fix $Q, \Sigma, \Gamma, \delta$, and F. $\square$

Remarks. 1) Because δ is a partial function, it may be undefined for certain arguments q_i, z_r, that is, $\delta(q_i, z_r)\uparrow$. In other words, a partial Turing program has no instruction $\delta(q_i, z_r) = (q_j, z_w, D)$ that would tell the machine what to do when in the state q_i it reads the symbol z_r. Thus, for such pairs, the machine halts. This always happens in final states (q_{yes}) and, for some Turing programs, also in some non-final states (q_{no}). 2) The interpretation of these two different ways of halting (i.e., what they tell us about the input word or the result) will depend on what purpose the machine will be used for. (We will see later that the Turing machine can be used for computing the function values, generating the sets, or recognizing the sets.)

Computation. What does a *computation* on a Turing machine look like? Recall (Sect. 5.2.4) that the internal configuration of an abstract computing machine describes all the relevant information the machine possesses at a particular step of the computation. We now define the internal configuration of a Turing machine.

Definition 6.2. (Internal Configuration) Let T be a basic TM and w an arbitrary input word. Start T on w. The **internal configuration** of T after a finite number of computational steps is the word $u q_i v$, where

- q_i is the current state of T;
- $uv \in \Gamma^*$ are the contents of T's tape up to (a) the rightmost non-blank symbol or (b) the symbol to the left of the window, whichever is rightmost. We assume that $v \neq \varepsilon$ in the case (a), and $v = \varepsilon$ in the case (b).
- T is scanning the leftmost symbol of v in the case (a), and the symbol $\sqcup$ in the case (b).

Not every sequence can be an internal configuration of T (given input w). Clearly, the configuration prior to the first step of the computation is $q_1 w$; we call it the *initial* configuration. After that, only sequences $u q_i v$ that can be *reached* from the initial configuration by executing the program δ are internal configurations of T. So, if $u q_i v$ is an internal configuration, then the *next* internal configuration can easily be constructed using the instruction $\delta(q_i, z_r)$, where z_r is the scanned symbol.

The computation of T on w is represented by a sequence of internal configurations starting with the initial configuration. Just as the computation may not halt, the sequence may also be infinite. (We will use internal configurations in Theorem 9.2.)

Example 6.1. (Addition on TM) Let us construct a Turing machine that transforms an input word $1^{n_1}01^{n_2}$ into $1^{n_1+n_2}$, where n_1, n_2 are natural numbers. For example, 111011 is to be transformed to 11111. Note that the input word can be interpreted as consisting of two unary-encoded natural numbers n_1, n_2 and the resulting word containing their sum. Thus, the Turing machine is to compute the function $\mathrm{sum}(n_1 + n_2) = n_1 + n_2$.

First, we give the intuitive description of the Turing program. If the first symbol of the input word is 1, then the machine deletes it (instruction 1), and then moves the window to the right over all the symbols 1 (instruction 2) until the symbol 0 is read. The machine then substitutes this symbol with 1 and halts (instruction 3). However, if the first of the input word is 0, then the machine deletes it and halts (instruction 4).

Formally, the Turing machine is $T = (Q, \Sigma, \Gamma, \delta, q_1, \sqcup, \{q_3\})$, where:

- $Q = \{q_1, q_2, q_3\}$;   // $s = 3$; q_1 is the initial state, q_3 is the final state (hence $F = \{q_3\}$);
- $\Sigma = \{0, 1\}$;
- $\Gamma = \{0, 1, \sqcup\}$;   // $t = 3$;
- the Turing program consists of the following instructions:

  1. $\delta(q_1, 1) = (q_2, \sqcup, \mathrm{Right})$;
  2. $\delta(q_2, 1) = (q_2, 1, \mathrm{Right})$;
  3. $\delta(q_2, 0) = (q_3, 1, \mathrm{Stay})$;
  4. $\delta(q_1, 0) = (q_3, \sqcup, \mathrm{Stay})$.

The state q_3 is final, because $\delta(q_3, z) \uparrow$ for all $z \in \Gamma$ (there are no instructions $\delta(q_3, z) = \ldots$).

For the input word 111011, the computation is illustrated in Fig. 6.4.

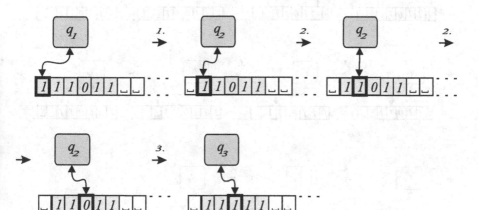

Fig. 6.4 Computing $3+2$ on a TM. Over the arrows are written the numbers of applied instructions

The corresponding sequence of internal configurations is:

$$q_1 111011 \rightarrow q_2 11011 \rightarrow 1q_2 1011 \rightarrow 11q_2 011 \rightarrow 11q_3 111.$$

If the input was 011, instruction 4 would execute, leaving the result 11. For the input 1110, the computation would proceed as in the figure above; only the result would be 111. $\square$

Example 6.2. (Addition on Another TM) Let us solve the same problem with another Turing machine that has a different Turing program. First, the window is moved to the right until $\sqcup$ is reached. Then the window is moved to the left (i.e., to the last symbol of the input word) and the symbol is deleted. If the deleted symbol is 0, the machine halts. Otherwise, the window keeps moving to the left and upon reading 0 the symbol 1 is written and the machine halts.

The Turing machine is $T' = (Q, \Sigma, \Gamma, \delta, q_1, \sqcup, \{q_5\})$, where:

- $Q = \{q_1, q_2, q_3, q_4, q_5\}$; $// s = 5$; q_1 is the initial and q_5 the final state ($F = \{q_5\}$);
- $\Sigma = \{0, 1\}$;
- $\Gamma = \{0, 1, \sqcup\}$; $// t = 3$;
- Turing program:

    1. $\delta(q_1, 1) = (q_2, 1, \text{Right})$ 6. $\delta(q_3, 0) = (q_5, \sqcup, \text{Stay})$
    2. $\delta(q_1, 0) = (q_2, 0, \text{Right})$ 7. $\delta(q_3, 1) = (q_4, \sqcup, \text{Left})$
    3. $\delta(q_2, 1) = (q_2, 1, \text{Right})$ 8. $\delta(q_4, 1) = (q_4, 1, \text{Left})$
    4. $\delta(q_2, 0) = (q_2, 0, \text{Right})$ 9. $\delta(q_4, 0) = (q_5, 1, \text{Stay})$
    5. $\delta(q_2, \sqcup) = (q_3, \sqcup, \text{Left})$

    The are no instructions of the form $\delta(q_5, z) = \ldots$, so the state q_5 is final.

For the input word 111011, the computation is illustrated in Fig. 6.5.

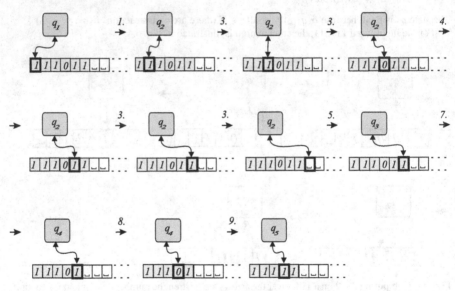

Fig. 6.5 Computing $3 + 2$ on a different TM. The numbers of applied instructions are on the arrows

The corresponding sequence of internal configurations is:

$q_1 111011 \rightarrow 1q_2 11011 \rightarrow 11q_2 1011 \rightarrow 111q_2 011 \rightarrow$
$\rightarrow 1110q_2 11 \rightarrow 11101q_2 1 \rightarrow 111011q_2 \rightarrow 11101q_3 1 \rightarrow$
$\rightarrow 1110q_4 1 \rightarrow 111q_4 01 \rightarrow 111q_5 11$ $\square$

6.1.2 Generalized Models

Turing considered different variants of the basic model; each is a generalization of the basic model in some respect. The variants differ from the basic model in their *external configurations*. For example, finite memory can be added to the control unit; the tape can be divided into parallel tracks, or it can become unbounded in both directions, or it can even be multi-dimensional; additional tapes can be introduced; and non-deterministic instructions can be allowed in Turing programs. The variants V of the basic model are:

- **Finite Storage TM.** This variant V has in its control unit a finite storage capable of memorizing $k \geqslant 1$ tape symbols and using them during the computation. The Turing program is formally $\delta_V : Q \times \Gamma \times \Gamma^k \to Q \times \Gamma \times \{\text{Left}, \text{Right}, \text{Stay}\} \times \Gamma^k$.

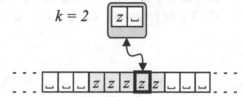

Fig. 6.6 Finite storage TM can store k tape symbols

- **Multi-track TM.** This variant V has the tape divided into $tk \geqslant 2$ *tracks*. On each track there are symbols from the alphabet Γ. The window displays tk-tuples of symbols, one symbol for each track. Formally the Turing program is the function $\delta_V : Q \times \Gamma^{tk} \to Q \times \Gamma^{tk} \times \{\text{Left}, \text{Right}, \text{Stay}\}$.

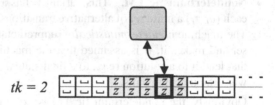

Fig. 6.7 Multi-track TM has tk tracks on its tape

- **Two-Way Unbounded TM.** This variant V has the tape unbounded in both directions. Formally, the Turing program is $\delta_V : Q \times \Gamma \to Q \times \Gamma \times \{\text{Left}, \text{Right}, \text{Stay}\}$.

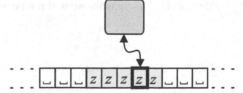

Fig. 6.8 Two-way TM has unbounded tape in both directions

- **Multi-tape TM.** This variant V has $tp \geqslant 2$ unbounded tapes. Each tape has its own window that is independent of other windows. Formally, the Turing program is $\delta_V : Q \times \Gamma^{tp} \to Q \times (\Gamma \times \{\text{Left}, \text{Right}, \text{Stay}\})^{tp}$.

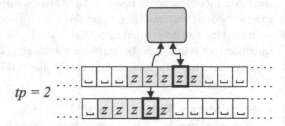

$tp = 2$

Fig. 6.9 Multi-tape TM has tp tapes with separate windows

- **Multidimensional TM.** This variant V has a d-dimensional tape, $d \geqslant 2$. The window can move in d dimensions, i.e., $2d$ directions $L_1, R_1, L_2, R_2, \ldots, L_d, R_d$. The Turing program is $\delta_V : Q \times \Gamma \to Q \times \Gamma \times \{L_1, R_1, L_2, R_2, \ldots, L_d, R_d, \text{Stay}\}$.

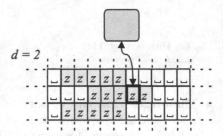

$d = 2$

Fig. 6.10 Multidimensional TM has a d-dimensional tape

- **Nondeterministic TM.** This variant V has a Turing program δ that assigns to each (q_i, z_r) a finite *set* of alternative transitions $\{(q_{j_1}, z_{w_1}, D_1), (q_{j_2}, z_{w_2}, D_2), \ldots\}$. The machine *nondeterministically* (unpredictably) chooses a transition from the set and makes it. It is assumed that the machine chooses one of the transitions that lead it to a solution (e.g., to a final state q_{yes}), *if* such transitions exist; otherwise, the machine halts.

Obviously, the nondeterministic TM is not a reasonable model of computation because it can foretell the future when choosing from alternative transitions. Nevertheless, it is a very useful tool, which makes it possible to define the minimum number of steps needed to compute the solution (when a solution exists). Again, this is important when we investigate the *computational complexity* of problem solving. For the *computability* on unrestricted models of computation, this is irrelevant. For this reason, we will not be using nondeterministic Turing machines.

6.1.3 Equivalence of Generalized and Basic Models

Although each generalized model V of the TM seems to be more powerful than the basic model T, it is not so; T can compute anything that V can compute. We will prove this by describing how T can simulate V. (The other way round is obvious as T is a special case of V.) In what follows we describe the main ideas of the proofs and leave the details to the reader as an exercise.

- **Simulation of a Finite Storage TM.** Let V be a finite storage TM with the storage capable of memorizing $k \geq 1$ tape symbols. The idea of the simulation is that T can memorize finite number of tape symbols by *encoding the symbols in the indexes of its states*. Of course, this may considerably enlarge the number of T's states, but recall that in the definition of the Turing machine there is no limitation of the number of states, as far as this number is finite.

  To implement the idea, we do the following: 1) We redefine the indexes of T's states by enforcing an internal structure on each index (i.e., by assuming that certain data is encoded in each index). 2) Using the redefined indexes, we are able to describe how T's instructions make sure that the tape symbols are memorized in the indexes of T's states. 3) Finally, we show that the redefined indexes can be represented by natural numbers. This shows that T is still the basic model of the Turing machine.

  The details are as follows:

  1. Let us denote by $[i, m_1, \ldots, m_k]$ any index that encodes k symbols $z_{m_1}, \ldots, z_{m_k}$ that were read from T's tape (not necessarily the last read symbols). Thus, the state $q_{[i,m_1,\ldots,m_k]}$ represents some usual T state q_i as well as the contents $z_{m_1}, \ldots, z_{m_k}$ of V's storage. Since at the beginning of the computation there are no memorized symbols, i.e., $z_{m_1} = \ldots = z_{m_k} = \sqcup$, and since $\sqcup = z_t$, where $t = |\Gamma|$, we see that the initial state is $q_{[1,t,\ldots,t]}$.
  2. The Turing program of T consists of two kinds of instructions. Instructions of the first kind *do not* memorize the symbol z_r that has been read from the current cell. Such instructions are of the form $\delta(q_{[i,m_1,\ldots,m_k]}, z_r) = (q_{[j,m_1,\ldots,m_k]}, z_w, D)$. Note that they leave the memorized symbols $z_{m_1}, \ldots, z_{m_k}$ unchanged. In contrast, the instructions of the second kind *do* memorize the symbol z_r. This symbol can substitute any of the memorized symbols (as if z_r had been written into any of the k locations of V's storage). Let ℓ, $1 \leq \ell \leq k$, denote which of the memorized symbols is to be substituted by z_r. The general form of the instruction is now $\delta(q_{[i,m_1,\ldots,m_k]}, z_r) = (q_{[j,m_1,\ldots,m_{\ell-1},r,m_{\ell+1},\ldots,m_k]}, z_w, D)$. After the execution of such an instruction, T will be in a new state that represents some usual state q_j as well as the new memorized symbols $z_{m_1}, \ldots, z_{m_{\ell-1}}, z_r, z_{m_{\ell+1}}, \ldots, z_{m_k}$.
  3. There are st^k indexes $[i, m_1, \ldots, m_k]$, where $1 \leq i \leq s = |Q|$ and $1 \leq m_\ell \leq t = |\Gamma|$ for $\ell = 1, \ldots, k$. We can construct a bijective function f from the set of all the indexes $[i, m_1, \ldots, m_k]$ onto the set $\{1, 2, \ldots, st^k\}$. (We leave the construction

of f as an exercise.) Using f, the states $q_{[i,m_1,\ldots,m_k]}$ can be renamed into the usual form $q_{f([i,m_1,\ldots,m_k])}$, where the indexes are natural numbers.

In summary, a finite storage TM V can be simulated by a basic Turing machine T if the indexes of T's states are appropriately interpreted.

- **Simulation of a Multi-track TM.** Let V be an tk-track TM, $tk \geqslant 2$. The idea of the simulation is that T considers tk-tuples of V's symbols as single symbols (see Fig. 6.11). Let V have on the ith track symbols from Γ_i, $i = 1, 2, \ldots, tk$. Then let T's tape alphabet be $\Gamma_1 \times \Gamma_2 \times \ldots \times \Gamma_{tk}$. If V has in its program δ_V an instruction $\delta_V(q_i, z_{r_1}, \ldots, z_{r_{tk}}) = (q_j, z_{w_1}, \ldots, z_{w_{tk}}, D)$, then let T have in its program δ_T the instruction $\delta_T(q_i, (z_{r_1}, \ldots, z_{r_{tk}})) = (q_j, (z_{w_1}, \ldots, z_{w_{tk}}), D)$. Then T simulates V.

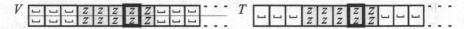

Fig. 6.11 Each pair of symbols in the V's window is considered as a single symbol by T

- **Simulation of a Two-Way Unbounded TM.** Let V be a two-way unbounded TM. Imagine that an arbitrary cell of V is chosen—denote it by c_0—and the tape is folded at the left border of c_0, so that the left part of the tape comes under the right part (see Fig. 6.12). After this, the cell c_{-1} is located under c_0, the cell c_{-2} under c_1, and so on. The result can be viewed as a one-way tape with two tracks. So, let us imagine that this is the tape of a two-track TM, which we denote by V'.

Now, the machine V' can simulate V as follows: 1) whatever V would do on the *right* part of its tape, V' does on the *upper* track of its tape; 2) whatever V would do on the *left* part of its tape, V' does on the *lower* track of its tape, while moving its window in the direction opposite to that of V's window; 3) when V would cross the left border of c_0, the machine V' passes to the opposite track in its first cell. To finish the proof recall from the previous simulation that V' can be simulated by a basic TM T. Thus, V can be simulated by T.

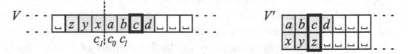

Fig. 6.12 The left part of V's tape is folded under the right one to obtain the tracks of V''s tape

- **Simulation of a Multi-tape TM.** Let V be a tp-tape TM, $tp \geqslant 2$. Note that after $k \geqslant 1$ steps the leftmost and rightmost window can be at most $2k + 1$ cells apart. (The distance is maximum when, in each step, the two windows move apart by two cells.) Other windows are between (or as far as) the two outermost windows.

Let us now imagine a two-way unbounded, $2tp$-track Turing machine V' (see Fig. 6.13). The machine V' can simulate V as follows. Each two successive tracks of V' describe the situation on one tape of V. That is, the situation on the ith tape of V is described by the tracks $2i-1$ and $2i$ of V'. The contents of the track $2i-1$ are the same as would be the contents of the ith tape of V. The track $2i$ is empty, except for one cell containing the symbol X. The symbol X is used to mark what V would see on its ith tape (i.e., where the window on the ith tape would be).

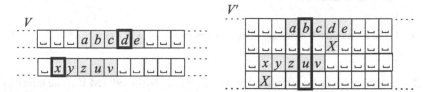

Fig. 6.13 Each V's tape is represented by two successive tracks of V''s tape

In one step, V would read tp symbols through its windows, write tp new symbols, move tp windows, and change the state. How can V' simulate all of this? The answer is by moving its single window to and fro and changing the contents of its tracks until they reflect the situation after V's step. Actually, this V' can do in *two sweeps*, first by moving its window from the leftmost X to the rightmost one, and then back to the leftmost X. When, during the *first* sweep, V' reads an X, it records the symbol above the X. In this way, V' records all the tp symbols that V would read through its windows. During the *second* sweep, V' uses information about V's Turing program: if V' detects an X, it substitutes the symbol above the X with another symbol (the same symbol that V would write on the corresponding tape) and, if necessary, moves the X to the neighboring cell (in the direction in which V would move the corresponding window). After the second sweep is complete, V' changes to the new state (corresponding to V's new state).

Some questions are still open. How does V' know that an outermost X has been reached so that the sweep is completed? The machine V' must have a counter (on additional track) that tells how many Xs are still to be detected in the current sweep. Before a sweep starts, the counter is set to tp, and during the sweep is decremented upon each detection of an X. When the counter reaches 0, the window is over an outermost X. This happens in finite time because the outermost Xs are at most $2k+1$ cells apart. Since each single move of V can be simulated by V' in finite time, every computation of V can also be simulated by V'.

The machine V' is a two-way unbounded, multi-track TM. However, from the previous simulations it follows that V' can be simulated by some basic model of the Turing machine T. Hence, V can be simulated by T.

- **Simulation of a Multidimensional TM.** Let V be a d-dimensional TM, $d \geqslant 2$. Let us call d-box the minimal rectangular d-polytope containing every nonempty cell of V's tape. For example, the 2-box is a rectangle (see Fig. 6.14) and the 3-box is a rectangular hexahedron. For simplicity we continue with $d = 2$. A 2-box contains rows (of cells) of equal length (the number of cells in the row).

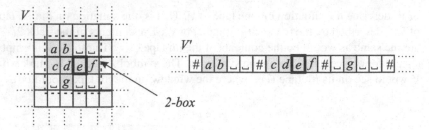

Fig. 6.14 The rows of the V's current 2-box are delimited by # on V''s tape

The machine V can be simulated by a two-way unbounded TM V'. The tape of V' contains finitely many rows of V's current 2-box, delimited by the symbol #. If V were to move its window *within the current 2-box*, then V' moves its window either to the neighboring cell in the same row (when V would move its window in the same row), or to the corresponding cell in the left/right neighboring row (when V would move its window to the upper/lower row). If, however, V would move its window *across the border of the current 2-box*, then V' either adds a new row before/after the existing rows (when V would cross the upper/lower border), or adds one empty cell to the beginning/end of each existing row and moves # as necessary (when V would cross the left/right border). We see that in order to extend a row by one cell, V' has to shift a part of the contents of its tape. V' can achieve this by moving the contents in a stepwise fashion, cell after cell. (Alternatively, V' can be supplied with a finite storage in its control unit. Using this, V' can shift the contents in a single sweep in a caterpillar-type movement.) The generalization to higher dimensions d is left to the reader as an exercise.

The machine V' is a two-way unbounded TM. As we know, V' can be simulated by a basic model of TM T.

- **Simulation of a Nondeterministic TM.**   Let V be a nondeterministic TM with a Turing program δ_V. The instructions are of the form

$$\delta_V(q_i, z_r) = \{(q_{j_1}, z_{w_1}, D_1), (q_{j_2}, z_{w_2}, D_2), \ldots, (q_{j_k}, z_{w_k}, D_k)\},$$

where $(q_{j_1}, z_{w_1}, D_1), (q_{j_2}, z_{w_2}, D_2), \ldots$ are alternative transitions and k depends on i and r. Call $|\delta_V(q_i, z_r)|$ the *indeterminacy of the instruction* $\delta_V(q_i, z_r)$, and $u \overset{\text{def}}{=} \max_{q_i, z_r} |\delta_V(q_i, z_r)|$ the *indeterminacy of the program* δ_V. We call a sequence of numbers $i_1, i_2, \ldots, i_\ell$, $1 \leqslant i_j \leqslant u$, the *scenario* of the execution of δ_V. We say that δ_V *executes alongside the scenario* $i_1, i_2, \ldots, i_\ell$ if the first instruction makes the i_1th transition, the second instruction makes the i_2th transition, and so on.

Let us now describe a three-tape (deterministic) TM V' that will simulate the machine V. The first tape contains the input word x (the same as V). The second tape is used for systematic generation of the scenarios of the execution of δ_V (in shortlex order). The third tape is used to simulate the execution of δ_V on x as if δ_V executed alongside the current scenario.

The machine V' operates in four steps: 1) V' generates the next scenario on the second tape. 2) V' clears the contents of the third tape and copies x from the first tape to the third one. 3) On the third tape, V' simulates the execution of the program δ_V alongside the current scenario; if, during the simulation, the scenario requires a nonexisting transition, V' returns to the first step. 4) If the current simulation halts in a state q_{yes} (i.e., V would halt on x in q_{yes}), the machine V' reports this (as would report V) and halts.

We see this: if, for arbitrary x, V halts in a final state q_{yes}, then it does so after executing a certain scenario. But V' generates the scenario in finite time and simulates V according to it. So, V' also halts on x and returns the same result as V.

It remains to simulate V' with T. But we have already proved that this is possible.

Remark. There is another useful view of the simulation. With each Turing program δ_V and input word x we can associate a directed tree, called the *decision tree of* δ_V for the input x. Vertices of the tree are the pairs $(q_i, z_r) \in Q \times \Gamma$ and there is an arc $(q_i, z_r) \rightsquigarrow (q_j, z_w)$ *iff* there is a possible transition from (q_i, z_r) to (q_j, z_w), i.e., $(q_j, z_w, D) \in \delta_V(q_i, z_r)$ for some D. The root of the tree is the vertex (q_1, a), where a is the first symbol of x. A scenario is a finite path from the root of the tree to an internal vertex. During the execution of δ_V, the nondeterministic TM V starts in the root and then *miraculously* chooses, at each vertex, the arc which is on some path to some final vertex (q_{yes}, z). In the case where none of the arcs leads towards a final vertex, the machine miraculously detects that and immediately halts. In contrast, the simulator T has no such capability, so it must systematically generate and check scenarios until it finds one ending in a vertex (q_{yes}, z). If there is no such scenario, T never halts.

6.1.4 Reduced Model

In Definition 6.1 (p. 102) of the basic TM, certain parameters are fixed; e.g., q_1 denotes the initial state; z_1, z_2, z_t denote the symbols $0, 1, \sqcup$, respectively; and the tape is a one-way unbounded single-track tape. We could also fix other parameters, e.g. Σ, Γ and F (with the exception of δ and Q, because fixing any of the two would result in a finite number of different Turing programs). We say that, by fixing these parameters, the basic TMs are *reduced*.

But why do that? The answer is that reduction simplifies many things, because reduced TMs differ only in their δs and Qs, i.e., in their Turing programs. So let us fix Γ and Σ, while fulfilling the condition $\{0, 1\} \subseteq \Sigma \subseteq \Gamma - \{\sqcup\}$ from Definition 6.1. We choose the simplest option, $\Sigma = \{0, 1\}$ and $\Gamma = \{0, 1, \sqcup\}$. No generality is lost by doing so, because any other Σ and Γ can be encoded by 0s and 1s. In addition, by merging the final states into one, say q_2, we can fix the set of final states to a singleton $F = \{q_2\}$. Then, a *reduced Turing machine* is a seven-tuple

$$T = (Q, \{0, 1\}, \{0, 1, \sqcup\}, \delta, q_1, \sqcup, \{q_2\}).$$

To obtain a particular reduced Turing machine, we must choose only Q and δ.

6.1.5 Equivalence of Reduced and Basic Models

Is the reduced model of TM less powerful than the basic one? The answer is no. The two models are equivalent, as they can simulate each other. Let us describe this.

Given an arbitrary reduced TM $R = (Q_R, \{0,1\}, \{0,1,\sqcup\}, \delta_R, q_{1R}, \sqcup, \{q_{2R}\})$, there is a basic TM $T = (Q_T, \Sigma_T, \Gamma_T, \delta_T, q_{1T}, \sqcup, F_T)$ capable of simulating R. Just take $Q_T := Q_R$, $\Sigma_T := \{0,1\}$, $\Gamma_T := \{0,1,\sqcup\}$, $\delta_T := \delta_R$, $q_{1T} := q_{1R}$, and $F_T := \{q_{2R}\}$.

Conversely, let $T = (Q_T, \Sigma_T, \Gamma_T, \delta_T, q_{1T}, \sqcup, F_T)$ be an arbitrary basic TM. We can describe a *finite-storge* TM $S = (Q_S, \{0,1\}, \{0,1,\sqcup\}, \delta_S, q_{1S}, \sqcup, \{q_{2S}\})$ that simulates T on an arbitrary input $w \in \Sigma_T^*$. Since S does not recognize T's tape symbols, we assume that w is binary-encoded, i.e., each symbol of w is replaced with its binary code. Let $n = \lceil \log_2 |\Gamma_T| \rceil$ be the length of this code. Thus, the binary input to S is of length $n|w|$. The machine S has storage in its control unit to record the code of T's current symbol (i.e., the symbol that would be scanned by T at that time). In addition, S's control unit has storage to record T's current state (i.e., the state in which T would be at that time). This storage is initiated to q_{1T}, T's initial state. Now S can start simulating T. It repeatedly does the following: (1) It reads and memorizes n consecutive symbols from its tape. (2) Suppose that the memorized T's current state is q_{iT} and the memorized symbols encode the symbol $z_r \in \Gamma_T$. If $\delta_T(q_{iT}, z_r) = (q_{jT}, z_w, D_{kT})$ is an instruction of δ_T, then S writes the code of z_w into n cells of its tape (thus replacing the code of z_r with the code of z_w), memorizes q_{jT}, and moves the window to the beginning of the neighboring group of n cells (if D_{kT} is Left or Right) or to the beginning of the current group of n cells (if $D_{kT} = $ Stay). As we have seen in Sect. 6.1.3, we can replace S with an equivalent TM R with no storage in its control unit. This is the sought-for reduced TM that is equivalent to T.

NB *The reduced model of the Turing machine enables us to identify Turing machines with their Turing programs. This can simplify the discussion.*

6.1.6 Use of Different Models

The computations on different models of the Turing machine can considerably differ in terms of time (i.e., the number of steps) and space (i.e., the number of visited cells). But this becomes important only when we are interested in the *computational complexity* of problem solving. As concerns the *general computability*, i.e., the computability on unrestricted models of computation, this is irrelevant.

So the question is: Are different models of the Turing machine of any use in *Computability Theory*? The answer is yes. The generalized models are useful when we try to prove the *existence* of a TM for solving a given problem. Usually, the construction of such a TM is easier if we choose a more versatile model of the Turing machine. In contrast, if we must prove the *nonexistence* of a TM for solving a given problem, then it is usually better to choose a more primitive model (e.g., the basic or reduced model).

Fortunately, by referring to *Computability Thesis*, we can avoid the cumbersome and error-prone constructing of TMs (see Sect. 5.3.5). We will do this only whenever the existence of a TM will be important (and not a particular TM).

6.2 Universal Turing Machine

Recall that Gödel enumerated formulas of *Formal Arithmetic* **A** and, in this way, enabled formulas to express facts about other formulas and, eventually, about themselves. As we have seen (Sect. 4.2.3), such a self-reference of formulas revealed the most important facts about formal axiomatic systems and their theories. Can a similar approach reveal important facts about Turing machines as well? The idea is this: If Turing machines were somehow enumerated (i.e., each TM described by a characteristic natural number, called the *index*), then each Turing machine T could compute with other Turing machines simply by including their indexes into T's input word. Of course, certain questions would immediately arise: How does one enumerate Turing machines with natural numbers? What kind of computing with indexes makes sense? How does one use computing with indexes to discover new facts about Turing machines? In this section we will explain how, in 1936, Turing answered these questions.

6.2.1 Coding and Enumeration of Turing Machines

In order to enumerate Turing machines, we must define how Turing machines will be encoded, that is, represented by words over some coding alphabet. The idea is that we only encode Turing *programs* δ, but in such way that the other components Q, Σ, Γ, F, which determine the particular Turing machine, can be restored from the program's code. An appropriate coding alphabet is $\{0, 1\}$, because it is included in the input alphabet Σ of every TM. In this way, every TM would also be able to read codes of other Turing machines and compute with them, as we suggested above. So, let us see how a Turing machine is encoded in the alphabet $\{0, 1\}$.

Let $T = (Q, \Sigma, \Gamma, \delta, q_1, \sqcup, F)$ be an arbitrary basic Turing machine. If

$$\delta(q_i, z_j) = (q_k, z_\ell, D_m)$$

is an instruction of its Turing program, we encode the instruction by the word

$$K = 0^i 10^j 10^k 10^\ell 10^m,$$

where $D_1 = \text{Left}$, $D_2 = \text{Right}$, and $D_3 = \text{Stay}$.

In this way, we encode each instruction of the program δ. Then, from the obtained codes $K_1, K_2, \ldots, K_r$ we construct the *code* $\langle T \rangle$ of the TM T as follows:

$$\langle T \rangle = 111 K_1 11 K_2 11 \dots 11 K_r 111. \tag{$*$}$$

We can interpret $\langle T \rangle$ to be the binary code of some natural number. Let us call this number the *index* of the Turing machine T (i.e., its program). Note, however, that some natural numbers are not indexes, because their binary codes are not structured as ($*$). To avoid this, we make the following **convention**: any natural number whose binary code is not of the form ($*$) is an index of a special Turing machine called the *empty* Turing machine. The program of this machine is everywhere undefined, i.e., for each possible pair of state and tape symbol. Thus, for every input, the empty Turing machine immediately halts, in 0 steps.

Consequently, the following proposition holds.

Proposition 6.1. *Every natural number is the index of exactly one Turing machine.*

Given an arbitrary index $\langle T \rangle$, we can easily restore from it the components Σ, Γ, Q and F of the corresponding basic TM $T = (Q, \Sigma, \Gamma, \delta, q_1, \sqcup, F)$; see Sect. 7.2. So, the above construction of $\langle T \rangle$ implicitly defines a *total* mapping $g : \mathbb{N} \to \mathcal{T}$, where $\mathcal{T}$ is the set of all basic Turing machines. Given an arbitrary $n \in \mathbb{N}$, $g(n)$ can be viewed as the nth basic TM and be denoted by T_n. By letting n run through $0, 1, 2, \dots$ we obtain the sequence $T_0, T_1, T_2, \dots$, i.e., an *enumeration* of all basic TMs. Later, on p. 129, g will be called the *enumeration function* of the set $\mathcal{T}$. Of course, corresponding to $T_0, T_1, T_2, \dots$ is the enumeration $\delta_0, \delta_1, \delta_2, \dots$ of Turing programs.

Remark. We could base the coding and enumeration of TMs on the *reduced* model. In this case, instructions $\delta(q_i, z_j) = (q_k, z_\ell, D_m)$ would be encoded by somewhat shorter words $K = 0^i 10^j 10^k 10^\ell 10^m$, since $j, \ell \in \{0, 1, 2\}$, and no restoration of Σ and Γ from $\langle T \rangle$ would be needed. However, due to the simulation, the programs of the reduced TMs would contain more instructions than the programs of the equivalent basic models. Consequently, their codes would be even longer.

Example 6.3. (TM Code) The code of the TM T in Example 6.1 (p. 105) is

$$\langle T \rangle = 111 \underbrace{01001001000100}_{K_1} 11 \underbrace{00100100100100}_{K_2} 11 \underbrace{001010001001000}_{K_3} 11 \underbrace{010100010001000}_{K_4} 111$$

and the corresponding index is 1075142408958020240455, a large number.

The code $\langle T' \rangle$ of the TM T' in Example 6.2 (p. 106) is

$$\langle T' \rangle = 111 \underbrace{01001001001000}_{K_1} 11 \underbrace{00101001010100}_{K_2} 11 \underbrace{00100100100100}_{K_3} 11 \underbrace{001010010100}_{K_4} 11$$
$$\underbrace{001000100100010}_{K_5} 11 \underbrace{00010100000100001000}_{K_6} 11 \underbrace{0001001000010000100010}_{K_7} 11 \underbrace{00001001000010010}_{K_8} 11$$
$$\underbrace{00001010000010010000}_{K_9} 111.$$

This code is different from $\langle T \rangle$ because T' has a different Turing program. The corresponding index is 13310153014029126947161548189999893572326199465677 (no need to remember it). $\quad\square$

Obviously, the indexes of Turing machines are huge numbers and hence not very practical. Luckily, we will almost never need their actual values, and they will not be operands of any arithmetic operation.

6.2.2 The Existence of a Universal Turing Machine

In 1936, using the enumeration of his machines, Turing discovered a seminal fact about Turing machines. We state the discovery in the following proposition.

Proposition 6.2. *There is a Turing machine that can compute whatever is computable by any other Turing machine.*

Proof. The idea is to construct a Turing machine U that is capable of simulating *any* other TM T. To achieve this, we use the method of proving by CT (see Sect. 5.3.5): (a) we describe the concept of the machine U and informally describe the algorithm executed by its Turing program, and (b) we refer to CT to prove that U exists.

(a) The concept of the machine U:

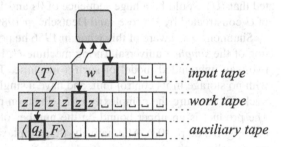

Fig. 6.15 Universal TM: tapes and their contents

- Tapes of the machine U (see Fig. 6.15):

  1. The first is the *input tape*. This tape contains an input word consisting of two parts: the code $\langle T \rangle$ of an arbitrary TM $T = (Q, \Sigma, \Gamma, \delta, q_1, \sqcup, F)$, and an arbitrary word w.
  2. The second is the *work tape*. Initially, it is empty. The machine U will use it in exactly the same way as T would use its own tape when given the input w.
  3. The third is the *auxiliary tape*. Initially, it is empty. The machine U will use it to record the current state in which the simulated T would be at that time, and for comparing this state with the final states of T.

- The Turing program of U should execute the following informal algorithm:

   1. Check whether the input word is $\langle T, w \rangle$, where $\langle T \rangle$ is a code of some TM. If it is not, halt.
   2. From $\langle T \rangle$ restore the set F and write the code $\langle q_1, F \rangle$ to the auxiliary tape.
   3. Copy w to the work tape and shift the window to the beginning of w.
   4. // Let the aux. tape have $\langle q_i, F \rangle$ and the work tape window scan a symbol z_r. If $q_i \in F$, halt.   // T would halt in a final state.
   5. On the input tape, search in $\langle T \rangle$ for the instruction beginning with "$\delta(q_i, z_r) =$"
   6. If not found, halt.   // T would halt in a non-final state.
   7. // Suppose that the found instruction is $\delta(q_i, z_r) = (q_j, z_w, D)$. On the work tape, write the symbol z_w and move the window in direction D.
   8. On the auxiliary tape, replace $\langle q_i, F \rangle$ by $\langle q_j, F \rangle$.
   9. Continue with step 4.

(b) The above algorithm can be executed by a human. So, according to the *Computability Thesis*, there is a Turing machine $U = (Q_U, \Sigma_U, \Gamma_U, \delta_U, q_{1U}, \sqcup, F_U)$ whose program δ_U executes this algorithm. We call U the *Universal Turing Machine*. □

Small Universal Turing Machines

The universal Turing machine U was actually described in detail. It was to be expected that $\langle U \rangle$ would be a huge sequence of 0s and 1s. Indeed, for example, the code of U constructed by Penrose[2] and Deutsch[3] in 1989 had about 5,500 bits.

Shannon[4] was aware of this when in 1956 he posed the problem of the construction of the *simplest* universal Turing machine U. He was interested in the simplest two-way unbounded model of such a machine. Thus, U was to be deterministic with no storage in its control unit, and have a single two-way infinite tape with one track. To measure the complexity of U Shannon proposed the product $|Q_U| \cdot |\Gamma_U|$. The product is an upper bound on the number of instructions in the program δ_U (see Fig. 6.3 on p. 103). Alternatively, the complexity of U could be measured more realistically by the number of *actual* instructions in δ_U.

Soon it became clear that there is a trade-off between $|Q_U|$ and $|\Gamma_U|$: the number of states can be decreased if the number of tape symbols is increased, and vice versa. So the researchers focused on different *classes* of universal Turing machines. Such a class is denoted by $\text{UTM}(s, t)$, for some $s, t \geq 2$, and by definition contains all the universal Turing machines with s states and t tape symbols (of the above model).

In 1996, Rogozhin[5] found universal Turing machines in the classes $\text{UTM}(2,18)$, $\text{UTM}(3,10)$, $\text{UTM}(4,6)$, $\text{UTM}(5,5)$, $\text{UTM}(7,4)$, $\text{UTM}(10,3)$, and $\text{UTM}(24,2)$. Of these, the machine $U \in \text{UTM}(4,6)$ has the smallest number of instructions: 22.

---

[2] Roger Penrose, 1931, British physicist, mathematician and philosopher.

[3] David Elieser Deutsch, 1953, British physicist.

[4] Claude Elwood Shannon, 1916–2001, American mathematician and electronics engineer.

[5] Yurii Rogozhin, 1949, Moldavian mathematician and computer scientist.

6.2.3 The Importance of the Universal Turing Machine

We can now upgrade Fig. 5.12 (p. 90) to Fig. 6.16. The existence of the universal TM indicated that it might be possible to design a general-purpose computing machine—something that is today called the general-purpose computer.

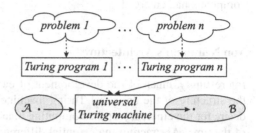

Fig. 6.16 The universal Turing machine can execute any Turing program and thus compute the solution of any problem

6.2.4 Practical Consequences: Data vs. Instructions

Notice that the machine U uses both the *program* of T and the input *data* to T as its own input *data*, i.e., as two input words $\langle T \rangle$ and w written in the *same* alphabet Σ_U. U interprets the word $\langle T \rangle$ as a sequence of instructions to be simulated (i.e., executed), and the word w as input to the simulated T. This consequence is one of the important discoveries of Turing.

Consequence 6.1. (Data vs. Instructions) *There is no a priori difference between data and instructions; the distinction between the two is established by their interpretation.*

6.2.5 Practical Consequences: General-Purpose Computer

Turing's proof of the existence of a universal Turing machine was a theoretical proof that *a general-purpose computing machine* is possible. This answers the question raised by Babbage a century earlier (see p. 6). Thus, the following practical consequence of Turing's discovery was evident.

Consequence 6.2. (General-Purpose Computer) *It is possible to construct a physical computing machine that can compute whatever is computable by any other physical computing machine.*

The construction of a general-purpose computing machine started at the beginning of the 1940s. After initial unsuccessful trials, which were mainly due to the teething troubles of electronics, researchers developed the first, increasingly efficient general-purpose computing machines, now called *computers*. These included ENIAC, EDVAC, and IAS, which were developed by research teams led by Mauchly,[6] Eckert,[7] von Neumann, and others. By the mid-1950s, a dozen other computers had emerged.

Von Neumann's Architecture

Interestingly, much of the development of early computers did not closely follow the structure of the universal Turing machine. The reasons for this were both the desire for the efficiency of the computing machine and the technological conditions of the time. Abstracting the essential differences between these computers and the universal TM, and describing the differences in terms of Turing machines, we find the following:

- Cells are now *enumerated*.
- The control unit does not access cells by a time-consuming movement of the window. Indeed, there is no window. Instead, the *control unit directly accesses an arbitrary cell in constant time* by using an additional component.
- The program is no longer in the control unit. Instead, the *program is written in cells* of the tape (as is the input data to the program).
- The control unit still executes the program in a stepwise fashion, but the *control unit has different duties*. Specifically, in each step, it typically does the following:

1. reads an instruction from a cell;
2. reads operands from cells;
3. executes the operation on the operands;
4. writes the result to a cell.

To do this, the control unit uses additional components: the *program counter* that describes from which cell the next instruction of the program will be read, *registers* which store operands, and a special register, called the *accumulator*, where the result of the operation is left.

Of course, due to these differences, terminological differences also arose. For example, *main memory* ($\approx$ tape), *program* ($\approx$ Turing program), *processor* ($\approx$ control unit), *memory location* ($\approx$ cell), and *memory address* ($\approx$ cell number). The general structure of these computers was called the *von Neumann architecture* (after an influential report on the logical design of the EDVAC computer, in which von Neumann described the key findings of its design team).

---

[6] John William Mauchly, 1907–1980, American physicist.

[7] John Adam Presper Eckert, Jr., 1919–1995, American engineer and computer scientist.

6.2.6 Practical Consequences: Operating System

What takes care of loading a program P, which is to be executed by a computer, into the memory? This is the responsibility of the *operating system* (*OS*). The operating system is a special program that is *resident* in memory. When it executes, it takes care of everything needed to execute any other program P. In particular:

1. It *reads* the program P and its input data from the computer's environment.
2. It *loads* P into the memory.
3. It *sets apart* additional memory space, which P will be using during its execution. This space contains the *data region* with P's input data and other global variables; the *runtime stack* for local variables of procedures and procedure-linkage information; and the *heap* for dynamic allocation of space when explicitly demanded by P.
4. It *initiates* P's execution by transferring control to it, i.e., by writing to the program counter the address of P's first instruction.
5. When P halts, it takes over and gives a chance to the next waiting program.

In time, additional goals and tasks were imposed on operating systems, mainly because of the desire to improve the efficiency of the computer and its user-friendliness. Such tasks include multiprogramming (i.e., supporting the concurrent execution of several programs), memory management (i.e., simulating a larger memory than available in reality), file system (i.e., supporting permanent data storage), input/output management (i.e., supporting communication with the environment), protection (i.e., protecting programs from other programs), security (i.e., protecting programs from the environment), and networking (i.e., supporting communication with other computers).

But there remains the question of what loads the very OS into the memory and starts it. For this, *hardware* is responsible. There is a small program, called the *bootstrap loader*, embedded in the hardware. This program

1. starts automatically when the computer is turned on,
2. loads the OS to the memory, and
3. transfers the control to the OS.

In terms of Turing machines, the bootstrap loader is the Turing program in the control unit of the universal Turing machine U. Its role, however, is to read the *simulator* (i.e., OS) into the control unit, thus turning U into a true universal TM. (The reading of finitely many data and hence the whole simulator into the control unit of a Turing machine is possible, as we described in Sect. 6.1.3.)

We conclude that a modern general-purpose computer can be viewed as a universal Turing machine.

6.2.7 Practical Consequences: RAM Model of Computation

After the first general-purpose computers emerged, researchers tried to abstract the essential properties of these machines in a suitable model of computation. They suggested several models of computation, of which we mention the *register machine* and its variants *RAM* and *RASP*.[8] All of these were proved to be equivalent to the Turing machine: what can be computed on the Turing machine, can also be computed on any of the new models, and vice versa. In addition, the new models displayed a property that was becoming increasingly important as solutions to more and more problems were attempted by the computers. Namely, these models proved to be more suitable for and realistic in estimating the computational resources (e.g., time and space) needed for computations on computers. This is particularly true of the RAM model. Analysis of the *computational complexity* of problems, i.e., an estimation of the amount of computing resources needed to solve a problem, initiated the *Computational Complexity Theory*, a new area of *Computability Theory*. So, let us describe the RAM model of computation.

While the Turing machine has *sequential access* to data on its tape, RAM accesses data *randomly*. There are several variants of RAM but they differ only in the level of detail in which they reflect the von Neumann architecture. We present a more detailed one in Fig. 6.17.

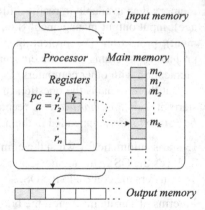

Fig. 6.17 RAM model of computation

Definition 6.3. (RAM) The **random access machine (RAM)** model of computation has several components: the *processor* with *registers*, two of which are the *program counter* and the *accumulator*; the *main memory*; the *input* and *output memory*; and a *program* with *input data*. Also the following holds:

---

[8] The *register machine* and its variants RAM (random access machine) and RASP (random access stored program) were gradually defined by Wang (1954), Melzak and Minsky (1961), Shepherdson and Sturgis (1963), Elgot and Robinson (1964), Hartmanis (1971), and Cook and Rechow (1973).

1. *a. Input* and *Output Memory:* During the computation, input data is read sequentially from the input memory, and the results are written to the output memory. Each memory is a sequence of equally sized *locations*. The location size is arbitrary, but finite. A location is empty or contains an integer.

   b. Main Memory: This is a potentially infinite sequence of equally sized locations $m_0, m_1, \ldots$ The index i is called the *address* of m_i. Each location is *directly accessible* by the processor: given an arbitrary i, reading from m_i or writing to m_i is accomplished in constant time.

   c. Registers: This is a sequence of locations $r_1, r_2, \ldots, r_n, n \geqslant 2$, in the processor. Registers are directly accessible. Two of them have special roles. *Program counter* $pc \, (= r_1)$ contains the address of the location in the main memory, which contains the instruction to be executed next. *Accumulator* $a \, (= r_2)$ is involved in the execution of each instruction. Other r_i are given roles as needed.

   d. Program: The program is a finite sequence of *instructions*. The details of the instructions are not very important as long as the RAM is of limited capability (see Sect. 5.2.4). So, it is assumed that the instructions are similar to the instructions of real computers. Thus, there are arithmetical, logic, input/output, and (un)conditional jump instructions. If $n = 2$, each instruction contains the information op about the operation and, depending on op, the information i about the operand. There may be additional modes of addressing, e.g., indirect (denoted by $*i$) and immediate (denoted by $=i$). Examples of instructions are:

read	(read data from input memory to accumulator a)
store i	($m_i := a$)
load i	($a := m_i$)
add i	($a := a + m_i$)
jmp i	($pc := i$)
jz i	(if $a = 0$ then $pc := i$)
jgz i	(if $a > 0$ then $pc := i$)
write	(write data from accumulator a to output memory)
halt	(halt)
load $*i$	($a := m_{m_i}$)
add $=i$	($a := a + i$)

2. Before the RAM is started, the following is done: (a) a program is loaded into the main memory (into successive locations starting with m_0); (b) input data are written to the input memory; (c) the output memory and registers are cleared.

3. From this point on, the RAM operates independently in a mechanical stepwise fashion as instructed by its program. Let $pc = k$ at the beginning of a step. (Initially, $k = 0$.) From the location m_k, the instruction I is read and started. At the same time pc is incremented. So, when I is completed, the next instruction to be executed is in m_{k+1}, *unless* one of the following holds: a) I was jmp i; b) I was jz i or jgz i and pc was assigned i; c) I was halt; d) I changed pc so that it contains an address outside the program. In c) and d) the program halts.   □

We now state the following important proposition.

Proposition 6.3. (RAM vs. TM) *The RAM and the Turing machine are equivalent: what can be computed on one of them can be computed on the other.*

Box 6.1 (Proof of Proposition 6.3). We show how TM and RAM simulate each other.

Simulation of a TM with a RAM. Let $T = (Q, \Sigma, \Gamma, \delta, q_1, \sqcup, F)$ be an arbitrary TM. The RAM will have its main memory divided into three parts. The first part, consisting of the first p locations $m_0, \ldots, m_{p-1}$, will contain the RAM's program. The second part will contain the Turing program δ. The third part will be the rest of the main memory; during the simulation, it will contain the same data as T would have on its tape. Two of RAM's registers will play special roles: r_3 will reflect the current state of T, and r_4 will contain the address of the location in the RAM's main memory corresponding to the cell under T's window.

The RAM is initialized as follows. Let us view δ as a table $\Delta = Q \times \Gamma$, where the component $\Delta[q, z] = (q', z', D)$ if $\delta(q, z) = (q', z', D)$ is an instruction of δ, and $\Delta[q, z] = 0$ if $\delta(q, z)\!\uparrow$. Since there are $d = |Q| \cdot |\Gamma|$ components in Δ, we can bijectively map them to the d locations $m_p, \ldots, m_{p+d-1}$. So, we choose a bijection $\ell : \Delta \to \{p, \ldots, p + d - 1\}$ and write each $\Delta[q, z]$ into the location $m_{\ell(q,z)}$. (A possible ℓ would map row after row of Δ into the memory.) The third part of the RAM's memory is cleared and T's input word is written to the beginning of it, i.e., into the locations $m_{p+d}, m_{p+d+1}, \ldots$. The registers r_3 and r_4 are set to q_1 and $p + d$, respectively.

Now, the simulation of T starts. Each step of T is simulated as follows. Based on the values $q = (r_3)$ and $z = m_{(r_4)}$, the RAM reads the value of $\delta(q, z)$ from $m_{\ell(q,z)}$. If this value is 0, the RAM halts because $\delta(q, z)\!\uparrow$. Otherwise, the value read is (q', z', D), so the RAM must simulate the instruction $\delta(q, z) = (q', z', D)$. This it can do in three steps: 1) $r_3 := q'$; 2) $m_{(r_4)} := z'$; 3) depending on D, it decrements r_4 ($D =$ Left), increments r_4 ($D =$ Right), or leaves r_4 unchanged ($D =$ Stay).

Remark. Note that the RAM could simulate any other TM. To do this, it would only change δ in the second part of the main memory and, of course, adapt the value of d.

Simulation of a RAM with a TM. Let R be an arbitrary RAM. R will be simulated by the following multi-tape TM T. For each R's register r_i there is a tape, called the r_i-tape, whose contents will be the same as the contents of r_i. Initially, it contains 0. There is also a tape, called the m-tape, whose contents will be the same as the contents of the R's main memory. If, for $i = 0, 1, 2, \ldots$, the location m_i would contain c_i, then the m-tape will have written $| \ 0 : c_0 \ | \ 1 : c_1 \ | \ 2 : c_2 \ | \ldots | \ i : c_i \ | \ldots$ (up to the last nonempty word).

T operates as follows. It reads from the pc-tape the value of the program counter, say k, and increments this value on the pc-tape. Then it searches the m-tape for the subsequence $| \ k : $. If the subsequence is found, T reads c_k, i.e., R's instruction I, and extracts from I both the information op about the operation and the information i about the operand. What follows depends on the addressing mode: (a) If I is op $= i$ (immediate addressing), then the operand is i. (b) If I is op i (direct addressing), then T searches the m-tape for the subsequence $| \ i : $. If the subsequence is found, then the operand is c_i. (c) If I is op $*i$ (indirect addressing), then, after T has found c_i, it searches the m-tape for the subsequence $| \ c_i : $. If the subsequence is found then the operand is c_{c_i}. When the operand if known, T executes op using the operand and the contents of the a-tape. $\square$

NB *General-purpose computers are capable of computing exactly what Turing machines can—assuming there are no limitations on the time and space consumed by the computations. Because of this, we will continue to investigate computability by using a Turing machine as the model of computation. The conclusions that we will come to will also hold for modern general-purpose computers.*

6.3 Use of a Turing Machine

In this section we will describe three elementary tasks for which we can use a Turing machine: 1) to compute the values of a function; 2) to generate elements of a set; and 3) to find out whether objects are members of a set.

6.3.1 Function Computation

A Turing machine is implicitly associated, for each natural $k \geqslant 1$, with a k-ary function mapping k words into one word. We will call it the k-*ary proper function* of the Turing machine. Here are the details.

Definition 6.4. (Proper Function) Let $T = (Q, \Sigma, \Gamma, \delta, q_1, \sqcup, F)$ be an arbitrary Turing machine and $k \geqslant 1$ a natural number. The k-**ary proper function** of T is a partial function $\psi_T^{(k)} : (\Sigma^*)^k \to \Sigma^*$, defined as follows:

If the input word to T consists of words $u_1, \ldots, u_k \in \Sigma^*$, then

$$\psi_T^{(k)}(u_1, \ldots, u_k) \stackrel{\text{def}}{=} \begin{cases} v, & \textit{if} \quad T \text{ halts in any state } \textit{and} \text{ the tape contains} \\ & \qquad \text{only the word } v \in \Sigma^*; \\ \uparrow, & \textit{else}, \text{ i.e., } T \text{ doesn't halt } \textit{or} \text{ the tape doesn't contain} \\ & \qquad \text{a word in } \Sigma^*. \end{cases}$$

If e is an index of T, we also denote the k-ary proper function of T by $\psi_e^{(k)}$. When k is known from the context or it is not important, we write ψ_T or ψ_e. The domain of ψ_e is also denoted by $\mathcal{W}_e$, i.e., $\mathcal{W}_e = \text{dom}(\psi_e) = \{x \mid \psi_e(x)\downarrow\}$.

So, given k words $u_1, \ldots, u_k$ written in the input alphabet of the Turing machine, we write the words to the tape of the machine, start it and wait until the machine halts and leaves a single word on the tape written in the same alphabet. If this does happen, and the resulting word is denoted by v, then we say that the machine has computed the value v of its k-ary proper function for the arguments $u_1, \ldots, u_k$.

The interpretation of the words $u_1, \ldots, u_k$ and v is left to us. For example, we can view the words as the encodings of natural numbers. In particular, we may use the alphabet $\Sigma = \{0, 1\}$ to encode $n \in \mathbb{N}$ by 1^n, and use 0 as a delimiter between the different encodings on the tape. For instance, the word 11101011001111 represents the numbers $3, 1, 2, 0, 4$. (The number 0 was represented by the empty word $\varepsilon = 1^0$.) In this case, $\psi_T^{(k)}$ is a k-ary numerical function with values represented by the words $1 \ldots 1$. When the function value is 0, the tape is empty (i.e., contains $1^0 = \varepsilon$). Another encoding will be given in Sect. 6.3.6.

TM as a Computer of a Function

Often we face the opposite task: Given a function $\varphi : (\Sigma^*)^k \to \Sigma^*$, find a TM capable of computing φ's values. Thus, we must find a TM T such that $\psi_T^{(k)} \simeq \varphi$. Depending on how powerful, if at all, such a T can be, i.e., depending on the extent to which φ can possibly be computed, we distinguish between three kinds of functions φ (in accordance with the formalization on p. 95).

Definition 6.5. Let $\varphi : (\Sigma^*)^k \to \Sigma^*$ be a function. We say that

φ is **computable**   *if*   there is a TM that can compute φ
                                            anywhere on $\mathrm{dom}(\varphi)$
                                     and $\mathrm{dom}(\varphi) = (\Sigma^*)^k$;

φ is **partial computable**   *if*   there is a TM that can compute φ
                                            anywhere on $\mathrm{dom}(\varphi)$;

φ is **incomputable**   *if*   there is *no* TM that can compute φ
                                            anywhere on $\mathrm{dom}(\varphi)$.

A TM that can compute φ anywhere on $\mathrm{dom}(\varphi)$ is also called the **computer** of φ.

Example 6.4. (Addition in Different Ways) The two Turing machines in Examples 6.1 (p. 105) and 6.2 (p. 106) are computers of the function $\mathrm{sum}(n_1, n_2) = n_1 + n_2$, where $n_1, n_2 \geqslant 0$.   □

A slight generalization defines (in accordance with Definition 5.3 on p. 96) what it means when we say that such a function is computable *on a set* $\mathcal{S} \subseteq (\Sigma^*)^k$.

Definition 6.6. Let $\varphi : (\Sigma^*)^k \to \Sigma^*$ be a function and $\mathcal{S} \subseteq (\Sigma^*)^k$. We say that

φ is **computable on** $\mathcal{S}$ *if*   there is a TM that can compute φ
                                            anywhere on $\mathcal{S}$;

φ is **partial computable on** $\mathcal{S}$ *if*   there is a TM that can compute φ
                                            anywhere on $\mathcal{S} \cap \mathrm{dom}(\varphi)$;

φ is **incomputable on** $\mathcal{S}$ *if*   there is *no* TM that can compute φ
                                            anywhere on $\mathcal{S} \cap \mathrm{dom}(\varphi)$.

6.3.2 Set Generation

When can elements of a set be enumerated? That is, when can elements of a set be listed in a finite or infinite sequence so that each and every element of the set sooner or later appears in the sequence? Moreover, when can the sequence be generated by an algorithm? These questions started the quest for the formalization of set generation. The answers were provided by Post, Church, and Turing.

Post's Discoveries

Soon after Hilbert's program appeared in the 1920s, Post started investigating the decidability of formal theories. In 1921, he proved that the *Propositional Calculus* **P** is a decidable theory by showing that there is a decision procedure (i.e., algorithm) which, for an arbitrary proposition of **P**, decides whether or not the proposition is provable in **P**. The algorithm uses truth-tables. Post then started investigating the decidability of the formal theory developed in *Principia Mathematica* (see Sect. 2.2.3). Confronted with *sets* of propositions, he realized that there is a strong similarity between the *process of proving* propositions in a formal theory and the *process of* *"mechanical, algorithmic generating"* of the elements of a set. Indeed, proving a proposition is, in effect, the same as "generating" a new element of the set of all theorems. This led Post to the questions,

> *What does it mean to "algorithmically generate" elements of a set?*
> *Can every countable set be algorithmically generated?*

Consequently, he started searching for a *model* that would formally define the intuitive concept of the "algorithmic generation of a set" (i.e., an algorithmic listing of all the elements of a set). Such a model is called the **generator** of a set.

In 1920–1921, Post developed *canonical systems* and *normal systems* and proposed these as generators. Informally, a canonical system consists of a symbol S, an alphabet Σ, and a finite set $\mathcal{P}$ of transformation rules, called productions. Starting with the symbol S, the system gradually transforms S through a sequence of intermediate words into a word over Σ. We say that the word has been generated by the canonical system. In each step of the generation, the last intermediate word is transformed by a production applicable to it. Since there may be several applicable productions, one must be selected. As different selections generally lead to different generated words, the canonical system can generate a *set* of words over Σ. Post also showed that productions can be simplified while retaining the generating power of the system. He called canonical systems with simplified productions *normal systems*. The set generated by a normal system he called the *normal set*.[9]

---

[9] Post described his ideas in a less abstract and more readable way than was usual practice at the time. The reader is advised to read his influential and informative paper from 1944 (see References). Very soon, Post's user-friendly style was adopted by other researchers. This speeded up exchange of the ideas and results and, hence, the development of *Computability Theory*.

Box 6.2 (Normal Systems).

A canonical system is a quintuple $(\mathcal{V}, \Sigma, \mathcal{X}, \mathcal{P}, S)$, where $\mathcal{V}$ is a finite set of symbols and $\Sigma \subset \mathcal{V}$. The symbols in Σ are said to be *final*, and the symbols in $\mathcal{V} - \Sigma$ are *non-final*. There is a distinguished non-final symbol, S, called the *start* symbol. $\mathcal{X}$ is a finite set of *variables*. A variable can be assigned a value that is an element of $\mathcal{V}^*$, i.e., any finite sequence of final and non-final symbols.

The remaining component of the quintuple is $\mathcal{P}$. This is a finite set of transformation rules called *productions*. A production describes the conditions under which, and the manner in which, subwords of a word can be used to build a new word. The general form of a production is

$$\alpha_0 x_1 \alpha_1 x_2 \dots \alpha_{n-1} x_n \alpha_n \to \beta_0 x_{i_1} \beta_1 x_{i_2} \dots \beta_{m-1} x_{i_m} \beta_m,$$

where $\alpha_i, \beta_j \in \mathcal{V}^*$ and $x_k \in \mathcal{X}$. When and how can a production p be applied to a word $v \in \mathcal{V}^*$? If each variable x_k of the left-hand side of p can be assigned a subword of v so that the left-hand side of p becomes equal to v, then v can be transformed into a word v' that is obtained from the right-hand side of p by substituting variables x_k with their assigned values. Note that p may radically change the word v: some subwords of v may disappear; new subwords may appear in v'; and all these constituents may be arbitrarily permuted in v'.

We write $S \xrightarrow{*}_{\mathcal{P}} v$ to denote that the start symbol S can be transformed to v in a finite number of applications of productions of $\mathcal{P}$. We say that S *generates* the word w if $S \xrightarrow{*}_{\mathcal{P}} w$ and $w \in \Sigma^*$. The set of words generated by $\mathcal{P}$ is denoted by $G(\mathcal{P})$, that is, $G(\mathcal{P}) = \{w \in \Sigma^* \mid S \xrightarrow{*}_{\mathcal{P}} w\}$.

Post proved that the productions of any canonical system can be substituted with productions of the form $\alpha_i x_k \to x_k \beta_j$. Canonical systems with such productions are said to be *normal*.

Post proved that the formal theory developed in *Principia Mathematica* can be represented as a normal system. Consequently, the set of theorems of *Principia Mathematica* is a normal set. Encouraged by this result, in 1921 he proposed the following formalization of the intuitive notion of set "generation":

Post Thesis. *A set* S *can be "generated"* $\longleftrightarrow$ S *is normal*

Post did not prove this proposition. He was not aware that it cannot be proved. The reasons are the same as those that, 15 years later, prevented the proving of the *Computability Thesis* (see Sect. 5.3). Namely, the proposition is a variant of the *Computability Thesis*.

In order to move on, he used the proposition as a working hypothesis (now called the *Post Thesis*), i.e., something that he will eventually prove. The thesis enabled him to progress in his research. Indeed, he made several important findings that were, 15 years later, independently and in different ways discovered by Gödel, Church and Turing. (In essence, these are the existence of undecidable sets and the existence of the universal Turing machine. We will come to these in the following sections.) Unfortunately, Post did not publish his results because he felt sure that he should prove his thesis first. As a byproduct of his attempts to do that, he proposed in 1936 a model of computation, which is now called the Post machine (see Sect. 5.2.3).

Church's Approach

In 1936, Church also became interested in the questions of set "generation." While investigating sets, whose elements are values of computable functions, he noticed: If a function g is computable on $\mathbb{N}$, then one can successively compute the values $g(0), g(1), g(2), \ldots$ and hence generate the set $\{g(i) \mid i \in \mathbb{N}\}$.

But, the opposite task is more interesting: Given a set S, find a computable function $g : \mathbb{N} \to S$ so that $\{g(i) \mid i \in \mathbb{N}\} = S$. If g exists, then S can be listed, i.e., all the elements of S are $g(0), g(1), g(2), \ldots$, and enumerated, i.e., an element $x \in S$ is said to be nth in order if n is the smallest $i \in \mathbb{N}$ for which $g(i) = x$. Such a g is said to be the *enumeration function* of the set S. (For example, the mapping g on p. 116 is the enumeration function of the set of all Turing programs.)

These ideas were also applied by Kleene. He imagined a function g that maps natural numbers into systems of equations $\mathscr{E}$, as defined by Herbrand and Gödel (see p. 76). If g is computable, then by computing $g(0), g(1), g(2), \ldots$ one generates systems $\mathscr{E}_0, \mathscr{E}_1, \mathscr{E}_2, \ldots$, each of which defines a computable function. Kleene then proved that there is *no* (total) computable function g on $\mathbb{N}$ such that g would generate *all* systems of equations that define computable functions. The reader may (correctly) suspect that Kleene's proof is connected with the diagonalizaton in Sect. 5.3.3.

Naturally, Post was interested in seeing how his normal systems compared with Church's generator (i.e., total computable functions). He readily proved the following theorem. (We omit the proof.)

Theorem 6.1 (Normal Set). *A set S is normal $\Longleftrightarrow S = \emptyset \vee S$ is the range of a computable function on $\mathbb{N}$.*

TM as a Generator

In addition to Post's normal systems and Church's computable functions, Turing machines can also be generators. A Turing machine that generates a set S will be denoted by G_S (see Fig. 6.18). The machine G_S writes to its tape, in succession, the elements of S and nothing else. The elements are delimited by the appropriate tape symbol in $\Gamma - \Sigma$ (e.g., #).

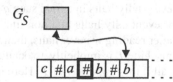

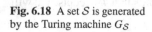
Fig. 6.18 A set S is generated by the Turing machine G_S

It turned out that the three generators are equivalent in their generating power. That is, if a set can be generated by one of them, it can be generated by any other. Because the Turing machine most convincingly formalized the basic notions of computation, we restate the *Post Thesis* in terms of this model of computation.

Post Thesis. (TM) *A set* S *can be "generated"* $\longleftrightarrow$ S *can be generated by a TM*

Due to the power of the Turing machine, the intuitive concept of set "generation" was finally formalized. Therefore, we will no longer use quotation marks.

Sets that can be algorithmically generated were called normal by Post. Today, we call them *computably enumerable* sets. Here is the official definition.

Definition 6.7. (c.e. Set) A set S is **computably enumerable** (in short **c.e.**)[10] if S can be generated by a TM.

From Theorem 6.1 and the above discussion we immediately deduce the following important corollary.

Corollary 6.1. *A set* S *is c.e.* $\Longleftrightarrow$ $S = \emptyset \vee S$ *is the range of a computable function on* $\mathbb{N}$

Remarks. 1) Note that the order in which the elements of a c.e. set are generated is not prescribed. So, any order will do. 2) An element may be generated more than once; what matters is that it be generated at least once. 3) Each element of a c.e. set is generated in finite time (i.e., a finite number of steps of a TM). This, however, does not imply that the *whole* set can be generated in finite time, as the set may be infinite.

6.3.3 Set Recognition

Let T be a Turing machine and w an arbitrary word. Let us write w as the input word to T's tape and start T. There are three possible outcomes. If T after reading w eventually halts in a final state q_{yes}, then we say that T *accepts* w. If T after reading w eventually halts in a non-final state q_{no}, we say that T *rejects* w. If, however, T after reading w never halts, then we say that T *does not recognize* w. Thus, a Turing machine T is implicitly associated with a set of all the words that T accepts. We will call it the *proper set* of T. Here is the definition.

---

[10] The older name is *recursively enumerable (r.e.)* set. See more about the renaming on p. 142.

Definition 6.8. (Proper Set) Let $T = (Q, \Sigma, \Gamma, \delta, q_1, \sqcup, F)$ be a Turing machine. The **proper set**[11] of T is the set $L(T) \overset{\text{def}}{=} \{w \in \Sigma^* \mid T \text{ accepts } w\}$.

Often we are confronted with the opposite task: Given a set S, find a Turing machine T such that $L(T) = S$. Put another way, we must find a Turing machine that accepts exactly the given set. Such a T is called the **acceptor** of the set S. However, we will see that the existence of an acceptor of S is closely connected with S's amenability to *set recognition*. So let us focus on this notion.

TM as a Recognizer of a Set

Informally, to *completely recognize a set* in a given environment, also called the *universe*, is to determine which elements of the universe are members of the set and which are not. Finding this out separately for each and every element of the universe is impractical because the universe may be too large. Instead, it suffices to exhibit an algorithm capable of deciding this for an arbitrary element of the universe.

What can be said about such an algorithm? Let us be given the universe $\mathcal{U}$, an arbitrary set $S \subseteq \mathcal{U}$, and an arbitrary element $x \in \mathcal{U}$. We ask: "Is x a member of the set S?", or, in short, $x \in ?S$. The answer is either YES or NO, as there is no third possibility besides $x \in S$ and $x \notin S$.

However, the answer may not be obvious (Fig. 6.19).

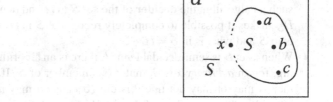

Fig. 6.19 Is x in S or in $\overline{S} = \mathcal{U} - S$? The border between S and $\overline{S}$ is not clear

So, let us focus on the construction of an algorithm A that will be capable of answering the question $x \in ?S$. First, recall the definition of the characteristic function of a set.

Definition 6.9. The **characteristic function** of a set S, where $S \subseteq \mathcal{U}$, is a function $\chi_S : \mathcal{U} \to \{0, 1\}$ defined by

$$\chi_S(x) \overset{\text{def}}{=} \begin{cases} 1 \, (\equiv \text{YES}), & \text{if } x \in S; \\ 0 \, (\equiv \text{NO}), & \text{if } x \notin S. \end{cases}$$

---

[11] Also called the *language* of the Turing machine T.

By definition, χ_S is *total* on $\mathcal{U}$, that is, $\chi_S(x)$ is defined for every $x \in \mathcal{U}$. If the sought-for algorithm A could compute χ_S's values, then it would answer the question $x \in ?S$ simply by computing the value $\chi_S(x)$. In this way the task of set recognition would be reduced to the task of function computation.

But how would A compute the value $\chi_S(x)$? The general definition of the characteristic function χ_S reveals nothing about how to compute χ_S and, consequently, how to construct A. What is more, the definition reveals nothing about the computability of the function χ_S. So, until $\mathcal{U}$ and S are defined in greater detail, nothing particular can be said about the design of A and the computability of χ_S.

Nevertheless, we *can* distinguish between three kinds of sets S, based on the extent to which the values of χ_S can possibly be computed on $\mathcal{U}$ (and, consequently, how S can possibly be recognized in $\mathcal{U}$).

Definition 6.10. Let $\mathcal{U}$ be the universe and $S \subseteq \mathcal{U}$ be an arbitrary set. We say that the set

S is **decidable** (or **computable**[12]) in $\mathcal{U}$ *if* χ_S is computable function on $\mathcal{U}$;

S is **semi-decidable** in $\mathcal{U}$ *if* χ_S is computable function on S;

S is **undecidable** (or **incomputable**) in $\mathcal{U}$ *if* χ_S is incomputable function on $\mathcal{U}$.

(Remember that $\chi_S : \mathcal{U} \to \{0,1\}$ is total.)

This, in turn, tells us how powerful the algorithm A can be:

- When a set S is decidable in $\mathcal{U}$, there exists an algorithm (Turing program) A capable of deciding, for *arbitrary* element $x \in \mathcal{U}$, whether or not $x \in S$. We call such an algorithm the **decider** of the set S in $\mathcal{U}$ and denote it by D_S. The decider D_S makes it possible to completely recognize S in $\mathcal{U}$, that is, to determine what is in S and what is in $\overline{S} = \mathcal{U} - S$.

- When a set S is semi-decidable in $\mathcal{U}$, there is an algorithm A capable of determining, for an *arbitrary* $x \in S$, that x *is* a member of S. If, however, in truth $x \in \overline{S}$, then A may or may not find this out (because it may not halt). We call such an algorithm the **recognizer** of the set S in $\mathcal{U}$ and denote it by R_S. The recognizer R_S makes it possible to completely determine what is in S, *but* it may or may not be able to completely determine what is in $\overline{S}$.

- When a set S is undecidable in $\mathcal{U}$, there is no algorithm A that is capable of deciding, for *arbitrary* element $x \in \mathcal{U}$, whether or not $x \in S$. In other words, *any* candidate algorithm A fails to decide, for *at least one* element $x \in \mathcal{U}$, whether or not $x \in S$ (because on such an input A never halts). This can happen for an x that is in truth a member of S or $\overline{S}$. In addition, there can be several such elements. Hence, we cannot completely determine what is in S, or in $\overline{S}$, or in both.

A word of caution. Observe that a decidable set is by definition also semi-decidable. Hence, a decider of a set is also a recognizer of the set. The inverse

---

[12] The older name is *recursive set*. We describe the reasons for a new name on p. 142.

does not hold in general; we will prove later that there are semi-decidable sets that
are not decidable.

Remarks. The adjectives *decidable* and *computable* will be used as synonyms and depend on
the context. For example, we will say "decidable set" in order to stress that the set is completely
recognizable in its universe. But we will say "computable set" in order to stress that the characteristic function of the set is computable on the universe. The reader should always bear in mind
the alternative adjective and its connotation. We will use the adjectives *undecidable* and *incomputable* similarly. Interestingly, the adjective *semi-decidable* will find a synonym in the adjective
computably enumerable (c.e.). In the following we explain why this is so.

6.3.4 Generation vs. Recognition

In this section we show that set generation and set recognition are closely connected tasks. We do this in two steps: first, we prove that every c.e. set is also semi-
decidable; then we prove that every semi-decidable set is also c.e. So, let $\mathcal{U}$ be the
universe and $S \subseteq \mathcal{U}$ be an arbitrary set.

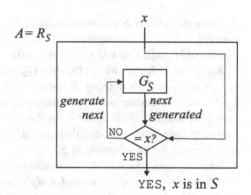

Fig. 6.20 R_S checks whether
G_S generated x

    Suppose that the set S is c.e. We can use the generator G_S to answer the question
"Is x a member of S?" for arbitrary $x \in \mathcal{U}$. To do this, we enhance G_S so that the
resulting algorithm A checks each generated word to see whether or not it is x (see
Fig. 6.20). If and when this happens, the algorithm A outputs the answer YES (i.e.,
$x \in S$) and halts; otherwise, the algorithm A continues generating and checking the
members of S. (Clearly, when in truth $x \notin S$ and S is infinite, A never halts.) It is
obvious, that A is R_S, the recognizer of the set S. Thus, S is semi-decidable.
    We have proven the following theorem.

Theorem 6.2. *A set S is c.e.* $\Longrightarrow$ *S is semi-decidable in* $\mathcal{U}$.

What about the other way round? Is a semi-decidable set also c.e.? In other words, given the recognizer R_S of S, can we construct the generator G_S of S? The answer is yes, but the proof is trickier.

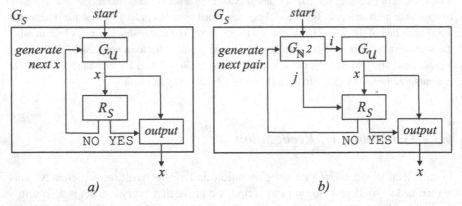

Fig. 6.21 a) A naive construction of G_S that works only for decidable sets S. b) An improved construction of G_S that works for any semi-decidable set S

Suppose that the set S is semi-decidable. The *naive* construction of G_S is as follows (see Fig. 6.21a). Assuming that $\mathcal{U}$ is c.e., G_S uses 1) the generator $G_\mathcal{U}$ to generate elements of $\mathcal{U}$ and 2) the recognizer R_S to find out, for each generated $x \in \mathcal{U}$, whether or not $x \in S$. If R_S answers YES, then G_S outputs (generates) x, and if the answer is NO, then G_S outputs nothing. *However*, there is a pitfall in this approach: If in truth $x \notin S$, then the answer NO is not guaranteed (because S is semi-decidable). So G_S may wait indefinitely long for it. In that case, the operation of G_S drags on for an infinitely long time, although there are still elements of $\mathcal{U}$ that should be generated and checked for membership in S.

This trap can be avoided by a technique called *dovetailing* (see Fig. 6.21b). The idea is to ensure that G_S waits for R_S's answer for only a finitely long time. To achieve this, G_S must allot to R_S a finite number of steps, say j, to answer the question $x \in ?S$. If R_S answers YES (i.e., $x \in S$) in exactly j steps, then G_S generates (outputs) x. Otherwise, G_S asks R_S to recognize, in the same controlled fashion, some other candidate element of $\mathcal{U}$. Of course, later G_S must start again the recognition of x, but this time allotting to R_S a larger number of steps.

To implement this idea, G_S must systematically label and keep track of each candidate element of $\mathcal{U}$ (e.g., with the natural number i) as well as of the currently allotted number of steps (e.g., with the natural number j). For this reason, G_S uses a generator $G_{\mathbb{N}^2}$, which systematically generates pairs $(i, j) \in \mathbb{N}^2$ in such a way that each pair is generated exactly once. The details are given in the proof of the next theorem (see Box 6.3).

Theorem 6.3. *Let the universe $\mathcal{U}$ be c.e. Then:*
$$A \text{ set } S \text{ is semi-decidable in } \mathcal{U} \implies S \text{ is c.e.}$$

Box 6.3 (Proof).

Recall that a c.e. set is the range of a computable function on $\mathbb{N}$ (Corollary 6.1, p. 130). Hence, there is a computable function $f : \mathbb{N} \to \mathcal{U}$ such that, for every $x \in \mathcal{U}$, there is an $i \in \mathbb{N}$ for which $x = f(i)$. Thus, the set $\mathcal{U}$ can be generated by successive computing of the values $f(0), f(1), f(2), \dots$ After $f(i)$ is computed, it is fed into the recognizer R_S for a *controlled recognition*, i.e., a finite number j of steps are allotted to R_S in which it tries to decide $x \in ?S$. (This prevents R_S from getting trapped in an endless computation of $\chi_S(x)$.) If the decision is not made in the jth step, it might be made in the following step. So R_S tries to recognize x later again, this time with $j + 1$ allotted steps.

Obviously, we need a generator capable of generating all the *pairs* (i, j) of natural numbers in such a way that each pair is generated exactly once. This can be done by a generator that generates pairs in the order of visiting dots (representing pairs) as described in Fig 6.22.

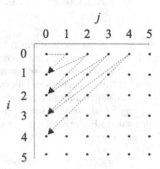

Fig. 6.22 The order of generating pairs $(i, j) \in \mathbb{N}^2$. Each pair is generated exactly once

The initial generated pairs are $(0,0), (0,1), (1,0), (0,2), (1,1), (2,0), (0,3), (1,2), (2,1), (3,0), \dots$

The generator must output pairs $(i, j) \in \mathbb{N}^2$ so that, for each $k = 0, 1, 2, 3, \dots$, it systematically generates all the pairs having $i + j = k$. These pairs correspond to the dots on the same diagonal in the table. It is not hard to conceive such a generator. The generator successively increments the variable k and, for each k, first outputs $(0, k)$ and then all the remaining pairs up to and including $(k, 0)$, where each pair is constructed from the previous one by incrementing its first component and decrementing its second component by 1.

This was but an intuitive description of the generator's algorithm (program). However, according to the *Computability Thesis*, there exists an actual Turing machine performing all the described tasks. We denote this Turing machine by $G_{\mathbb{N}^2}$. It is easy to see that, for arbitrary $i, j \in \mathbb{N}$, the machine $G_{\mathbb{N}^2}$ generates every (i, j) exactly once.

How will $G_{\mathbb{N}^2}$ be used? With the pair generator $G_{\mathbb{N}^2}$, the function f, and the recognizer R_S, we can now construct an improved generator G_S of the set S. (See Fig. 6.21b.) This generator repeats the following four steps: 1) it demands from $G_{\mathbb{N}^2}$ the next pair (i, j); 2) it generates an element $x \in \mathcal{U}$ by computing $x := f(i)$; 3) it demands from R_S the answer to the question $x \in ?S$ in exactly j steps; 4) if the answer is YES (i.e., $x \in S$), then G_S outputs (generates) x; otherwise it generates nothing and returns to 1). $\qquad\square$

6.3.5 The Standard Universes Σ^* and $\mathbb{N}$

In the rest of the book, the universe $\mathcal{U}$ will be either Σ^*, the set of all the words over the alphabet Σ, or $\mathbb{N}$, the set of all natural numbers. In what follows we show that both Σ^* and $\mathbb{N}$ are c.e. sets. This will closely link the tasks of set generation and set recognition.

Theorem 6.4. Σ^* and $\mathbb{N}$ are c.e. sets.

Proof. We intuitively describe the generators of the two sets.

a) The generator G_{Σ^*} will output words in the *shortlex* order (i.e., in order of increasing length, and, in the case of equal length, in lexicographical order; see Appendix A, p. 303). For example, for $\Sigma = \{a, b, c\}$, the first generated words are

$$\varepsilon,$$
$$a,b,c,$$
$$aa,ab,ac,ba,bb,bc,ca,cb,cc,$$
$$aaa,aab,aac,aba,abb,abc,aca,acb,acc,baa,bab,bac,bba,bbb,bbc,bca,bcb,bcc,caa,cab,cac,cba,cbb,cbc,cca,ccb,ccc,$$
$$\vdots$$

To generate the words of length $\ell + 1$, G_{Σ^*} does the following: for each previously generated word w of length ℓ, it outputs the words ws for each symbol $s \in \Sigma$.

b) The generator $G_{\mathbb{N}}$ will generate binary representations of natural numbers $n = 0, 1, \dots$. To achieve this, it operates as follows. First, it generates the two words of length $\ell = 1$, that is, 0 and 1. Then, to generate all the words of length $\ell + 1$, it outputs, for each previously generated word w of length ℓ *which starts with* 1, the words $w0$ and $w1$. (In this way the words with leading 0s are not generated.) For example, the first generated binary representations and the corresponding natural numbers are:

$$0, 1, \qquad\qquad\qquad\qquad\qquad 0, 1,$$
$$10, 11, \qquad\qquad\qquad\qquad\qquad 2, 3,$$
$$100, 101, 110, 111, \qquad\qquad\qquad 4, 5, 6, 7,$$
$$1000, 1001, 1010, 1011, 1100, 1101, 1110, 1111 \qquad 8, 9, 10, 11, 12, 13, 14, 15,$$
$$\vdots \qquad\qquad\qquad\qquad\qquad\qquad \vdots$$

The existence of the Turing machines G_{Σ^*} and $G_{\mathbb{N}}$ is assured by the *Computability Thesis*. □

Combining Theorems 6.2, 6.3 and 6.4 we find that when the universe is Σ^* or $\mathbb{N}$, set generation and set recognition are closely linked tasks.

Corollary 6.2. *Let the universe $\mathcal{U}$ be Σ^* or $\mathbb{N}$. Then:*
A set S is semi-decidable in $\mathcal{U} \Longleftrightarrow S$ is c.e.

Remark. Since in the following the universe will be either Σ^* or $\mathbb{N}$, the corollary allows us to use the adjectives *semi-decidable* and *computably enumerable* as synonyms. Similarly to the pairs of adjectives *decidable–computable* and *undecidable–incomputable*, we will use them in accordance with our wish to stress the amenability of a set of interest to recognition or generation.

6.3.6 Formal Languages vs. Sets of Natural Numbers

In this section we show that it is not just by chance that both Σ^* and $\mathbb{N}$ are c.e. sets. This is because there is a bijective function from Σ^* to $\mathbb{N}$. We prove this in Box 6.4.

Theorem 6.5. *There is a bijection* $f : \Sigma^* \to \mathbb{N}$.

Box 6.4 (Proof).

First we prove the theorem for the alphabet $\Sigma = \{0,1\}$ and then for a general Σ.

a) Let $\Sigma = \{0,1\}$. We define the operation $\#_2$ as follows: If $u \in \{0,1\}^*$ is the binary code of a natural number, then let $\#_2(u)$ be that number. For example, $\#_2(101) = 5$. Now, let $w \in \{0,1\}^*$ be an arbitrary word. We associate w with a natural number $f(w)$, where

$$f(w) \stackrel{\text{def}}{=} \#_2(1w) - 1.$$

For example, $f(\varepsilon) = 0; f(0) = 1; f(1) = 2; f(00) = 3; f(01) = 4; f(10) = 5; f(11) = 6; f(000) = 7$. Next, we prove that $f : \{0,1\}^* \to \mathbb{N}$ is bijective. The function f is injective: If $w_1 \neq w_2$, then $1w_1 \neq 1w_2$, so $\#_2(1w_1) - 1 \neq \#_2(1w_2) - 1$, and $f(w_1) \neq f(w_2)$. The function f is surjective: If n is an arbitrary natural number, then $n = f(w)$, where w is obtained from the binary code of $n+1$ by canceling the leading symbol (which is 1). For example, for $n = 4$, the binary code of $n+1$ is 101, so $w = 01$. Thus, $4 = f(01)$.

b) Let Σ be an arbitrary alphabet. Let us generalize the above function f. We write $b = |\Sigma|$ and define the operation $\#_b$ as follows: If $u \in \Sigma^*$ is the code of a natural number in a positional number system with base b, then let $\#_b(u)$ be that number. For example, if $b = 3$, then $\#_3(21) = 2 \cdot 3^1 + 1 \cdot 3^0 = 7$. Consider the word $u = 00\ldots0 \in \Sigma^*$ and write $\ell = |u|$. If we append to the left-hand side of u the symbol 1, we find that $\#_b(1u) = \#_b(100\ldots0) = b^\ell$. In canonical (i.e., shortlex) order there are $b^0 + b^1 + \ldots + b^{\ell-1}$ words preceding the word u. Assuming that the searched-for function f maps these words in the same order into numbers $0, 1, \ldots$, then u must be mapped into the number $\sum_{i=0}^{\ell-1} b^i$. Now, let $w \in \Sigma^*$ be an arbitrary word. The function f must map w into the number $f(w)$, where $f(w) = \#_b(1w) - b^\ell + \sum_{i=0}^{\ell-1} b^i$. After rearranging the right-hand side, we obtain

$$f(w) = \#_b(1w) - b^\ell + \frac{b^\ell - 1}{b - 1}, \quad \text{where } \ell = |w|.$$

Take, for example, $\Sigma = \{0,1,2\}$. Then $w = 21$ is mapped to $f(21) = \#_3(121) - 3^2 + \frac{8}{2} = 16 - 9 + 4 = 11$. The function $f : \Sigma^* \to \mathbb{N}$ is bijective. We leave the proof of this as an exercise. $\square$

A subset of the set Σ^* is said to be a *formal language* over the alphabet Σ. The above theorem states that every formal language $S \subseteq \Sigma^*$ is associated with exactly one set $f(S) \subseteq \mathbb{N}$ of natural numbers—and vice versa. How can we use this? The answer is given in the following remark.

NB *When a property of sets is independent of the nature of their elements, we are allowed to choose whether to study the property using formal languages or sets of natural numbers. The results will apply to the alternative, too. For Computability Theory, three properties of this kind are especially interesting: the decidability, semi-decidability, and undecidability of sets. We will use the two alternatives based on the context and the ease and clarity of the presentation.*

6.4 Chapter Summary

In addition to the basic model of the Turing machine there are several variants. Each is a generalization of the basic model in some respect. Nevertheless, the basic model is capable of computing anything that can be computed by any other variant.

Turing machines can be encoded by words consisting of 0s and 1s. This enables the construction of the universal Turing machine, a machine that is capable of simulating any other Turing machine. Thus, the universal Turing machine can compute anything that can be computed by any other Turing machine. Practical consequences of this are the existence of the general-purpose computer and the operating system.

RAM is a model of computation that is equivalent to the Turing machine but is more appropriate in *Computational Complexity Theory*, where time and space are bounded computational resources.

Turing machines can be enumerated and generated. This allows us to talk about the nth Turing machine, for any natural n. The Turing machine can be used as a computer (to compute values of a function), or as a generator (to generate elements of a set), or as a recognizer (to find out which objects are members of a set and which are not).

A function $\varphi : \mathcal{A} \to \mathcal{B}$ is said to be partial computable on $\mathcal{A}$ if there exists a Turing machine capable of computing the function's values wherever the function is defined. A partial computable (p.c.) function $\varphi : \mathcal{A} \to \mathcal{B}$ is said to be computable on $\mathcal{A}$ if it is defined for every member of $\mathcal{A}$. A partial function is incomputable if there exists no Turing machine capable of computing the function's values wherever the function is defined.

A set is decidable in a universe if the characteristic function of the set is computable on the universe. A set is undecidable in a universe if the characteristic function of the set is incomputable on the universe. A set is semi-decidable if the characteristic function of the set is computable on this set. A set is computably enumerable (c.e.) if there exists a Turing machine capable of generating exactly the members of the set.

There is a bijective function between the sets Σ^* and $\mathbb{N}$; so instead of studying the decidability of formal languages, we can study the decidability of sets of natural numbers.

Problems

6.1. Given Q and δ, how many reduced TMs $T = (Q, \{0,1\}, \{0,1,\sqcup\}, \delta, q_1, \sqcup, \{q_2\})$ are there?

6.2. Informally design an algorithm that restores Σ, Γ, Q, and F from the code $\langle T \rangle$ of a basic TM.

6.3. Informally design the following Turing machines:

(a) a basic TM that accepts the language $\{0^n 1^n \mid n \geqslant 1\}$;
   [*Hint.* Repeatedly replace the currently leftmost 0 by X and the currently rightmost 1 by Y.]

(b) a TM that accepts the language $\{0^n 1^n 0^n \mid n \geqslant 1\}$;

(c) a basic TM that accepts an input if its first symbol does not appear elsewhere on the input;
   [*Hint.* Use a finite storage TM.]

(d) a TM that recognizes the set of words with an equal number of 0s and 1s;

(e) a TM that decides whether or not a binary input greater than 2 is a prime;
   [*Hint.* Use a multi-track TM. Let the input be on the first track. Starting with number 1, repeatedly generate on the second track the next larger number less than the input; for each generated number copy the input to the third track and then subtract the second-track number from the third-track number as many times as possible.]

(f) a TM that recognizes the language $\{wcw \mid w \in \{0,1\}^*\}$;
   [*Hint.* Check off the symbols of the input by using a multi-track TM.]

(g) a TM that recognizes the language $\{ww \mid w \in \{0,1\}^*\}$;
   [*Hint.* Locate the middle of the input.]

(h) a TM that recognizes the language $\{ww^R \mid w \in \{0,1\}^*\}$ of palindromes;
   [*Hint.* Use a multi-tape TM.]

(i) a TM that shifts its tape contents by three cells to the right;
   [*Hint.* Use a finite storage TM to perform caterpillar-type movement.]

(j) Adapt the TM from (g) for moving its tape contents by three cells to the left.

6.4. Prove: A nondeterministic d-dimensional tp-tape tk-track TM can be simulated by a basic TM.

6.5. Prove:

(a) A set is c.e. *iff* it is generated by a TM.
(b) If a set is c.e. then there is a generator that generates each element in the set exactly once.
(c) A set is computable *iff* it is generated by a TM in shortlex order.
(d) Every c.e. set is accepted by a TM with only two nonaccepting states and one accepting state.

6.6. Informally design TMs that compute the following functions (use any version of TM):

(a) $\text{const}_j^k(n_1, \ldots, n_k) \overset{\text{def}}{=} j$, for $j \geqslant 0$ and $k \geqslant 1$;

(b) $\text{add}(m,n) \overset{\text{def}}{=} m + n$;

(c) $\text{mult}(m,n) \overset{\text{def}}{=} mn$;
   [*Hint:* start with $0^m 10^n$, put 1 after it, and copy 0^n onto the right end m times.]

(d) $\text{power}(m,n) \overset{\text{def}}{=} m^n$;

(e) $\text{fact}(n) \overset{\text{def}}{=} n!$

(f) $\text{tower}(m,n) \overset{\text{def}}{=} m^{m^{\cdot^{\cdot^{\cdot^{m}}}}} \Big\} n$ levels

(g) $\text{minus}(m,n) \overset{\text{def}}{=} m \dot{-} n = \begin{cases} m - n & \text{if } m \geqslant n; \\ 0 & \text{otherwise.} \end{cases}$

(h) $\text{div}(m,n) \overset{\text{def}}{=} m \div n = \lfloor \frac{m}{n} \rfloor$;

(i) $\mathrm{floorlog}(n) \overset{\text{def}}{=} \lfloor \log_2 n \rfloor$

(j) $\log^*(n) \overset{\text{def}}{=}$ the smallest k such that $\overbrace{\log(\log(\cdots(\log(n))\cdots))}^{k\ \text{times}} \leqslant n$;

(k) $\gcd(m,n) \overset{\text{def}}{=}$ greatest common divisor of m and n;

(l) $\mathrm{lcm}(m,n) \overset{\text{def}}{=}$ least common multiplier of m and n;

(m) $\mathrm{prime}(n) \overset{\text{def}}{=}$ the nth prime number;

(n) $\pi(x) \overset{\text{def}}{=}$ the number of primes not exceeding x;

(o) $\phi(n) \overset{\text{def}}{=}$ the number of positive integers not exceeding n relatively prime to n (*Euler function*);

(p) $\max^k(n_1,\ldots,n_k) \overset{\text{def}}{=} \max\{n_1,\ldots,n_k\}$, for $k \geqslant 1$;

(r) $\mathrm{neg}(x) \overset{\text{def}}{=} \begin{cases} 0 & \text{if } x \geqslant 1; \\ 1 & \text{if } x = 0. \end{cases}$

(s) $\mathrm{and}(x,y) \overset{\text{def}}{=} \begin{cases} 1 & \text{if } x \geqslant 1 \wedge y \geqslant 1; \\ 0 & \text{otherwise.} \end{cases}$

(t) $\mathrm{or}(x,y) \overset{\text{def}}{=} \begin{cases} 1 & \text{if } x \geqslant 1 \vee y \geqslant 1; \\ 0 & \text{otherwise.} \end{cases}$

(u) $\text{if-then-else}(x,y,z) \overset{\text{def}}{=} \begin{cases} y & \text{if } x \geqslant 1; \\ z & \text{otherwise.} \end{cases}$

(v) $\mathrm{eq}(x,y) \overset{\text{def}}{=} \begin{cases} 1 & \text{if } x = y; \\ 0 & \text{otherwise.} \end{cases}$

(w) $\mathrm{gr}(x,y) \overset{\text{def}}{=} \begin{cases} 1 & \text{if } x > y; \\ 0 & \text{if } x \leqslant y. \end{cases}$

(x) $\mathrm{geq}(x,y) \overset{\text{def}}{=} \begin{cases} 1 & \text{if } x \geqslant y; \\ 0 & \text{if } x < y. \end{cases}$

(y) $\mathrm{ls}(x,y) \overset{\text{def}}{=} \begin{cases} 1 & \text{if } x < y; \\ 0 & \text{if } x \geqslant y. \end{cases}$

(z) $\mathrm{leq}(x,y) \overset{\text{def}}{=} \begin{cases} 1 & \text{if } x \leqslant y; \\ 0 & \text{if } x > y. \end{cases}$

6.7. Prove: A set is c.e. if it is the domain of a p.c. function.

Definition 6.11. (Pairing Function) A **pairing function** is any computable bijection $f : \mathbb{N}^2 \to \mathbb{N}$ whose inverse functions f_1^{-1}, f_2^{-1}, defined by $f(f_1^{-1}(n), f_2^{-1}(n)) = n$, are computable. So, $f(i,j) = n$ iff $f_1^{-1}(n) = i$ and $f_2^{-1}(n) = j$. The **standard** (or Cantor) pairing function $p : \mathbb{N}^2 \to \mathbb{N}$ is defined by

$$p(i,j) \overset{\text{def}}{=} \frac{1}{2}(i+j)(i+j+1)+i.$$

6.8. Prove: The function $p(i,j)$ is a pairing function with inverse functions

$$p_1^{-1}(n) = i = n - \frac{1}{2}w(w+1) \quad \text{and}$$
$$p_2^{-1}(n) = j = w - i$$

where

$$w = \left\lfloor \frac{\sqrt[2]{8n+1}-1}{2} \right\rfloor.$$

6.9. What is the connection between p and the generation of pairs in Fig. 6.22 (p.135)?

6.10. We can use pairing functions to define bijective functions that map $\mathbb{N}^k$ onto $\mathbb{N}$.

(a) Describe how we can use p (from Problem 6.8) to define and compute a bijection $p^{(3)} : \mathbb{N}^3 \to \mathbb{N}$?

(b) Can you find the corresponding inverse functions $p_1^{(3)-1}, p_2^{(3)-1}, p_3^{(3)-1}$?

(c) How would you generalize to bijections from $\mathbb{N}^k$ onto $\mathbb{N}$, where $k \geqslant 3$?

[*Hint.* Define $p^{(k)} : \mathbb{N}^k \to \mathbb{N}$ by $p^{(k)}(i_1, \ldots, i_k) = p(p^{(k-1)}(i_1, \ldots, i_{k-1}), i_k)$ for $k \geqslant 3$, and $p^{(2)} = p$.]

6.11. Informally design the following generators (use any version of TM):

(a) $G_{\mathbb{N}^3}$, the generator of 3-tuples $(i_1, i_2, i_3) \in \mathbb{N}^3$.

(b) $G_{\mathbb{N}^k}$, the generator of k-tuples $(i_1, \ldots, i_k) \in \mathbb{N}^k$ (for $k \geqslant 3$);

(c) $G_{\mathbb{Q}}$, the generator of rational numbers $\frac{i}{j}$.

Bibliographic Notes

- Turing introduced his machine in [178]. Here, he also described in detail the construction of a universal Turing machine.
- The search for the smallest universal Turing machine was initiated in Shannon [148]. The existence and implementation of several small universal Turing machines is described in Rogozhin [140].
- The RAM model was formally considered in Cook and Reckhow [27]. In describing the model RAM we leaned on Aho et al. [5]. The influential von Neumann's report on the design of the EDVAC computer is [183].
- The concept of the *computably enumerable* (*c.e.*) set was formally defined in Kleene [79] (where he used the term recursively enumerable). Fifteen years before Kleene, Post discovered an equivalent concept of the *generated set*, but it remained unpublished until Post [122].
- The most significant works of Turing, along with key commentaries from leading scholars in the field, are collected in Cooper and van Leeuwen [30]. A comprehensive biography of Alan Turing is Hodges [69].

The following excellent books were consulted and are sources for the subjects covered in Part II of this book:

- General monographs on *Computability Theory* from the 1950s are Kleene [82] and Davis [34]. The first one contains a lot of information about the reasons for searching for the models of computation.
- In the mid-1960s, a highly influential monograph on *Computability Theory* was Rogers [139]. The treatise contains all the results up to the mid-1960s, presented in a less formal style. Due to this, the book remained among the main sources of the *Computability Theory* for the next twenty years.
- Some of the influential monographs from the 1970s and 1980s are Machtey and Young [98], Hopcroft and Ullman [72], Cutland [32], and Lewis and Papadimitriou [96].
- In the mid-1980s, the classic monograph of Rogers was supplemented and substituted by Soare [164]. This is a collection of the results in *Computability Theory* from 1931 to the mid-1980s. An influential monograph of that decade was also Boolos and Jeffrey [16].
- In the mid-1990s, computability theorists defined a uniform terminology (see below). Some of the influential general monographs are Davis [38] and Kozen [85].
- The 2000s were especially fruitful and bore Boolos et al. [15], Shen and Vereshchagin [149], Cooper [29], Harel and Feldman [62], Kozen [86], Sipser [157], Rich [136], Epstein [43], and Rosenberg [141].
- The first half of the 2010s brought the monographs of Enderton [42], Weber [186], and Soare [169].

- The conceptual foundations of *Computability Theory* are discussed in light of our modern understanding in Copeland et al. [31]. An overview of *Computability Theory* until the end of the millennium is Griffor [59].
- Many of the recent general books on *Computational Complexity Theory* offer the basics of *Computability Theory*. These include Goldreich [57, 58] and Homer and Selman [71]. Two recent monographs that investigate the impact of randomness on computation and complexity are Nies [113] and Downey and Hirschfeldt [39]. For an intuitive introduction to this area, see Weber [186, Sect. 9.2].

Remark. (On the new terminology in *Computability Theory*.) The terminology in the papers of the 1930s is far from uniform. This is not surprising, though, as the theory was just born. Initially, *partial recursive function* was used for any function constructed using the rules of Gödel and Kleene (see p. 73). However, after *Computability Thesis* was accepted, the adjective "partial recursive" expanded to eventually designate any computable function, *regardless of the model of computation on which the function was defined*. At the same time, the Turing machine became widely accepted as the most convincing model of computation. But the functions computable on Turing machines do not exhibit recursiveness (i.e., self-reference) as explicitly as the functions computable on Gödel-Kleene's model of computation. (We will explain this in the *Recursion Theorem*, Sect. 7.4.) Consequently, the adjective "partial recursive" lost its sense and was groundless when the subjects under discussion were Turing-computable functions. A more appropriate adjective would be "computable" (which was used by Turing himself). Nevertheless, with time the whole research field took the name *Recursion Theory*, in spite of the fact that its prime interest has always been computability (and not only recursion).

In 1996, Soare[13] proposed corrections to the terminology so that notions and concepts would regain their original meanings, as intended by the first researchers (see Soare [165]). In summary:

- the term *computable* and its variants should be used in connection with notions: computation, algorithm; Turing machine, register machine; function (defined on one of these models), set (generated or recognized on one of these models); relative computability;
- the term *recursive* and its variants should be used in connection with notions: recursive (inductive) definition; general recursive function (Herbrand-Gödel), recursive and partial recursive function (Gödel-Kleene) and some other notions from the theory of recursive functions.

---

[13] Robert Irving Soare, 1940, American mathematician.

Chapter 7
The First Basic Results

*Recursion is a method of defining objects in which the object
being defined is applied within its own definition.*

Abstract In the previous chapters we have defined the basic notions and concepts of a theory that we are interested in, *Computability Theory*. In particular, we have rigorously defined its basic notions, i.e., the notions of algorithm, computation, and computable function. We have also defined some new notions, such as the decidability and semi-decidability of a set, that will play key roles in the next chapter (where we will further develop *Computability Theory*). As a side product of the previous chapters we have also discovered some surprising facts, such as the existence of the universal Turing machine. It is now time to start using this apparatus and deduce the first theorems of *Computability Theory*. In this chapter we will first prove several simple but useful theorems about decidable and semi-decidable sets and their relationship. Then we will deduce the so-called *Padding Lemma* and, based on it, introduce the extremely important concept of the index set. This will enable us to deduce two influential theorems, the *Parametrization Theorem* and the *Recursion Theorem*. We will not be excessively formal in our deductions; instead, we will equip them with meaning and motivation wherever appropriate.

7.1 Some Basic Properties of Semi-decidable (C.E.) Sets

For starters, we prove in this section some basic properties of semi-decidable sets. In what follows, $\mathcal{A}$, $\mathcal{B}$ and $\mathcal{S}$ are sets.

Theorem 7.1. *$\mathcal{S}$ is decidable $\Longrightarrow$ $\mathcal{S}$ is semi-decidable*

Proof. This is a direct consequence of Definition 6.10 (p. 132) of decidable and semi-decidable sets. □

© Springer-Verlag Berlin Heidelberg 2015
B. Robič, *The Foundations of Computability Theory*,
DOI 10.1007/978-3-662-44808-3_7

Theorem 7.2. *S is decidable $\Longrightarrow$ $\overline{S}$ is decidable*

Proof. The decider $D_{\overline{S}}$ starts D_S and reverses its answers.   □

The next theorem is due to Post and is often used.

Theorem 7.3. (Post's Theorem) *S and $\overline{S}$ are semi-decidable $\Longleftrightarrow$ S is decidable*

Proof. Let S and $\overline{S}$ be semi-decidable sets. Then, there are recognizers R_S and $R_{\overline{S}}$. Every $x \in \mathcal{U}$ is a member of either S or $\overline{S}$; the former situation can be detected by R_S and the latter by $R_{\overline{S}}$. So let us combine the two recognizers into the following algorithm: 1) Given $x \in \mathcal{U}$, simultaneously start R_S and $R_{\overline{S}}$ on x and wait until one of them answers YES. 2) If the YES came from R_S, output YES (i.e., $x \in S$) and halt; otherwise (i.e., the YES came from $R_{\overline{S}}$) output NO (i.e., $x \notin S$) and halt. This algorithm decides, for arbitrary $x \in \mathcal{U}$, whether or not x is in S. Thus, the algorithm is D_S, the decider for S.   □

Theorem 7.4. *S is semi-decidable $\Longleftrightarrow$ S is the domain of a computable function*

Proof. If S is semi-decidable, then it is (by Definition 6.10 on p. 132) the domain of the characteristic function χ_S, which is computable on S. Inversely, if S is the domain of a computable function φ, then the characteristic function, defined by $\chi_S(x) = 1 \Leftrightarrow \varphi(x)\downarrow$, is computable on S (see Definition 6.6 on p. 126). Hence, S is semi-decidable.   □

Theorem 7.5.

A and B are semi-decidable $\Longrightarrow$ $A \cup B$ and $A \cap B$ are semi-decidable
A and B are decidable $\Longrightarrow$ $A \cup B$ and $A \cap B$ are decidable

Proof. a) Let A and B be semi-decidable sets. Their characteristic functions χ_A and χ_B are computable on A and B, respectively. The function $\chi_{A \cup B}$, defined by

$$\chi_{A \cup B}(x) \stackrel{\text{def}}{=} \begin{cases} 1, & \text{if } \chi_A(x) = 1 \vee \chi_B(x) = 1, \\ 0, & \text{otherwise} \end{cases}$$

is the characteristic function of the set $A \cup B$ and is computable on $A \cup B$. Hence, $A \cup B$ is semi-decidable. Similarly, the function $\chi_{A \cap B}$, defined by

$$\chi_{A \cap B}(x) \stackrel{\text{def}}{=} \begin{cases} 1, & \text{if } \chi_A(x) = 1 \wedge \chi_B(x) = 1, \\ 0, & \text{otherwise} \end{cases}$$

is the characteristic function of the set $A \cap B$ and is computable on $A \cap B$. So, $A \cap B$ is semi-decidable.

b) Let A and B be decidable sets. Then, χ_A and χ_B are computable functions on $\mathcal{U}$. Also, the functions $\chi_{A \cup B}$ and $\chi_{A \cap B}$ are computable on $\mathcal{U}$. Consequently, $A \cup B$ and $A \cap B$ are decidable sets.   □

7.2 Padding Lemma and Index Sets

We have seen in Sect. 6.2.1 that each natural number can be viewed as the index of exactly one Turing machine (see Fig. 7.1). Specifically, given an arbitrary $e \in \mathbb{N}$, we can find the corresponding Turing machine $T_e = (Q, \Sigma, \Gamma, \delta, q_1, \sqcup, F)$ by the following algorithm:

> **if** the binary code of e is of the form $111 K_1 11 K_2 11 \ldots 11 K_r 111$, where each
> K is of the form $0^i 10^j 10^k 10^\ell 10^m$ for some i, j, k, ℓ, m,
> **then** determine $\delta, Q, \Sigma, \Gamma, F$ by inspecting $K_1, K_2, \ldots, K_r$ and taking into account
> that $K = 0^i 10^j 10^k 10^\ell 10^m$ encodes the instruction $\delta(q_i, z_j) = (q_k, z_\ell, D_m)$;
> **else** T_e is the empty Turing machine.

But remember that the Turing machine T_e is implicitly associated, for each $k \geqslant 1$, with a k-ary partial computable function $\psi_e^{(k)}$, the k-ary *proper function of T_e* (see Sect. 6.3.1). Consequently, each $e \in \mathbb{N}$ is the index of exactly one k-ary partial computable function for any fixed $k \geqslant 1$. See Fig. 7.1.

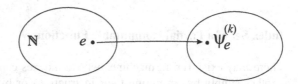

Fig. 7.1 Each natural number e is the index of exactly one partial computable function (denoted by $\psi_e^{(k)}$)

Padding Lemma

What about the other way round? Is each Turing machine represented by exactly one index? The answer is no; a Turing machine has several indexes. To see this, let T be an arbitrary Turing machine and

$$\langle T \rangle = 111 K_1 11 K_2 11 \ldots 11 K_r 111$$

its code. Here, each subword K encodes an instruction of T's program δ. Let us now permute the subwords $K_1, K_2, \ldots, K_r$ of $\langle T \rangle$. Of course, we get a different code, but notice that the new code still represents the same Turing program δ, i.e., the same *set* of instructions, and hence the same Turing machine, T. Thus, T has several different indexes (at least $r!$, where r is the number of instructions in δ).

Still more: we can insert into $\langle T \rangle$ new subwords $K_{r+1}, K_{r+2}, \ldots$, where each of them represents a *redundant* instruction, i.e., an instruction which will *never* be executed. By such *padding* we can construct an unlimited number of new codes each of which describes a different Turing program. But notice that, when started, each of these programs *behaves* (i.e., executes) in the same way as the T's program δ. In other words, each of the constructed indexes defines the same partial computable function, the T's proper function. Formally: if e is the index of T and x an arbitrary index constructed from e as described, then $\psi_x \simeq \psi_e$. (See Fig. 7.2.)

We have just proved the so-called *Padding Lemma*.

Lemma 7.1. (Padding Lemma) *A partial computable function has countably infinitely many indexes. Given one of them, the others can be generated.*

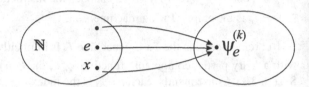

Fig. 7.2 Each p.c. function (denoted by $\psi_e^{(k)}$) has countably infinitely many indexes

Index Set of a Partial Computable Function

There may exist Turing machines whose indexes cannot be be transformed one to another simply by permuting their instructions or by instruction padding, and yet the machines compute the same partial computable function (i.e., solve the same problem). For instance, we described two such machines in Examples 6.1 and 6.2 (see pp. 105,106). The machines compute the sum of two integers in two different ways. In other words, the machines have different programs but they compute the same partial computable function $\psi^{(2)} : \mathbb{N}^2 \to \mathbb{N}$, that is, the function $\psi^{(2)}(m,n) = m+n$. We say that these machines are equal in their *global behavior* (because they compute the same function) but they differ in their *local behavior* (as they do this in different ways).

All of this leads in a natural way to the concept of the *index set of a partial computable function*. Informally, the index set of a p.c. function consists of all the indexes of all the Turing machines that compute this function. (See Fig. 7.3.) So let φ be an arbitrary p.c. function. Then there exists at least one Turing machine T capable of computing the values of φ. Let e be the index of T. Then $\psi_e \simeq \varphi$, where ψ_e is the proper function of T (having the same number of arguments as φ, i.e., the same k-arity). Now let us collect *all* the indexes of *all* the Turing machines that compute the function φ. We call this set the *index set* of φ and denote it by $\text{ind}(\varphi)$.

Definition 7.1. (Index Set) The **index set** of a p.c. function φ is the set $\operatorname{ind}(\varphi) \overset{\text{def}}{=} \{x \in \mathbb{N} \mid \psi_x \simeq \varphi\}$.

Informally, $\operatorname{ind}(\varphi)$ contains all the (encoded) Turing programs that compute φ. Taking into account the *Computability Thesis*, we can say that $\operatorname{ind}(\varphi)$ contains all the (encoded) algorithms that compute (values of) the function φ. There are countably infinitely many algorithms in $\operatorname{ind}(\varphi)$. Although they may differ in their local behavior, they all exhibit the same global behavior.

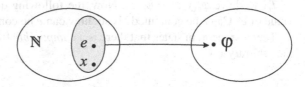

Fig. 7.3 Index set $\operatorname{ind}(\varphi)$ of a p.c. function φ

Index Set of a Semi-decidable Set

We can now in a natural way define one more useful notion. Let S be an arbitrary semi-decidable set. Of course, its characteristic function χ_S is partial computable. But now we know that the index set $\operatorname{ind}(\chi_S)$ of the function χ_S contains all the (encoded) algorithms that are capable of computing χ_S. In other words, $\operatorname{ind}(\chi_S)$ consists of all the (encoded) *recognizers* of the set S. For this reason we also call $\operatorname{ind}(\chi_S)$ the *index set of the set* S and denote it by $\operatorname{ind}(S)$.

7.3 Parametrization (s-m-n) Theorem

Consider an arbitrary multi-variable partial computable function φ. Select some of its variables and assign arbitrary values to them. This changes the role of these variables, because the fixed variables become *parameters*. In this way, we obtain a new function ψ of the rest of the variables. The *Parametrization Theorem*, which is the subject of this section, tells us that the index of the new function ψ depends only on the index of the original function φ and its parameters; what is more, the index of ψ can be computed, for any φ and parameters, with a computable function s.

There are important practical consequences of this theorem. First, since the index of a function represents the Turing program that computes the function's values, the *Parametrization Theorem* tells us that the parameters of φ can always be *incorporated into* the program for φ to obtain the program for ψ. Second, since parameters are natural numbers, they can be viewed as indexes of Turing programs, so the incorporated parameters can be interpreted as *subprograms* of the program for ψ.

7.3.1 Deduction of the Theorem

First, we develop the basic form of the theorem. Let $\varphi_x(y,z)$ be an arbitrary partial computable function. Its values can be computed by a Turing machine T_x. Let us pick an arbitrary natural number—call it $\bar{y}$—and substitute each occurrence of y in the expression of $\varphi_x(y,z)$ by $\bar{y}$. We say that the variable y has been changed to the *parameter* $\bar{y}$. The resulting expression represents a new function $\varphi_x(\bar{y},z)$ of one variable z. Note, that $\varphi_x(\bar{y},z)$ is a partial computable function (otherwise $\varphi_x(y,z)$ would not be partial computable). Therefore, there is a Turing machine—call it T_e—that computes $\varphi_x(\bar{y},z)$. (At this point e is not known.) Since ψ_e is the proper function of T_e, we have $\varphi_x(\bar{y},z) = \psi_e(z)$. Now, the following questions arise: "What is the value of e? Can e be computed? If so, how can e be computed?"

The next theorem states that there is a *computable* function, whose value is e, for the arbitrary $x,\bar{y}$.

Theorem 7.6. (Parametrization Theorem) *There is injective computable function* $s: \mathbb{N}^2 \to \mathbb{N}$ *such that, for every* $x,\bar{y} \in \mathbb{N}$,

$$\varphi_x(\bar{y},z) = \psi_{s(x,\bar{y})}(z).$$

Proof idea. Let $x,\bar{y} \in \mathbb{N}$ be given. First, we conceive a Turing machine T that operates as follows: Given an arbitrary z as the input, T inserts $\bar{y}$ before z and then starts simulating Turing machine T_x on the inputs $\bar{y},z$. Obviously, T computes the value $\varphi_x(\bar{y},z)$. We then show that such a T can be *constructed* for *any* pair $x,\bar{y}$. Therefore, the index of the constructed T can be denoted by $s(x,\bar{y})$, where s is a computable function mapping $\mathbb{N}^2$ on $\mathbb{N}$. The details of the proof are given in Box 7.1.

Box 7.1 (Proof of the *Parametrization Theorem*).

Let us describe the actions of the machine T_e that will entail the equality $\varphi_x(\bar{y},z) = \psi_e(z)$ for every $x,\bar{y} \in \mathbb{N}$. Initially, there is only z written on T_e's tape. Since $\bar{y}$ will also be needed, T_e prepares it on its tape. To do this, it shifts z to the right for $\bar{y}$ cells, and writes unary-encoded $\bar{y}$ to the emptied space. After this, T_e starts simulating T_x's program on inputs $\bar{y}$ and z and thus computing the value $\varphi_x(\bar{y},z)$. Consequently, $\varphi_x(\bar{y},z) = \psi_e(z)$.

But where is the function s? It follows from the above description that the program P_e of the machine T_e is a sequence $P_1;P_2$ of two programs P_1 and P_2, such that P_1 inserts $\bar{y}$ on the tape and then leaves the control to P_2, which is responsible for computing the values of the function $\varphi_x(y,z)$. Hence, $\langle P_e \rangle = \langle P_1;P_2 \rangle$. Now we see the role of s: The function s must compute $\langle P_1;P_2 \rangle$ for arbitrary x and $\bar{y}$. To do this, s must perform the following tasks (informally):

1. Construct the code $\langle P_1 \rangle$ from the parameter $\bar{y}$.
   (Note that P_1 is simple: it must shift the contents of the tape $\bar{y}$ times to the right and write the symbol 1 to each of the emptied cells.)
2. Construct the code $\langle P_2 \rangle$ from the index x.

3. Construct the code $\langle P_1 ; P_2 \rangle$ from $\langle P_1 \rangle$ and $\langle P_2 \rangle$.
   (To do this, take the word $\langle P_1 \rangle \langle P_2 \rangle$ and change it so that, instead of halting, P_1 starts the first instruction of P_2.)
4. Compute the index e from $\langle P_1 ; P_2 \rangle$ and return $s(x, \bar{y}) := e$.

All the tasks are computable. So, s is a computable function. It is injective, too. □

Next, we generalize the theorem. Let the function φ_x have $m \geqslant 1$ parameters and $n \geqslant 1$ variables. That is, the function is $\varphi_x(\bar{y}_1, \ldots, \bar{y}_m, z_1, \ldots, z_n)$. The proof of the following generalization of the *Parametrization Theorem* uses the same ideas as the previous one, so we omit it.

Theorem 7.7. (s-m-n Theorem) *For arbitrary $m, n \geqslant 1$ there is an injective computable function $s_n^m : \mathbb{N}^{m+1} \to \mathbb{N}$ such that, for every $x, \bar{y}_1, \ldots, \bar{y}_m \in \mathbb{N}$,*

$$\varphi_x(\bar{y}_1, \ldots, \bar{y}_m, z_1, \ldots, z_n) = \psi_{s_n^m(x, \bar{y}_1, \ldots, \bar{y}_m)}(z_1, \ldots, z_n).$$

Summary. If the variables $y_1, \ldots, y_m$ of a function φ_x are assigned fixed values $\bar{y}_1, \ldots, \bar{y}_m$, respectively, then we can build the values in T_x's Turing program and obtain a Turing program for computing the function ψ of the rest of the variables. The new program is represented by the index $s_n^m(x, \bar{y}_1, \ldots, \bar{y}_m)$, where s_n^m is an injective and computable function. It can be proved that s_n^m is a primitive recursive function.

7.4 Recursion (Fixed-Point) Theorem

The construction of the universal Turing machine brought, as a byproduct, the cognizance that a Turing machine can compute with other Turing machines simply by manipulating their indexes and simulating programs represented by the indexes. But, can a Turing machine manipulate its own index and consequently compute with itself? Does the question make sense? The answer to both questions is yes, as will be explained in this section. Namely, we have seen that there is another model of computation, the (partial) recursive functions of Gödel and Kleene (see Sect. 5.2.1), that allows functions to be defined and hence computed by referring to themselves, i.e., recursively (see Box 5.1). Since the Turing machine and recursive functions are equivalent models of computation, it is reasonable to ask whether Turing machines also allow for, and can make sense of, recursiveness (i.e., self-reference). That this is actually so follows from the *Recursion Theorem* that was proved by Kleene in 1938. The theorem is also called the *Fixed-Point Theorem*.

7.4.1 Deduction of the Theorem

1. Let $i \in \mathbb{N}$ be an arbitrary number and $f : \mathbb{N} \to \mathbb{N}$ an arbitrary computable function.

2. **Definition.** Given i and f, let $T^{(i)}$ be a TM which performs the following steps:

   a. $T^{(i)}$ has two input data: i (as picked above) and an arbitrary x;
   b. $T^{(i)}$ interprets input i as an index and constructs the TP of T_i;
   c. $T^{(i)}$ simulates T_i on input i;   // i.e., $T^{(i)}$ tries to compute $\psi_i(i)$
   d. if the simulation halts, then $T^{(i)}$ performs steps (e)–(g):   // i.e., if $\psi_i(i) \downarrow$
   e.   it applies the function f to the result $\psi_i(i)$;   // i.e., $T^{(i)}$ computes $f(\psi_i(i))$
   f.   it interprets $f(\psi_i(i))$ as an index and constructs the TP of $T_{f(\psi_i(i))}$;
   g.   it simulates $T_{f(\psi_i(i))}$ on input x.   // In summary: $T^{(i)}$ computes $\psi_{f(\psi_i(i))}(x)$

   We have seen that $T^{(i)}$ on input i, x computes the value $\psi_{f(\psi_i(i))}(x)$. Thus, the proper function of $T^{(i)}$ depends on input i.

3. But, the steps a–g are only a definition of the machine $T^{(i)}$. (We can define whatever we want.) So the question arises: "For which $i \in \mathbb{N}$ does $T^{(i)}$ actually *exist*?" And what is more: "For which $i \in \mathbb{N}$ can $T^{(i)}$ be *constructed*?"

   Proposition. $T^{(i)}$ *exists for every* $i \in \mathbb{N}$. *There is an algorithm that constructs* $T^{(i)}$ *for arbitrary i.*

   Proof. We have already proved: *i)* every natural number is an index of a TM; *ii)* every index can be transformed to a Turing program; and *iii)* every TM can be simulated by another TM. Therefore, each of the steps (a)–(g) is *computable*. Consequently, there is, for every $i \in \mathbb{N}$, a Turing machine that executes steps (a)–(g). We call this machine $T^{(i)}$.

   Can $T^{(i)}$ be constructed by an *algorithm*? To answer in the affirmative, it suffices to describe the algorithm with which we construct the program of $T^{(i)}$. Informally, we do this as follows: We write the program of $T^{(i)}$ as a sequence of five calls of subprograms, a call for each of the tasks (b), (c), (e), (f), (g). We need only three different subprograms: (1) to convert an index to the corresponding Turing program (for steps b, f); (2) to simulate a TM (for steps c, g); and (3) to compute the value of the function f (for step e). We already know that subprograms (1) and (2) can be constructed (see Sect. 6.2). The subprogram (3) is also at our disposal because, by assumption, f is computable. Thus, we have informally described an algorithm for constructing the program of $T^{(i)}$. □

4. According to the *Computability Thesis*, there is a Turing machine T_j which does the same thing as the informal algorithm in the above proof, i.e., T_j constructs $T^{(i)}$ for arbitrary $i \in \mathbb{N}$. But this means that T_j computes a computable function that maps $\mathbb{N}$ into the set of all Turing machines (or rather, their indexes). In other words, T_j's proper function ψ_j is total.

   Consequence. *There is a computable function* $\psi_j : \mathbb{N} \to \mathbb{N}$ *such that* $T^{(i)} = T_{\psi_j(i)}$.

Note, that ψ_j does not execute steps a–g; it only returns the index of $T^{(i)}$ which does execute these steps.

5. The unary proper function of $T_{\psi_j(i)}$ is denoted by $\psi_{\psi_j(i)}$. Applied to x, it computes the same value as $T_{\psi_{j(i)}} = T^{(i)}$ in steps a–g, that is, the value $\psi_{f(\psi_i(i))}(x)$. Consequently,

$$\psi_{\psi_j(i)}(x) = \psi_{f(\psi_i(i))}(x).$$

6. Recall that i is arbitrary. So, let us take $i := j$. After substituting i by j, we obtain

$$\psi_{\psi_j(j)}(x) = \psi_{f(\psi_j(j))}(x)$$

and, introducing $n := \psi_j(j)$, we finally obtain

$$\psi_n(x) = \psi_{f(n)}(x).$$

This equation was our goal. Recall that in the equation f is an *arbitrary* computable function. We have proven the *Recursion Theorem*.

Theorem 7.8. (Recursion Theorem) *For every computable function f there is a natural number n such that $\psi_n \simeq \psi_{f(n)}$. The number n can be computed from the index of the function f.*

7.4.2 Interpretation of the Theorem

What does the *Recursion Theorem* tell us? First, observe that from $\psi_n \simeq \psi_{f(n)}$ it does not necessarily follow that $n = f(n)$. For example, f defined by $f(k) = k+1$ is a computable function, yet there is no natural number n such that $n = f(n)$. Rather, the equality $\psi_n \simeq \psi_{f(n)}$ states that two partial computable functions, whose indexes are n and $f(n)$, are equal (i.e., ψ_n and $\psi_{f(n)}$ have equal domains and return equal values).

Second, recall that the indexes represent Turing programs (and hence Turing machines). So we can interpret f as a transformation of a Turing program represented by some $i \in \mathbb{N}$ to a Turing program represented by $f(i)$. Since f is supposed to be total, it represents a transformation that modifies *every* Turing program in some way. Now, in general, the original program (represented by i) and the modified program (represented by $f(i)$) are not equivalent, i.e., they compute different proper functions ψ_i and $\psi_{f(i)}$. This is where the *Recursion Theorem* makes an entry; it states that there exists a program (represented by some n) which is an exception to this rule. Specifically:

If a transformation f modifies every Turing program, then some Turing program n is transformed into an equivalent Turing program $f(n)$.

In other words, if a transformation f modifies every Turing machine, there is always some Turing machine T_n for which the modified machine $T_{f(n)}$ computes the same function as T_n. Although the Turing programs represented by n and $f(n)$ may completely differ in their instructions, they nevertheless return equal results, i.e., compute the same function $\psi\,(\simeq \psi_n \simeq \psi_{f(n)})$. We have said that such machines differ in their *local behavior* but are equal in their *global behavior* (see p. 146).

Can we find out the value of n, i.e., the (index of the) Turing program which is modified by f into the equivalent program? According to step 6 of the deduction we have $n = \psi_j(j)$, while according to step 3 j can be computed and is dependent on the function f only. Hence, n depends only on f. Since f is by assumption computable, n can be computed.

7.4.3 Fixed Points of Functions

The *Recursion Theorem* is also called the *Fixed-Point Theorem*. The reason is this: The number n for which $\psi_n = \psi_{f(n)}$ is for obvious reasons called the *fixed point of the function f*. Using this terminology we can restate the theorem as follows:

Theorem 7.9. (Fixed-Point Theorem) *Every computable function has a fixed point.*

We will use this version of the *Recursion Theorem* in Sect. 9.3.

The following question immediately arises: Is the number n, which we constructed from the function f, the only fixed point of f? The answer is *no*. Actually, there are countably infinitely many others.

Theorem 7.10. *A computable function has countably infinitely many fixed points.*

Proof. Assume that there exists a computable function f with only *finitely many* fixed points. Denote the finite set of f's fixed points by $\mathcal{F}$. Choose any partial computable function ψ_e with the property that none of its indexes is in $\mathcal{F}$, i.e., $\psi_e \not\simeq \varphi_x$ for every $x \in \mathcal{F}$. (In short, $\mathrm{ind}(\psi_e) \cap \mathcal{F} = \emptyset$.) Now comes the tricky part: Let $g : \mathbb{N} \to \mathbb{N}$ be a function that is implicitly defined by

$$\varphi_{g(x)} \simeq \begin{cases} \psi_e, & \text{if } x \in \mathcal{F}; \\ \varphi_{f(x)}, & \text{otherwise.} \end{cases}$$

Next we determine two relevant properties of g (we can do this even though g is not explicitly defined):

1. The function g is *computable*. The algorithm to compute $g(x)$ for an arbitrary x has two steps: (a) Decide $x \in ?\mathcal{F}$. (This can always be done because $\mathcal{F}$ is finite.) (b) If $x \in \mathcal{F}$, then $g(x) := e$; otherwise, compute $f(x)$ and assign it to $g(x)$. (This can always be done because f is computable.)

2. The function g has *no fixed point*. If $x \in \mathcal{F}$, then $\varphi_{g(x)} \simeq \psi$ (by definition) and $\psi \not\simeq \varphi_x$ (property of ψ), so $\varphi_{g(x)} \not\simeq \varphi_x$. This means that none of the elements of $\mathcal{F}$ is a fixed point of g. On the other hand, if $x \notin \mathcal{F}$, then $\varphi_{g(x)} \simeq \varphi_{f(x)}$ (by definition) and $\varphi_{f(x)} \not\simeq \varphi_x$ (as x is not a fixed point of f), which implies that $\varphi_{g(x)} \not\simeq \varphi_x$. This means that none of the elements of $\overline{\mathcal{F}}$ is a fixed point of g. In summary, the function g has no fixed point.

We have found that g is a computable function with no fixed points. But this contradicts the *Fixed-Point Theorem*. As a consequence, we must throw away our initial assumption. ☐

Consequently, we can upgrade the interpretation of the *Fixed-Point Theorem*: If a transformation f modifies every Turing program, then countably infinitely many Turing programs are transformed into equivalent Turing programs.

7.4.4 Practical Consequences: Recursive Program Definition

The *Recursion Theorem* allows a partial computable function to be defined with its own index, as described in the following schematic definition:

$$\varphi_i(x) \stackrel{\text{def}}{=} \underbrace{[\ldots i \ldots x \ldots]}_{P}$$

On the left-hand side there is a function φ_i which is being defined, and on the right-hand side there is a Turing program P, i.e., a sequence $[\ldots i \ldots x \ldots]$ of instructions, describing how to compute the value of φ_i at the argument x. Consider the variable i in P. Obviously, i is also the index of the function being defined, φ_i. But, an index of a function is just a description of the program computing that function (see Sect. 6.2.1). Thus, it looks as if the variable i in P describes the very program P, and hence the index i of the function being defined. But this is a circular definition, because the object being defined (i.e., i) is a part of the definition!

On the other hand, we know that partial recursive functions, i.e., the Gödel-Kleene model of computation, allow constructions, and hence definitions, of functions in a *self-referencing* manner by using the rule of primitive recursion (see Box 5.1 on p. 74). Since a Turing machine is an equivalent model of computation, we expect that it, too, allows self-reference. To prove that this is indeed so, we need both the *Recursion Theorem* and the *Parametrization Theorem*. Let us see how.

Notice first that the variables i and x have different roles in P. While x represents an *arbitrary* natural number, i.e., the *argument* of the function being defined, the variable i represents an unknown but *fixed* natural number, i.e., the *index* of this function. Therefore, i can be treated as a (yet unknown) *parameter* in P. Now, viewing the program P as a function of one variable x and one parameter i, we can apply the *Parametrization Theorem* and move the parameter i to the index of some new function, $\varphi_{s(i)}(x)$. Observe that this function is still defined by P:

$$\varphi_{s(i)}(x) \stackrel{\text{def}}{=} \underbrace{[\ldots i \ldots x \ldots]}_{P} \tag{$*$}$$

The index $s(i)$ describes a program computing $\varphi_{s(i)}(x)$, but it does not appear in the program P. Due to the *Parametrization Theorem*, the function s is computable. But then the *Recursion Theorem* tells us that there exists a natural number n such that $\varphi_{s(n)} \simeq \varphi_n$. So, let us take $i := n$ in the definition (∗). Taking into account that $\varphi_{s(n)} \simeq \varphi_n$, we obtain the following definition:

$$\varphi_n(x) \stackrel{\text{def}}{=} [\ldots n \ldots x \ldots] \qquad\qquad (**)$$

Recall that n is a particular natural number that is dependent on the function s. Consequently, φ_n is a particular partial computable function. This function has been defined in (∗∗) with its own index.

We have proved that there are partial functions computable according to Turing that can be defined recursively. This finding is most welcome, because the Turing machine as the model of computation does not exhibit the capability of supporting self-reference and recursiveness as explicitly as the Gödel-Kleene model of computation (i.e., the partial recursive functions).

7.4.5 Practical Consequences: Recursive Program Execution

Suppose that a partial computable function is defined recursively. Can the function be computed by a Turing machine? If so, what does the computation look like?

Let us be given a function defined by $\varphi(x) \stackrel{\text{def}}{=} x!$, where $x \in \mathbb{N}$. How can $\varphi(x)$ be defined and computed recursively by a Turing program δ? For starters, recall that $\varphi(x)$ can be recursively defined by the following system of equations:

$$\varphi(x) = x \cdot \varphi(x-1), \qquad \text{if } x > 0 \qquad\qquad (*)$$
$$\varphi(0) = 1$$

For example, $\varphi(4) = 4 \cdot \varphi(3) = 4 \cdot 3 \cdot \varphi(2) = 4 \cdot 3 \cdot 2 \cdot \varphi(1) = 4 \cdot 3 \cdot 2 \cdot 1 \cdot \varphi(0) = 4 \cdot 3 \cdot 2 \cdot 1 \cdot 1 = 24$.

The definition (∗) uncovers the actions that should be taken by the Turing program δ in order to compute the value $\varphi(x)$. Let $\delta(a)$ denote an instance of the execution of δ on input a. We call $\delta(a)$ the *activation* of δ on a; in this case, a is called the *actual parameter* of δ. To compute the value $\varphi(a)$, the activation $\delta(a)$ should do, in succession, the following:

1. *activate* $\delta(a-1)$, i.e., call δ on $a-1$ to compute $\varphi(a-1)$;
2. multiply its actual parameter $(= a)$ and the result r $(= \varphi(a-1))$ of $\delta(a-1)$;
3. return the product $(=$ the result r of $\delta(a))$ to the caller.

In this fashion the Turing machine should execute every $\delta(a)$, $a = x, x-1, \ldots, 2, 1$. The exception is the activation $\delta(0)$, whose execution is trivial and immediately returns the result $r = 1$. Returns to whom? To the activation $\delta(1)$, which activated $\delta(0)$ and has, since then, been waiting for $\delta(0)$'s result. This will enable $\delta(1)$ to resume execution at step 2 and then, in step 3, return its own result. Similarly, the following should hold in general: Every *callee* $\delta(a-1)$ should return its result r ($= \varphi(a-1)$) to the awaiting *caller* $\delta(a)$, thus enabling $\delta(a)$ to continue its execution.

We now describe how, in principle, activations are represented in a Turing machine, what actions the machine must take to start a new activation of its program and how it collects the awaited result. For each activation $\delta(a), a = x, \ldots, 2, 1$, the following *runtime actions* must be taken:

a. *Before control is transferred to $\delta(a)$, the activation record of $\delta(a)$ is created.*

The activation record (AR) of $\delta(a)$ is a finite sequence of tape cells set aside to hold the information which will be used or computed by $\delta(a)$. In the case of $\varphi(x) = x!$, the AR of $\delta(a)$ contains the actual parameter a and an empty field r for the result ($= \varphi(a)$) of $\delta(a)$. We denote the AR of $\delta(a)$ by $[a, r]$. (See Fig. 7.4 (a).)

b. $\delta(a)$ *activates* $\delta(a-1)$ *and waits for its result.*

To do this, $\delta(a)$ creates, next to its AR, a new AR $[a-1, r]$ to be used by $\delta(a-1)$. Note that $a-1$ is a value and r is empty. Then, $\delta(a)$ moves the tape window to $a-1$ of $[a-1, r]$ and changes the state of the control unit to the initial state q_1. In this way, $\delta(a)$ stops its own execution and invokes a new instance of δ. Since the input to the invoked δ will be $a-1$ from $[a-1, r]$, the activation $\delta(a-1)$ will start. (See Fig. 7.4 (b).)

c. $\delta(a-1)$ *computes its result, stores it in its AR, and returns the control to $\delta(a)$.*

The details are as follows. After $\delta(a-1)$ has computed its result $\varphi(a-1)$, it writes it into r of $[a-1, r]$. Then, $\delta(a-1)$ changes the state of the control unit to some previously designated state, say q_2, called the *return state*. In this way, $\delta(a-1)$ informs its caller that it has terminated. (See Fig. 7.4 (c).)

d. $\delta(a)$ *resumes its execution. It reads $\delta(a-1)$'s result from $\delta(a-1)$'s AR, uses it in its own computation, and stores the result into its AR.*

Specifically, when the control unit has entered the return state, $\delta(a)$ moves the tape window to $[a-1, r]$ of $\delta(a-1)$, copies r ($= \varphi(a-1)$) into its own $[a, r]$, and deletes $[a-1, r]$. Then $\delta(a)$ continues executing the rest of its program (steps 2 and 3) and finally writes the result ($= \varphi(a)$) into r of its AR. (See Fig. 7.4 (d).)

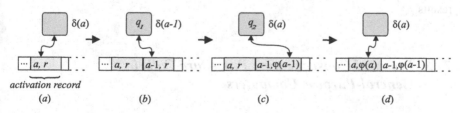

Fig. 7.4 (a) An activation $\delta(a)$ and its activation record. (b) $\delta(a-1)$ starts computing $\varphi(a-1)$. (c) $\delta(a)$ reads $\delta(a-1)$'s result. (d) $\delta(a)$ computes $\varphi(a)$ and stores it into its activation record

Example 7.1. (Computation of $\varphi(x) = x!$ on TM) Figure 7.5 shows the steps in the computation of the value $\varphi(4) = 4! = 24$. □

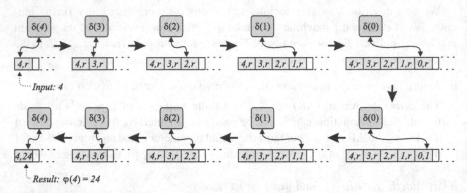

Fig. 7.5 Recursive computation of the value $\varphi(4) = 4!$ on a Turing machine

Clearly, δ must take care of all runtime actions described in *a, b, c, d*. Consequently, δ will consist of two sets of instructions:

- the instructions implementing the actions explicitly dictated by the algorithm (∗);
- the instructions implementing the *runtime actions a, b, c, d*, implicit in (∗).

To prove that there exists a TM that operates as described above, we refer to the *Computability Thesis* and leave the construction of the corresponding program δ as an exercise to the reader.

Looking at Fig. 7.5, we notice that adding and deleting of ARs on the tape of a Turing machine follow the rules characteristic of a data structure called the *stack*. The first AR that is created on the tape is at the *bottom* of the stack, and the rightmost existing AR is at the *top* of the stack. A new AR can be created next to the top AR; we say that it is *pushed* onto the stack. An AR can be deleted from the stack only if it is at the top of the stack; we say that it is *popped* off the stack. We will use these mechanisms shortly.

Later, we will often say that a TM T *calls* another TM T' to do a task for T. The call-and-return mechanism described above can readily be used to make T's program δ call the program δ' of T', pass actual parameters to δ', and collect δ''s results.

7.4.6 Practical Consequences: Procedure Calls in General-Purpose Computers

The mechanisms described in the previous section are used in general-purpose computers to handle procedure calls during program execution.

The Concept of the Procedure

In general-purpose computers we often use *procedures* to decompose large programs into components. The idea is that procedures have other procedures do subtasks. A procedure (*callee*) executes its task when it is *called* (invoked) by another procedure (*caller*). The caller can also be the operating system. A callee may return one or several values to its caller. Each procedure has its *private storage*, where it can access its private (*local*) variables that are needed to do its task.

High-level procedural programming languages support procedures. They make *linkage conventions* in order to define and implement in standard ways the key mechanisms for procedure management. These mechanisms are needed to:

- *invoke* a procedure and map its actual parameters to the callee's private space, so that the callee can use them as its input;
- *return* control to the caller after callee's termination, so that the caller can continue its execution immediately after the point of the call.

Most languages allow a procedure to return one or more values to the caller.

After a program (source code) is written, it must be compiled. One of the tasks of the compiler is to *embed* into the generated code all the *runtime* algorithms and data structures that are necessary to implement the call-and-return behavior implicitly dictated by the source code.

The code is then linked into the executable code, which is ready to be loaded into the computer's main memory for execution.

When the program is started, its call-and-return behavior can be modelled with a *stack*. In the simplest case, the caller pushes the return address onto the stack and transfers the control to the callee; when the callee returns, it pops the return address off the stack and transfers the control back to the caller to the return address. In reality, besides the return address there is more information passed between the two procedures.

The Role of the Compiler

We now see what the compiler must do. First, it must embed in the generated code runtime algorithms that will correctly push onto and pop off the stack the information that must be transferred between a caller and callee via the stack. Second, it must establish a data structure that will contain this information. The structure is called the *activation record* (AR). This must be implemented as a private block of memory associated with a specific invocation of a specific procedure. Consequently, the AR has a number of fields, each for one of the following data:

- *return address*, where the caller's execution will resume after the callee's termination;
- *actual parameters*, which are input data to the callee;
- *register contents*, which the caller preserves to resume its execution after return;
- *return values*, which the callee returns to the caller;

- *local variables*, which are declared and used by the callee;
- *addressability information*, which allows the callee to access non-local variables.

The Role of the Operating System

As described in Sect. 6.2.5, modern general-purpose computers and their operating systems allocate to each executable program, prior to its execution, a memory space that will be used by the program during its execution. Part of this memory is the *runtime stack*. During the execution of the program, an AR is pushed onto (popped off) the runtime stack each time a procedure is called (terminated). The current top AR contains the information associated with the currently executing procedure. In case the size of the runtime stack exceeds the limits set by the operating system, the operating system suspends the execution of the program and takes over. This can happen, for example, in a recursive program when a recursive procedure keeps invoking itself (because it does not meet its recursion-termination criterion).

Execution of Recursive Programs

So how would the program for computing the function $\varphi(x) = x!$ execute on a general-purpose computer? First, we write the program in a high-level procedural programming language, such as C or Java. In this we follow the recursive definition $(*)$ of the function $\varphi(x)$. We obtain a source program similar to P below. Of course, P is recursive (i.e., self-referencing) because the activation $P(x)$ starts a new activation $P(x-1)$, i.e., a new instance of P with a decreased actual parameter. The invocations of P come to an end when the recursion-termination criterion ($x = 1$) is satisfied.

```
program P(x : integer) return integer;
  {
    if x > 0 return x * P(x − 1) else
    if x = 0 return 1
    else Error(illegal_input)
  }
```

7.5 Chapter Summary

A decidable set is also semi-decidable. If a set is decidable, so is its complement. If a set and its complement are semi-decidable, they are both decidable. A set is semi-decidable if and only if it is the domain of a computable function. If two sets are decidable, their union and intersection are decidable. If two sets are semi-decidable, their union and intersection are semi-decidable.

The *Padding Lemma* states that a partial computable function has countably infinitely many indexes; given one of them, the other indexes of the function can be generated. The *index set* of a partial computable function contains all of its indexes, that is, all the encoded Turing programs that compute the function.

The *Parametrization (s-m-n) Theorem* states that the parameters of a function can be built in the function's Turing program to obtain a new Turing program computing the function without parameters. The new Turing program can be algorithmically constructed from the old program and the parameters only. The generalization of the theorem is called the *s-m-n Theorem*.

The *Recursion Theorem* states that if a transformation transforms every Turing program, then some Turing program is transformed into an equivalent Turing program. The *Recursion Theorem* allows a Turing-computable function to be defined recursively, i.e., with its own Turing program. The Turing machine model of computation supports the recursive execution of Turing programs. This lays theoretical grounds for the recursive execution of programs in general-purpose computers and, more generally, for the call-and-return mechanism that allows the use of procedures in programs.

Problems

7.1. Show that:

(a) all tasks in the proof of the *Parametrization Theorem* are computable;

    [*Hint*. See Box 7.1.]

(b) the function s in the *Parametrization Theorem* is injective.

7.2. Prove the following consequences of the *Parametrization Theorem*:

(a) there is a computable function f such that $\mathrm{rng}(\psi_{f(e)}) = \mathrm{dom}(\psi_e)$;

    [*Hint*. Suppose that there existed a p.c. function defined by $\varphi(e,x) \stackrel{\text{def}}{=} \begin{cases} x & \text{if } x \in \mathrm{dom}(\psi_e); \\ \uparrow & \text{otherwise.} \end{cases}$

    Viewing $\varphi(e,x)$ as a function $\lambda x.\varphi(e,x)$ of x, we would have $\mathrm{rng}(\varphi(e,x)) = \mathrm{dom}(\psi_e)$. But $\varphi(e,x)$ exists. It is computed by a TM that simulates UTM on e,x and outputs x if UTM halts.]

(b) there is computable function g such that $\mathrm{dom}(\psi_{g(e)}) = \mathrm{rng}(\psi_e)$;

(c) a set is c.e. *iff* it is the range of a p.c. function.

7.3. Prove the following consequences of the *Recursion Theorem*:

(a) there is a TM that on any input outputs its own code;

    [*Hint*. Prove that there is an $n \in \mathbb{N}$ such that ψ_n is the constant function with value n.]

(b) for every computable function f, there is an n such that $\mathcal{W}_n = \mathcal{W}_{f(n)}$ (recall: $\mathcal{W}_i = \mathrm{dom}(\psi_i)$);

(c) for every p.c. function $\varphi(e,x)$, there is an $n \in \mathbb{N}$ such that $\varphi(n,x) = \psi_n(x)$;

(d) there is an $n \in \mathbb{N}$ such that $\mathcal{W}_n = \{n^2\}$.

7.4. We say that a TM T_1 "calls" another TM T_2 to perform a subtask for T_1. We view this as the program δ_1 calling the program (procedure) δ_2. Explain in detail how such a procedure call can be implemented on Turing machines.

    [*Hint*. Merge δ_1 and δ_2 into δ_1', which will use activation records to activate δ_2 (see Sect. 7.4.5).]

Bibliographic Notes

- Theorem 7.3 was proved in Post [123].
- The *Parametrization Theorem* and *Recursion Theorem* were first proved in Kleene [80]. An extensive description of both theorems and their variants as well as their applications is in Odifreddi [116].
- For runtime environments, activation records and procedure activation in general-purpose computers, see Aho et al. [6, Chap. 7] and Cooper and Torczon [28, Chap. 6].

Chapter 8
Incomputable Problems

A problem is unsolvable if there is no single procedure that can
construct a solution for an arbitrary instance of the problem.

Abstract After the basic notions of computation had been formalized, a close link
between computational problems—in particular, decision problems—and sets was
discovered. This was important because the notion of the the set was finally settled,
and sets made it possible to apply diagonalization, a proof method already discovered by Cantor. Diagonalization, combined with self-reference, made it possible to
discover the first incomputable problem, i.e., a decision problem called the *Halting
Problem*, for which there is no single algorithm capable of solving every instance of
the problem. This was simultaneously and independently discovered by Church and
Turing. After this, *Computability Theory* blossomed, so that in the second half of
the 1930s one of the main questions became, "Which computational problems are
computable and which ones are not?" Indeed, using various proof methods, many
incomputable problems were discovered in different fields of science. This showed
that incomputability is a constituent part of reality.

8.1 Problem Solving

In previous chapters we have discussed how the values of functions can be computed, how sets can be generated, and how sets can be recognized. All of these
are *elementary* computational tasks in the sense that they are all closely connected
with the computational model, i.e., the Turing machine. However, in practice we are
also confronted with other kinds of problems that require certain computations to
yield their solutions.[1] All such problems we call *computational problems*. Now the
following question immediately arises: "Can we use the accumulated knowledge
about how to solve the three elementary computational tasks to solve other kinds of
computational problems?" In this section we will explain how this can be done.

---

[1] Obviously, we are not interested in psychological, social, economic, political, philosophical and
other related problems whose solutions require reasoning beyond (explicit) computation.

© Springer-Verlag Berlin Heidelberg 2015   161
B. Robič, *The Foundations of Computability Theory*,
DOI 10.1007/978-3-662-44808-3_8

8.1.1 Decision Problems and Other Kinds of Problems

Before we start searching for the answer to the above question, we must define precisely what we mean by other "kinds" of computational problems. It is a well-known fact, based upon our everyday experience, that there is a myriad of different computational problems. By grouping all the "similar" problems into classes, we can try to put this jungle of computational problems in order (a rather simple one, actually). The "similarity" between problems can be defined in several ways. For example, we could define a class of all the problems asking for numerical solutions, and another class of all the other problems. However, we will proceed differently. We will define two problems to be *similar* if their solutions are "equally simple." Now, "equally simple" is a rather fuzzy notion. Fortunately, we will not need to formalize it. Instead, it will suffice to define the notion informally. So let us define the following four *kinds (i.e., classes) of computational problems*:

- **Decision** problems (also called **yes/no** problems). The solution of a decision problem is the answer YES or NO. The solution can be represented by a *single bit* (e.g., 1 = YES, 0 = NO).
  Examples: Is there a prime number in a given set of natural numbers? Is there a Hamiltonian cycle in a given graph?

- **Search** problems. The solution of a search problem is an element of a given set S such that the element has a given property P. The solution is an *element of a set*.
  Examples: Find the largest prime number in a given set of natural numbers. Find the shortest Hamiltonian cycle in a given weighted graph.

- **Counting** problems. The solution of a counting problem is the number of elements of a given set S that have a given property P. The solution is a *natural number*.
  Examples: How many prime numbers are in a given set of natural numbers? How many Hamiltonian cycles are in a given graph?

- **Generating** problems (also called **enumeration** problems). The solution of a generating problem is a list of elements of a given set S that have a given property P. The solution is a *sequence of elements of a set*.
  Examples: List all the prime numbers in a given set of natural numbers. List all the Hamiltonian cycles of a given graph.

Which of these kinds of problems should we focus on? Here we make a pragmatic choice and focus on the *decision* problems, because these problems ask for the simplest possible solutions, i.e., solutions representable by a single bit. Our choice does not imply that other kinds of computational problems are not interesting—we only want to postpone their treatment until the decision problems are better understood.

8.1.2 Language of a Decision Problem

In this subsection we will show that there is a close link between *decision problems* and *sets*. This will enable us to *reduce* the questions about decision problems to questions about sets. We will uncover the link in four steps.

1.   Let $\mathcal{D}$ be a decision problem.

2. In practice we are usually confronted with a particular *instance d* of the problem $\mathcal{D}$. The instance d can be obtained from $\mathcal{D}$ by replacing the variables in the definition of $\mathcal{D}$ with actual data. Thus, the problem $\mathcal{D}$ can be viewed as the set of all the possible instances of this problem. We say that an instance $d \in \mathcal{D}$ is *positive* or *negative* if the answer to d is YES or NO, respectively.

   Example 8.1. (Problem Instance) Let $\mathcal{D}_{Prime} \equiv$ "Is n a prime number?" be a decision problem. If we replace the variable n by a particular natural number, say 4, we obtain the instance $d \equiv$ "Is 4 a prime number?" of $\mathcal{D}_{Prime}$. This instance is negative because its solution is the answer NO. In contrast, since 2009 we have known that the solution to the instance $d \equiv$ "Is $2^{43,112,609} - 1$ a prime number?" is YES, so the instance is positive.   □

   So:
                        Let d be an instance of $\mathcal{D}$.

3. Any instance of a decision problem is either positive or negative (due to the *Law of Excluded Middle*; see Sect. 2.1.2). So the answer to the instance d is either YES or NO. But how can we *compute* the answer, say on the Turing machine? In the natural-language description of d there can be various data, e.g., numbers, matrices, graphs. But in order to compute the answer on a machine—be it a modern computer or an abstract model such as the Turing machine—we must rewrite these data in a form that is understandable to the machine. Since any machine uses its own alphabet Σ (e.g., $\Sigma = \{0, 1\}$), we must choose a function that will transform every instance of $\mathcal{D}$ into a word over Σ. Such a function is called the *coding function* and will be denoted by *code*. Therefore, code : $\mathcal{D} \to \Sigma^*$. We will usually write $\langle d \rangle$ instead of the longer code(d). There are many different coding functions. From the point of view of *Computability Theory*, we can choose any of them as long as the function is *computable* on $\mathcal{D}$ and *injective*. These requirements[2] are natural because we want the coding process to halt and we do not want a word to encode different instances of $\mathcal{D}$.

   Example 8.2. (Instance Code) The instances of the problem $\mathcal{D}_{Prime} \equiv$ "Is n a prime number?" can be encoded by the function code : $\mathcal{D}_{Prime} \to \{0, 1\}^*$ that rewrites n in its binary representation; e.g., code("Is 4 a prime number?") = 100 and code("Is 5 a prime number?") = 101. □

   So:
                        Let code : $\mathcal{D} \to \Sigma^*$ be a coding function.

---

[2] In *Computational Complexity Theory* we also require that the coding function does not produce unnecessarily long codes.

4. After we have encoded d into $\langle d \rangle$, we could start searching for a Turing machine capable of computing the answer to d when given the input $\langle d \rangle$. However, we proceed differently: We gather the *codes of all the positive instances* of $\mathcal{D}$ in a *set* and denote it by $L(\mathcal{D})$. This is a subset of Σ^*, so it is a formal language (see Sect. 6.3.6). Here is the definition.

Definition 8.1. (Language of a Decision Problem) The **language of a decision problem** $\mathcal{D}$ is the set $L(\mathcal{D}) \stackrel{\text{def}}{=} \{\langle d \rangle \in \Sigma^* \mid d \text{ is a positive instance of } \mathcal{D}\}$.

Example 8.3. (Language of a Decision Problem)   The language of the decision problem $\mathcal{D}_{Prime}$ is the set $L(\mathcal{D}_{Prime}) = \{10,11,101,111,1011,1101,10001,10011,10111,110101,\ldots\}$. □

5. Now the following is obvious:

$$\text{An instance } d \text{ of } \mathcal{D} \text{ is positive} \iff \langle d \rangle \in L(\mathcal{D}). \tag{$*$}$$

What did we gain by this? The equivalence ($*$) tells us that computing the answer to d can be substituted with deciding whether or not $\langle d \rangle$ is in $L(\mathcal{D})$. (See Fig. 8.1.) Thus we have found a connection between decision problems and sets:

Solving a decision problem $\mathcal{D}$ can be reduced to recognizing the set $L(\mathcal{D})$ in Σ^*.

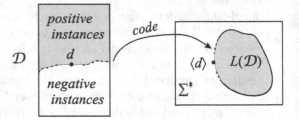

Fig. 8.1 The answer to the instance d of a decision problem $\mathcal{D}$ can be obtained by determining where the word $\langle d \rangle$ is relative to the set $L(\mathcal{D})$

The connection ($*$) is important because it enables us to apply (when solving decision problems) all the theory developed to recognize sets. Recall that recognizing a set is determining what is in the set and what is in its complement (Sect. 6.3.3). Now let us see what the recognizability of $L(\mathcal{D})$ in Σ^* tells us about the solvability of the decision problem $\mathcal{D}$:

- *Let $L(\mathcal{D})$ be decidable.* Then there is a decider $D_{L(\mathcal{D})}$ that, for arbitrary $\langle d \rangle \in \Sigma^*$, answers the question $\langle d \rangle \in ? \; L(\mathcal{D})$. Because of ($*$), the answer tells us whether d is a positive or a negative instance of $\mathcal{D}$. Consequently, there is an algorithm that, for the arbitrary instance $d \in \mathcal{D}$, decides whether d is positive or negative. The algorithm is $D_{L(\mathcal{D})}(\text{code}(\cdot))$, the composition of the coding function and the decider of $L(\mathcal{D})$.

- *Let $L(\mathcal{D})$ be semi-decidable.* Then, for arbitrary $\langle d \rangle \in L(\mathcal{D})$, the recognizer $R_{L(\mathcal{D})}$ answers the question $\langle d \rangle \in? L(\mathcal{D})$ with YES. However, if $\langle d \rangle \notin L(\mathcal{D})$, $R_{L(\mathcal{D})}$ may or may not return NO in finite time. So: There is an algorithm that for the arbitrary *positive* instance $d \in \mathcal{D}$ finds that d *is* positive. The algorithm is $R_{L(\mathcal{D})}(\text{code}(\cdot))$.

- *Let $L(\mathcal{D})$ be undecidable.* Then there is no algorithm capable of answering, for arbitrary $\langle d \rangle \in \Sigma^*$, the question $\langle d \rangle \in? L(\mathcal{D})$. Because of (∗), there is no algorithm capable of deciding, for the arbitrary instance $d \in \mathcal{D}$, whether d is positive or negative.

So we can extend our terminology about sets (Definition 6.10) to decision problems.

Definition 8.2. Let $\mathcal{D}$ be a decision problem. We say that the problem

$$\mathcal{D} \text{ is } \textbf{decidable (or computable)} \quad \textit{if} \quad L(\mathcal{D}) \text{ is decidable set;}$$
$$\mathcal{D} \text{ is } \textbf{semi-decidable} \quad \textit{if} \quad L(\mathcal{D}) \text{ is semi-decidable set;}$$
$$\mathcal{D} \text{ is } \textbf{undecidable (or incomputable)} \quad \textit{if} \quad L(\mathcal{D}) \text{ is undecidable set.}$$

NB *Instead of a(n) (un)decidable problem we can say (in)computable problem. But bear in mind that the latter notion is more general: it can be used with all kinds of computational problems, not only with the decision problems (Sects. 9.3 and 9.5). So, the (un)decidable problem is the same as the (in)computable decision problem. The term (un)solvable from this section's motto is even more general: it addresses all kinds of computational and non-computational problems. (See Fig. 8.2.)*

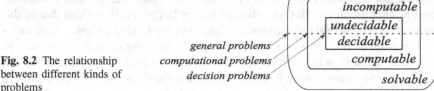

Fig. 8.2 The relationship between different kinds of problems

8.1.3 Subproblems of a Decision Problem

Often we encounter a decision problem that is a special version of another, more general decision problem. Is there any connection between the languages of the two problems? Is there any connection between the decidabilities of the two problems? Let us start with a definition.

Definition 8.3. (Subproblem) A decision problem $\mathcal{D}_{Sub}$ is the **subproblem** of a decision problem $\mathcal{D}_{Prob}$ if $\mathcal{D}_{Sub}$ is obtained from $\mathcal{D}_{Prob}$ by imposing additional restrictions on (some of) the variables of $\mathcal{D}_{Prob}$.

A restriction can be put in various ways: a variable can be assigned a particular value; or it can be restricted to take only the values from a given set of values; or a relation between a variable and another variable can be imposed. For instance, if $\mathcal{D}_{Prob}$ has the variables x and y, then we might want to impose the equality $x = y$. (Actually, we will do this in the next section.)

It should be obvious that the following holds:

$$\mathcal{D}_{Sub} \text{ is a subproblem of } \mathcal{D}_{Prob} \implies L(\mathcal{D}_{Sub}) \subseteq L(\mathcal{D}_{Prob})$$

It is also easy to prove the next theorem.

Theorem 8.1. *Let $\mathcal{D}_{Sub}$ be a subproblem of a decision problem $\mathcal{D}_{Prob}$. Then:*
$$\mathcal{D}_{Sub} \text{ is undecidable} \implies \mathcal{D}_{Prob} \text{ is undecidable}$$

Proof. Let $\mathcal{D}_{Sub}$ be undecidable and suppose that $\mathcal{D}_{Prob}$ is decidable. Then we could use the algorithm for solving $\mathcal{D}_{Prob}$ to solve $\mathcal{D}_{Sub}$, so $\mathcal{D}_{Sub}$ would be decidable. This is a contradiction. □

8.2 There Is an Incomputable Problem — Halting Problem

We have just introduced the notions of the decidable, semi-decidable, and undecidable decision problems. But the attentive reader has probably noticed that at this point we actually *do not know* whether there exists any semi-decidable or any undecidable problem. If there is no such problem, then the above definition is in vain, so we should throw it away, together with our attempt to explore semi-decidable and undecidable problems. (What is more, in this case the definition of semi-decidable and undecidable *sets* would also be superfluous.) In other words, the recently developed part of *Computability Theory* should be abandoned. Therefore, the important question at this point is this: "Is there any decision problem that is semi-decidable or undecidable?"

To prove the existence of an undecidable decision problem, we should find a decision problem $\mathcal{D}$ such that the set $L(\mathcal{D})$ is undecidable. But how can we find such a $\mathcal{D}$ (if there is one at all)? In 1936, Turing succeeded in this. (Independently and at the same time, Church also found such a problem.) How did he do that?

Turing was well aware of the fact that difficulties in obtaining computational results are caused by those Turing machines that may not halt on some input data. It would therefore be beneficial, he reckoned, if we could *check, for any Turing machine T and any input word w, whether or not T halts on w.* If such a checking were possible, then, given an arbitrary pair (T, w), we would first check the pair (T, w) and then, depending on the outcome of the checking, we would either start T on w, or try to improve T so that it would halt on w, too. This led Turing to define the following decision problem.

Definition 8.4. (Halting Problem) The **Halting Problem** $\mathcal{D}_{Halt}$ is defined by
$\mathcal{D}_{Halt} \equiv$ "Given a Turing machine T and a word $w \in \Sigma^*$, does T halt on w?"

Turing then proved the following theorem.

Theorem 8.2. *The Halting Problem $\mathcal{D}_{Halt}$ is undecidable.*

Before we go to the proof, we introduce two sets that will play an important role in the proof and, indeed, in the rest of the book. The sets are called the *universal* and *diagonal languages*, respectively.

Definition 8.5. (Universal Language $\mathcal{K}_0$) The **universal language**, denoted[3] $\mathcal{K}_0$, is the language of the *Halting Problem*, that is,

$$\mathcal{K}_0 \stackrel{\text{def}}{=} L(\mathcal{D}_{Halt}) = \{\langle T, w \rangle \mid T \text{ halts on } w\}$$

The second language, $\mathcal{K}$, is obtained from $\mathcal{K}_0$ by imposing the restriction $w = \langle T \rangle$.

Definition 8.6. (Diagonal Language $\mathcal{K}$) The **diagonal language**, denoted $\mathcal{K}$, is defined by
$$\mathcal{K} \stackrel{\text{def}}{=} \{\langle T, T \rangle \mid T \text{ halts on } \langle T \rangle\}$$

Observe that $\mathcal{K}$ is the language $L(\mathcal{D}_H)$ of the decision problem

$\mathcal{D}_H \equiv$ "Given a Turing machine T, does T halt on $\langle T \rangle$?"

The problem $\mathcal{D}_H$ is a *subproblem* of $\mathcal{D}_{Halt}$, since it is obtained from $\mathcal{D}_{Halt}$ by restricting the variable w to $w = \langle T \rangle$.

We can now proceed to the proof of Theorem 8.2.

---

[3] The reasons for the notation $\mathcal{K}_0$ are historical. Namely, Post proved that this language is *Turing-complete*. We will explain the notion of Turing-completeness in Part III of the book; see Sect. 14.1.

Proof. (of the theorem)   The plan is to prove (in a lemma) that $\mathcal{K}$ is an undecidable set; this will then imply that $\mathcal{K}_0$ is also an undecidable set and, consequently, that $\mathcal{D}_{Halt}$ is an undecidable problem. The lemma is instructive because it applies a cunningly defined Turing machine to its own code.

Lemma 8.1. (Undecidability of $\mathcal{K}$) *The set $\mathcal{K}$ is undecidable.*

Proof. (of the lemma) *Suppose that $\mathcal{K}$ is a decidable set.* Then there must exist a decider $D_{\mathcal{K}}$ that, for arbitrary T, answers the question $\langle T,T \rangle \in? \mathcal{K}$ with

$$D_{\mathcal{K}}(\langle T,T \rangle) = \begin{cases} \text{YES, if } T \text{ halts on } \langle T \rangle; \\ \text{NO, \ if } T \text{ does not halt on } \langle T \rangle. \end{cases}$$

Now we construct a new Turing machine S. The intention is to construct S in such a way that, when given as input its *own* code $\langle S \rangle$, S will expose the incapability of $D_{\mathcal{K}}$ in predicting whether or not S will halt on $\langle S \rangle$. The machine S is depicted in Fig. 8.3.

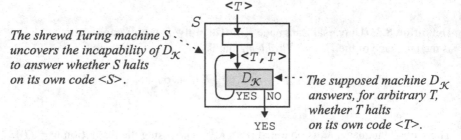

The shrewd Turing machine S uncovers the incapability of $D_{\mathcal{K}}$ to answer whether S halts on its own code <S>.

The supposed machine $D_{\mathcal{K}}$ answers, for arbitrary T, whether T halts on its own code <T>.

Fig. 8.3 *S exposes the incapability of the decider $D_{\mathcal{K}}$ to predict the halting of S on $\langle S \rangle$*

The machine S operates as follows. The input to S is the code $\langle T \rangle$ of an arbitrary Turing machine T. The machine S doubles $\langle T \rangle$ into $\langle T,T \rangle$, sends this to the decider $D_{\mathcal{K}}$, and starts it. The decider $D_{\mathcal{K}}$ eventually halts on $\langle T,T \rangle$ and answers either YES or NO to the question $\langle T,T \rangle \in? \mathcal{K}$. If $D_{\mathcal{K}}$ has answered YES, then S asks $D_{\mathcal{K}}$ again the same question. If, however, $D_{\mathcal{K}}$ has answered NO, then S outputs its own answer YES and *halts*.

But S is *shrewd*: if given as input its *own* code $\langle S \rangle$, it puts the supposed decider $D_{\mathcal{K}}$ in insurmountable trouble. Let us see why. Given the input $\langle S \rangle$, S doubles it into $\langle S,S \rangle$ and hands it over to $D_{\mathcal{K}}$, which in finite time answers the question $\langle S,S \rangle \in? \mathcal{K}$ with either YES or NO. Now let us analyze the consequences of each of the answers:

a. Suppose that $D_{\mathcal{K}}$ has answered with $D_{\mathcal{K}}(\langle S,S \rangle) =$ YES. Then S repeats the question $\langle S,S \rangle \in? \mathcal{K}$ to $D_{\mathcal{K}}$, which in turn stubbornly repeats its answer $D_{\mathcal{K}}(\langle S,S \rangle) =$ YES. It is obvious that S cycles and will *never* halt. But note that, at the same time, $D_\mathcal{K}$ repeatedly predicts just the opposite, i.e., that S *will* halt on $\langle S \rangle$. We must conclude that in the case (a) the supposed decider $D_\mathcal{K}$ fails to compute the correct answer.

b. *Suppose that $D_\mathcal{K}$ has answered with $D_\mathcal{K}(\langle S, S \rangle)$ = NO. Then S returns to the environment its own answer and halts.* But observe that just before that $D_\mathcal{K}$ has computed the answer $D_\mathcal{K}(\langle S, S \rangle)$ = NO and thus predicted that S will *not* halt on $\langle S, S \rangle$. We must conclude that in the case (b) the supposed decider $D_\mathcal{K}$ fails to compute the correct answer.

Thus, the supposed decider $D_\mathcal{K}$ is unable to correctly decide the question $\langle S, S \rangle \in ? \mathcal{K}$. This contradicts our supposition that $\mathcal{K}$ is a decidable set and $D_\mathcal{K}$ its decider. We must conclude that $\mathcal{K}$ is an undecidable set. *The lemma is proved.*

Because $\mathcal{K}$ is undecidable, so is the problem $\mathcal{D}_H$. But $\mathcal{D}_H$ is a subproblem of the *Halting Problem $\mathcal{D}_{Halt}$*, so the *Halting Problem* is undecidable too. This completes the proof of Theorem 8.2. □

As noted, the sets $\mathcal{K}_0$ and $\mathcal{K}$ are called the universal and diagonal languages, respectively. The reasons for such a naming are explained in the following subsection.

8.2.1 Consequences: The Basic Kinds of Decision Problems

Now we know that besides decidable problems there also exist undecidable problems. What about semi-decidable problems? Do they exist? The answer is clearly *yes*, because every decidable set is by definition semi-decidable. So the real question is: Are there *undecidable* problems that are *semi-decidable*? That is, are there decision problems such that only their positive instances are guaranteed to be solvable? The answer is *yes*. Let us see why.

Theorem 8.3. *The set $\mathcal{K}_0$ is semi-decidable.*

Proof. We must prove that there is a recognizer for $\mathcal{K}_0$ (see Sect. 6.3.3). Let us conceive the following machine: Given an arbitrary input $\langle T, w \rangle \in \Sigma^*$, the machine must simulate T on w, and if the simulation halts, the machine must return YES and halt. So, if such a machine exists, it will answer YES *iff* $\langle T, w \rangle \in \mathcal{K}_0$. But we already know that such a machine exists: it is the *Universal Turing machine U* (see Sect. 6.2.2). Hence, $\mathcal{K}_0$ is semi-decidable. □

Theorems 8.2 and 8.3 imply that $\mathcal{K}_0$ is an undecidable semi-decidable set.

The proof of the Theorem 8.3 has also revealed that $\mathcal{K}_0$ is the proper set of the Universal Turing machine. This is why $\mathcal{K}_0$ is called the *universal language*.

What about the set $\overline{\mathcal{K}_0}$, the complement of $\mathcal{K}_0$?

Theorem 8.4. *The set $\overline{\mathcal{K}_0}$ is not semi-decidable.*

Proof. If $\overline{\mathcal{K}_0}$ were semi-decidable, then both $\overline{\mathcal{K}_0}$ and $\mathcal{K}_0$ would be semi-decidable. Then $\mathcal{K}_0$ would be decidable because of Theorem 7.3 (p. 144). But this would contradict Theorem 8.2. So, the set $\overline{\mathcal{K}_0}$ is not semi-decidable.   □

Similarly, we prove that the undecidable set $\mathcal{K}$ is semi-decidable, and that $\overline{\mathcal{K}}$ is not semi-decidable.

We have proved the existence of undecidable semi-decidable sets (e.g., $\mathcal{K}_0$ and $\mathcal{K}$) and the existence of undecidable sets that are not even semi-decidable (e.g., $\overline{\mathcal{K}_0}$ and $\overline{\mathcal{K}}$). We now know that the class of all the sets partitions into three non-empty subclasses, as depicted in Fig. 8.4. (Of course, we continue to talk about sets that are subsets of Σ^* or $\mathbb{N}$, as explained in Sect. 6.3.5.)

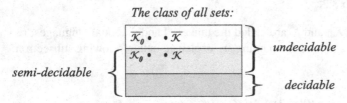

Fig. 8.4 The three main classes of sets according to their decidability

The Class of All the Decision Problems

We can view sets as languages of decision problems. Then, $\mathcal{K}_0$ and $\mathcal{K}$ (Fig. 8.4) are the languages of decision problems $\mathcal{D}_{Halt}$ and $\mathcal{D}_H$, respectively. What about $\overline{\mathcal{K}_0}$ and $\overline{\mathcal{K}}$? These are the languages of decision problems $\mathcal{D}_{\overline{Halt}}$ and $\mathcal{D}_{\overline{H}}$, respectively, where $\mathcal{D}_{\overline{Halt}} \equiv$ "Given a Turing machine T and a word $w \in \Sigma^*$, does T *never* halt on w?" and $\mathcal{D}_{\overline{H}} \equiv$ "Given a Turing machine T, does T *never* halt on $\langle T \rangle$?". Now we see (Fig. 8.5) that *the class of all the decision problems* partitions into two non-empty subclasses: the class of decidable problems and the class of undecidable problems. There is also a third class, the class of semi-decidable problems, which contains all the decidable problems and some, but not all, of the undecidable problems.

In other words, a decision problem $\mathcal{D}$ can be of one of the three kinds:

- $\mathcal{D}$ is *decidable*. This means that there is an algorithm D that can solve an *arbitrary* instance $d \in \mathcal{D}$. Such an algorithm D is called the *decider of the problem* $\mathcal{D}$.
- $\mathcal{D}$ is *semi-decidable undecidable*. This means that no algorithm can solve an arbitrary instance $d \in \mathcal{D}$, but that there is an algorithm R that can solve an arbitrary *positive* $d \in \mathcal{D}$. The algorithm R is called the *recognizer of the problem* $\mathcal{D}$.
- $\mathcal{D}$ is *not semi-decidable*. Now, for any algorithm there exists a positive instance and a negative instance of $\mathcal{D}$ such that the algorithm cannot solve either of them.

The class of all decision problems:

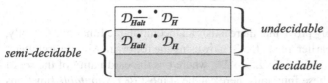

Fig. 8.5 The three main classes of decision problems according to their decidability

8.2.2 Consequences: Complementary Sets and Decision Problems

From the previous theorems it follows that there are only three possibilities for the decidability of a set S and its complement $\overline{S}$:

1. S and $\overline{S}$ are decidable (e.g., ▲, △ in Fig. 8.6);
2. S and $\overline{S}$ are undecidable; one is semi-decidable and the other is not (e.g., ○, ●);
3. S and $\overline{S}$ are undecidable and neither is semi-decidable (e.g., □,■).

The class of all sets:

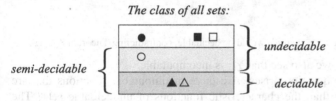

Fig. 8.6 Decidability of complementary sets

Formally, two complementary sets S and $\overline{S}$ are associated with two complementary decision problems $\mathcal{D}_S$ and $\overline{\mathcal{D}}_S$, where $\overline{\mathcal{D}}_S \stackrel{\text{def}}{=} \mathcal{D}_{\overline{S}}$. Of course, for the decidability of these two problems there are also only three possibilities (Fig. 8.7).

The class of all decision problems:

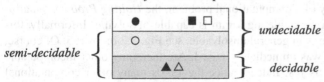

Fig. 8.7 Decidability of complementary decision problems

Remark. Because of the situation ●,○, two undecidable problems can differ in their undecidability. Does this mean that ● is more difficult than ○? Are there different *degrees* of undecidability? The answer is *yes*. Indeed, there is much to be said about this phenomenon, but we postpone the treatment of it until Part III.

8.2.3 Consequences: There Is an Incomputable Function

We are now able to easily prove that there exists an incomputable function. Actually, we foresaw this much earlier, in Sect. 5.3.3, by using the following simple argument. There are c different functions $f : \Sigma^* \to \Sigma^*$, where c is the cardinality of the set of real numbers. Among these functions there can be at most $\aleph_0$ *computable* functions (as there are at most $\aleph_0$ different Turing machines that can compute functions' values). Since $\aleph_0 < c$, there must be functions $f : \Sigma^* \to \Sigma^*$ that are not computable. However, this proof is not constructive: it neither constructs a particular incomputable function nor shows how such a function could be constructed, at least in principle. In the intuitionistic view, this proof is not acceptable (see Sect. 2.2.2).

But now, when we know that $\mathcal{K}_0$ is undecidable, it is an easy matter to exhibit a particular incomputable function. This is the characteristic function $\chi_{\mathcal{K}_0} : \Sigma^* \to \{0,1\}$, where

$$\chi_{\mathcal{K}_0}(x) = \begin{cases} 1 \, (\equiv \text{YES}), & \text{if } x \in \mathcal{K}_0; \\ 0 \, (\equiv \text{NO}), & \text{if } x \notin \mathcal{K}_0. \end{cases}$$

Corollary 8.1. (Incomputable Function) *The characteristic function $\chi_{\mathcal{K}_0}$ is incomputable.*

Proof. If $\chi_{\mathcal{K}_0}$ were computable, then $\mathcal{K}_0$ would be decidable, contradicting Theorem 8.2. □

By the same argument we also see that $\chi_{\mathcal{K}}$ is incomputable.

Now the following question arises: Are there incomputable functions that are defined less generally than the characteristic functions of undecidable sets? The answer is yes. We will define such a function, the so-called *Busy Beaver function*, in the next section.

8.3 Some Other Incomputable Problems

We have seen that the formalization of the basic notions of computability led to the surprising discovery of a computational problem, the *Halting Problem*, an arbitrary instance of which no single algorithm is capable of solving. Informally, this means that this problem is in general unsolvable; see Fig. 8.2 on p. 165. Of course, the following question was immediately raised: Are there any other incomputable problems? The answer is *yes*. Indeed, since the 1940s many other computational problems were proved to be incomputable. The first of these problems were somewhat unnatural, in the sense that they referred to the properties and the operations of models of computation. After 1944, however, more realistic incomputable problems were (and are still being) discovered in different fields of science and in other nonscientific fields. In this section we will list some of the known incomputable problems, grouped by the fields in which they occur. (We will postpone the discussion of the methods for proving their incomputability to the next section.)

8.3.1 Problems About Turing Machines

Halting of Turing Machines. We already know two incomputable problems about Turing machines: the *Halting Problem* $\mathcal{D}_{Halt}$ and its subproblem $\mathcal{D}_H$ that asks whether a Turing machine halts on its own code. But there are other incomputable problems about the halting of Turing machines.

▲ HALTING OF TURING MACHINES

Let T and T' be arbitrary Turing machines and M be a fixed Turing machine. Let $w \in \Sigma^*$ be an arbitrary word. Questions:

$$\mathcal{D}_{Halt} \equiv \text{``Does } T \text{ halt on } w\text{?''}$$
$$\mathcal{D}_H \equiv \text{``Does } T \text{ halt on } \langle T \rangle\text{?''}$$
$$\text{``Does } T \text{ halt on empty input?''}$$
$$\text{``Does } T \text{ halt on every input?''}$$
$$\text{``Does } M \text{ halt on } w\text{?''}$$
$$\text{``Do } T \text{ and } T' \text{ halt on the same inputs?''}$$

These problems are all undecidable; no algorithm can solve any of them in general.

Properties of Turing Machine Languages. Recall from Sect. 6.3.3 that $L(T)$, the proper set of a Turing machine T, contains exactly those words in Σ^* on which T halts in the accepting state. So, which Turing machines have empty and which ones nonempty proper sets? Next, by Definition 6.10 (p. 132), $L(T)$ is decidable if we can algorithmically decide, for an arbitrary word, whether or not the word is in $L(T)$. Which Turing machines have decidable languages and which ones do not?

▲ PROPERTIES OF TM LANGUAGES

Let T be an arbitrary Turing machine. Question:

$$\mathcal{D}_{Emp} \equiv \text{``Is the language } L(T) \text{ empty?''}$$
$$\text{``Is the language } L(T) \text{ decidable?''}$$

These problems are undecidable; no algorithm can solve any of them for arbitrary T.

Busy Beaver Problems. Informally, a busy beaver is the most productive Turing machine of its kind. Let us define the kind of Turing machines we are interested in. Let $n \geqslant 1$ be a natural number. Define $\mathcal{T}_n$ to be the set of all the two-way unbounded Turing machines $T = (Q, \Sigma, \Gamma, \delta, q_1, \sqcup, F)$, where $Q = \{q_1, \ldots, q_{n+1}\}$, $\Sigma = \{0,1\}$, $\Gamma = \{0,1,\sqcup\}$, $\delta : Q \times \Gamma \to Q \times \{1\} \times \{\text{Left}, \text{Right}\}$ and $F = \{q_{n+1}\}$. Informally, $\mathcal{T}_n$ contains all the Turing machines that have 1) the tape unbounded in both ways; 2) $n \geqslant 1$ non-final states (including the initial state q_1) and one final state q_{n+1}; 3) tape symbols $0, 1, \sqcup$ and input symbols $0, 1$; 4) instructions that write to a cell only the symbol 1 and move the window either to the left or right. We see that such TMs have unbounded space on their tape for writing the symbol 1, and they do not waste time either with writing symbols other than 1 or with leaving the window as it is.

It can be shown that there are finitely many Turing machines in $\mathcal{T}_n$. (See Problem 8.3, p. 189.) Now, let us pick an arbitrary $T \in \mathcal{T}_n$ and start it on an empty input (i.e., with the tape containing $\sqcup$ in each cell.) If T halts, we define its *score* $\sigma(T)$ to be the number of symbols 1 that remain on the tape after halting. If, however, T does not halt, we let $\sigma(T)$ be undefined. In this way we have defined a partial function $\sigma : \mathcal{T}_n \to \mathbb{N}$, where

$$\sigma(T) \stackrel{\text{def}}{=} \begin{cases} k & \text{if } T \text{ halts on empty input and leaves on its tape } k \text{ symbols } 1; \\ \uparrow & \text{otherwise.} \end{cases}$$

We will say that a Turing machine $T \in \mathcal{T}_n$ is a *stopper* if it halts on an empty input. Intuitively, we expect that, in general, different stoppers in $\mathcal{T}_n$—*if they exist*—attain different scores. Since there are only finitely many stoppers in $\mathcal{T}_n$, at least one of them must attain the highest possible score in $\mathcal{T}_n$. But, what if there were *no* stoppers in $\mathcal{T}_n$? Then, the highest score would not be defined for $\mathcal{T}_n$ (i.e., for that n). Can this happen at all? The answer is no; in 1962 Radó[4] proved that for *every* $n \geqslant 1$ there *exists* a stopper in $\mathcal{T}_n$ that attains, among all the stoppers in $\mathcal{T}_n$, the maximum value of σ. Such a stopper is called the *n-state Busy Beaver* and will be denoted by *n-BB*. Consequently, the function $s : \mathbb{N} \to \mathbb{N}$, defined by

$$s(n) \stackrel{\text{def}}{=} \sigma(n\text{-}BB)$$

is well defined (i.e., $s(n)\downarrow$ for every $n \in \mathbb{N}$). We call it the *Busy Beaver function*.

At last we can raise the following question: Given an arbitrary Turing machine of the above kind (i.e., a machine in $\bigcup_{i \geqslant 1} \mathcal{T}_i$), can we algorithmically decide whether or not the machine is an *n*-state Busy Beaver for some *n*? Interestingly, we cannot.

▲ BUSY BEAVER PROBLEM
   Let $T \in \bigcup_{i \geqslant 1} \mathcal{T}_i$ be an arbitrary Turing machine. Question:

$$\mathcal{D}_{BB} \equiv \text{``Is } T \text{ a Busy Beaver?''}$$

The problem is undecidable. There is no algorithm that can solve it for arbitrary T.

What about the Busy Beaver function s? Is it computable? The answer is *no*.

▲ BUSY BEAVER FUNCTION

The Busy Beaver function is incomputable.

So there is no algorithm capable of computing, for arbitrary $n \geqslant 1$, the score of the *n*-state Busy Beaver.

---

[4] Tibor Radó, 1895–1965, Hungarian-American mathematician.

8.3.2 Post's Correspondence Problem

This problem was defined and proved to be undecidable by Post in 1946 as a result of his research of normal systems (see Sect. 6.3.2). It is important because it was one of the first realistic undecidable problems to be discovered and because it was used to prove the undecidability of several other decision problems. The problem can be defined as follows (Fig. 8.8).

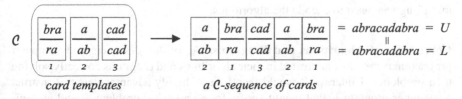

Fig. 8.8 A positive instance of *Post's Correspondence Problem*

Let $\mathcal{C}$ be a finite set of elements, called *card templates*. Each card template is divided into an upper and lower half. Each half contains a word over some alphabet Σ. We call the two words the upper and the lower word of the card template. For each card template there is a stack with a potentially infinite number of exact copies of the template, called the *cards*. Cards from any stack can be put one after another in an arbitrary order to form a finite sequence of cards, called the *C-sequence*. Each $\mathcal{C}$-sequence defines two compound words, called *sentences*. The upper sentence U is the concatenation of all the upper words in the $\mathcal{C}$-sequence. Similarly, the lower sentence L is the concatenation of all the lower words in the $\mathcal{C}$-sequence. Now the question is: Is there a $\mathcal{C}$-sequence such that $U = L$?

▲ POST'S CORRESPONDENCE PROBLEM
   Let $\mathcal{C}$ be a finite set of card templates, each with an unlimited number of copies. Question:

$$\mathcal{D}_{PCP} \equiv \text{``Is there a finite C-sequence such that } U = L?\text{''}$$

This problem is undecidable; no algorithm can solve it for arbitrary $\mathcal{C}$.

8.3.3 Problems About Algorithms and Computer Programs

Termination of Algorithms and Programs. Algorithm designers and computer programmers are usually interested in whether their algorithms and programs always terminate (i.e., do not get trapped into endless cycling). Since the Turing machine formalizes the notion of algorithm, we can easily restate two of the above halting problems into the following undecidable problems.

▲ TERMINATION OF ALGORITHMS (PROGRAMS)

Let A be an arbitrary algorithm and d be arbitrary input data. Questions:

$$\mathcal{D}_{Term} \equiv \text{"Does an algorithm A terminate on every input data?"}$$
$$\text{"Does an algorithm A terminate on input data d?"}$$

Both problems are undecidable. Consequently, there is no computer program capable of checking, for an arbitrary computer program P, whether or not P eventually terminates (even if the input data is fixed). This holds irrespectively of the programming language used to encode the algorithms.

Correctness of Algorithms and Programs. Algorithm designers and computer programmers are most interested in their algorithms and programs correctly solving their problems of interest. It would therefore be highly advantageous to construct a computer program V that would verify, for an arbitrary problem $\mathcal{P}$ and an arbitrary algorithm A, whether or not A correctly solves $\mathcal{P}$. In order to be read by such a would-be verifier, the problem $\mathcal{P}$ and the algorithm A must be appropriately encoded. So let $code(\mathcal{P})$ denote the word encoding the problem $\mathcal{P}$, and let $code(A)$ denote the computer program describing the algorithm A.

▲ CORRECTNESS OF ALGORITHMS (PROGRAMS)

Let $\mathcal{P}$ be an arbitrary computational problem and A an arbitrary algorithm. Question:

$$\mathcal{D}_{Corr} \equiv \text{"Does the algorithm code(A) correctly solve the problem code(P)?"}$$

The problem is undecidable; there is no algorithm (verifier) capable of solving it for arbitrary $\mathcal{P}$ and A. This is true irrespective of the programming language used to write the algorithms. Indeed, the problem remains undecidable even if $code(A)$ is allowed to use only the most elementary data types (e.g., integers, character strings) and perform only the most elementary operations on them. This is bad news for those who work in the field of program verification.

Shorter Equivalent Programs. Sometimes, programmers try to shorten their programs in some way. They may want to reduce the number of program statements, or the total number of symbols, or the number of variables in their programs. In any case, they use some reasonable measure of program length. While striving for a shorter program they want to retain its functionality; that is, a shortened program should be functionally equivalent to the original one. Here, we define two programs as being *equivalent* if, for every input, they return equal results. So, the question is: Is there a shorter equivalent program? In general, this is an undecidable problem.

▲ EXISTENCE OF SHORTER EQUIVALENT PROGRAMS

Let $code(A)$ be a program describing an algorithm A. Question:

"Given a program code(A), is there a shorter equivalent program?"

This is an undecidable problem; no algorithm can solve it in general. This holds for any reasonable definition of program length and for any programming language used to code algorithms.

8.3.4 Problems About Programming Languages and Grammars

The syntax of a language—be it natural, programming, or purely formal—is described by means of a *grammar*. This can be viewed as a Post canonical system (see Box 6.2 on p. 128) whose productions have been restricted (i.e., simplified) in any of several possible ways. For example, natural languages, such as English, developed complex syntaxes that demand the so-called *context-sensitive grammars* to deal with. In contrast, programming languages, which are artificial, have their syntaxes defined by simpler grammars, the so-called *context-free grammars* (CFGs). This is because such grammars simplify the recognition (parsing) of computer programs (i.e., checking whether the programs are syntactically correct). This in turn allows us to construct simpler and faster parsers (and compilers). Finally, the basic building blocks of computer programs, the so-called *tokens*, are words with even simpler syntax. Tokens can be described and recognized with the so-called *regular grammars*. In summary, a grammar G can be used in several ways: to *define* a language, which we will denote by $L(G)$; or to *generate* (some) elements of $L(G)$; or to *recognize* the language $L(G)$. For example, given a string w of symbols (be it a natural-language sentence or a computer program), we can answer the question $w \in ?L(G)$ by trying to generate w by G. In particular, the parser tries to generate by G the given computer program. When G is CFG, we say that $L(G)$ is a *context-free language* (CFL); and when G is regular, $L(G)$ is said to be a *regular language*. Many problems about grammars and their languages were defined in the fields of programming language design and translation. Some of these are incomputable.

Ambiguity of CFG Grammars. Programming language designers are only interested in grammars that do not allow a computer program be parsed in two different ways. If this happened, it would mean that the structure and meaning of the program can be viewed in two different ways. Which of the two was intended by the programmer? Such a grammar would be *ambiguous*. So the detection of ambiguous grammars is an important practical problem.

▲ AMBIGUITY OF CFG GRAMMARS

Let G be a context-free grammar. Question:

"Is there a word that can be generated by G in two different ways?"

The problem is undecidable; there is no algorithm capable of solving it for arbitrary G. As a consequence, programming-language designers must invent and apply, for different Gs, different approaches to prove that the Gs are unambiguous. To ease this, they additionally restrict feasible CFGs and deal with, for example, the so-called LL(k) grammars. Since CFGs are a subclass of the class of context-sensitive grammars, the above decision problem is a subproblem of the problem "Is there a word that can be generated by a context-sensitive grammar G in two different ways?" Hence, the latter is undecidable as well.

Equivalence of CFG Grammars. Sometimes a programming language designer wants to improve a grammar at hand while retaining the old generated language. So, after improving a grammar G_1 to a grammar G_2, the designer asks whether or not $L(G_2) = L(G_1)$. In other words, he or she asks whether or not G_1 and G_2 are *equivalent* grammars.

▲ EQUIVALENCE OF CFG GRAMMARS
   Let G_1 and G_2 be CFGs. Question:

$$\text{"Do } G_1 \text{ and } G_2 \text{ generate the same language?"}$$

This problem is undecidable; no algorithm can solve it for arbitrary G_1 and G_2. Language designers must invent, for different pairs G_1, G_2, different approaches to prove their equivalence. As above, the equivalence of context-sensitive grammars is an undecidable problem too.

Some Other Properties of CFGs and CFLs. There are other practical problems about context-free grammars and languages that turned out to be undecidable. Some are listed below. We leave it to the reader to interpret each of them as a problem of programming language design.

▲ OTHER PROPERTIES OF CFGs AND CFLs
   Let G and G' be arbitrary CFGs, and let C and R be an arbitrary CFL and a regular language, respectively. As usual, Σ is the alphabet. Questions:

   "Is $L(G) = \Sigma^*$?"
   "Is $L(G)$ regular?"
   "Is $R \subseteq L(G)$?"
   "Is $L(G) = R$?"
   "Is $\overline{L(G)}$ CFL?"
   "Is $L(G) \subseteq L(G')$?"
   "Is $L(G) \cap L(G') = \emptyset$?"
   "Is $L(G) \cap L(G')$ CFL?"
   "Is C ambiguous CFL?"
   "Is there a palindrome in $L(G)$?"

Each of these problems is undecidable. No algorithm can solve it in general. Of course, these problems are undecidable for context-sensitive G, G' and C.

Remark. As we saw, all the above undecidability results extend to *context-sensitive* grammars and languages. This was bad news for researchers in linguistics, such as Chomsky,[5] who did not expect a priori limitations on the mechanical, algorithmic processing of natural languages.

8.3.5 Problems About Computable Functions

For many properties about computable functions the following holds: the property depends neither on *how* the function's values are computed (i.e., with what program or algorithm) nor on *where* the computation is performed (i.e., on what computer or model of computation). Such a property is intrinsic to the function itself, that is, it belongs to the function as a correspondence $\varphi : A \to B$ between two sets. For instance, totality is such a property, because whether or not $\varphi : A \to B$ is total depends *only* on the definition of φ, A, and B, and it has nothing to do with the actual computation of φ's values. Deciding whether or not an arbitrary computable function has such an intrinsic property is usually an undecidable problem.

▲ INTRINSIC PROPERTIES OF COMPUTABLE FUNCTIONS
  Let $\varphi : A \to B$ and $\psi : A \to B$ be arbitrary computable functions. Questions:

$$\mathcal{D}_{\mathcal{K}_1} \equiv \text{``Is } \mathrm{dom}(\varphi) \text{ empty?''}$$
$$\mathcal{D}_{\mathcal{F}in} \equiv \text{``Is } \mathrm{dom}(\varphi) \text{ finite?''}$$
$$\mathcal{D}_{\mathcal{I}nf} \equiv \text{``Is } \mathrm{dom}(\varphi) \text{ infinite?''}$$
$$\mathcal{D}_{\mathcal{C}of} \equiv \text{``Is } A - \mathrm{dom}(\varphi) \text{ finite?''}$$
$$\mathcal{D}_{\mathcal{T}ot} \equiv \text{``Is } \varphi \text{ total?''}$$
$$\mathcal{D}_{\mathcal{E}xt} \equiv \text{``Can } \varphi \text{ be extended to total computable function?''}$$
$$\mathcal{D}_{\mathcal{S}ur} \equiv \text{``Is } \varphi \text{ surjective?''}$$
$$\text{``Is } \varphi \text{ defined at } x?''$$
$$\text{``Is } \varphi(x) = y \text{ for at least one } x?''$$
$$\text{``Is } \mathrm{dom}(\varphi) = \mathrm{dom}(\psi)?''$$
$$\text{``Is } \varphi \simeq \psi?''$$

All of the above problems are undecidable; no algorithm can solve any of them.

Remark. We can now understand the difficulties, described in Sect. 5.3.3, that would have arisen if we had used only *total* functions to formalize the basic notions of computation: the basic notions of computation would have been defined in terms of an undecidable mathematical notion (i.e., the property of being a total function).

---

[5] Avram Noam Chomsky, 1928, American linguist and philosopher.

8.3.6 Problems from Number Theory

Solvability of Diophantine Equations. Let $p(x_1, x_2, \ldots, x_n)$ be a polynomial with unknowns x_i and integer coefficients. A *Diophantine equation* is the equation $p(x_1, x_2, \ldots, x_n) = 0$, for some $n \geqslant 1$, where only integer solutions $x_1, x_2, \ldots, x_n$ are sought. For example, the Diophantine equation $2x^4 - 4xy^2 + 5yz + 3z^2 - 6 = 0$ asks for all the triples $(x, y, z) \in \mathbb{Z}^3$ that satisfy it. Depending on the polynomial $p(x_1, x_2, \ldots, x_n)$, the Diophantine equation may or may not have a solution. For example, $6x + 8y - 24 = 0$ and $x^4 z^2 + y^4 z^2 - 2z^4 = 0$ both have a solution, while $6x + 8y - 25 = 0$ and $x^2 + y^2 - 3 = 0$ do not. Since Diophantine equations find many applications in practice, it would be beneficial to devise an algorithm that would decide, for arbitrary $p(x_1, x_2, \ldots, x_n)$, whether or not the Diophantine equation $p(x_1, x_2, \ldots, x_n) = 0$ has a solution. This is known as *Hilbert's tenth problem*, the tenth in his list of 23 major problems that were unresolved in 1900.

▲ SOLVABILITY OF DIOPHANTINE EQUATIONS
Let $p(x_1, x_2, \ldots, x_n)$ be an arbitrary polynomial with unknowns $x_1, x_2, \ldots, x_n$ and rational integer coefficients. Question:

"Does a Diophantine equation $p(x_1, x_2, \ldots, x_n) = 0$ have a solution?"

This problem is undecidable. There is no algorithm capable of solving it for an arbitrary polynomial p. This was proved in 1970 by Matiyasevič,[6] who built on the results of Davis,[7] Putnam[8] and Robinson.[9]

8.3.7 Problems from Algebra

Mortal Matrix Problem. Let $m, n \geqslant 1$ and let $\mathcal{M} = \{M_1, \ldots, M_m\}$ be a set of $n \times n$ matrices with integer entries. Choose an arbitrary finite sequence $i_1, i_2, \ldots, i_k$ of numbers i_j, where $1 \leqslant i_j \leqslant m$, and multiply the matrices in this order to obtain the product $M_{i_1} \times M_{i_2} \times \ldots \times M_{i_k}$. Can it happen that the product is equal to O, the zero matrix of order $n \times n$? The question is known as the *Mortal Matrix Problem*.

▲ MORTAL MATRIX PROBLEM
Let $\mathcal{M}$ be a finite set of $n \times n$ matrices with integer entries. Question:

"Can the matrices of $\mathcal{M}$ be multiplied in some order, possibly with repetition, so that the product is the zero matrix O?"

The problem is undecidable. No algorithm can solve it for arbitrary $\mathcal{M}$.

---

[6] Juri Vladimirovič Matiyasevič, 1947, Russian mathematician and computer scientist.

[7] Martin David Davis, 1928, American mathematician.

[8] Hilary Whitehall Putnam, 1926, American philosopher, mathematician and computer scientist.

[9] Julia Hall Bowman Robinson, 1919–1985, American mathematician.

Word Problems. Let Σ be an arbitrary alphabet. A *word on* Σ is any finite sequence of the symbols of Σ. As usual, Σ^* is the set of all the words on Σ, including the empty word ε. If $u, v \in \Sigma^*$, then uv denotes the word that is the concatenation (i.e., juxtaposition) of the words u and v. The *rule over* Σ is an expression $x \to y$, where $x, y \in \Sigma^*$. Let $\mathcal{R}$ be an arbitrary finite set of rules over Σ. The pair $(\Sigma, \mathcal{R})$ is called the *semi-Thue system.*[10] In a semi-Thue system $(\Sigma, \mathcal{R})$ we can investigate whether and how a word of Σ^* can transform, using only the rules of $\mathcal{R}$, into another word of Σ^*. Given two words $u, v \in \Sigma^*$, a *transformation* of u into v in the semi-Thue system $(\Sigma, \mathcal{R})$ is a finite sequence of words $w_1, \ldots, w_n \in \Sigma^*$ such that 1) $u = w_1$ and $w_n = v$ and 2) for each $i = 1, \ldots n - 1$ there exists a rule in $\mathcal{R}$, say $x_i \to y_i$, such that $w_i = px_is$ and $w_{i+1} = py_is$, where the prefix p and suffix s are words in Σ^*, possibly empty. We write $u \xrightarrow{*} v$ to assert that there exists a transformation of u into v in the semi-Thue system $(\Sigma, \mathcal{R})$. Then the *word problem for semi-Thue system* $(\Sigma, \mathcal{R})$ is the problem of determining, for arbitrary words $u, v \in \Sigma^*$, whether or not $u \xrightarrow{*} v$. We can impose additional requirements on semi-Thue systems and obtain Thue systems. Specifically, a *Thue system* is a semi-Thue system $(\Sigma, \mathcal{R})$ in which $x \to y \in \mathcal{R} \iff y \to x \in \mathcal{R}$. Thus, in a Thue system we have $u \xrightarrow{*} v \iff v \xrightarrow{*} u$.

▲ WORD PROBLEM FOR SEMI-GROUPS
   Let $(\Sigma, \mathcal{R})$ be an arbitrary Thue system and $u, v \in \Sigma^*$ be arbitrary words. Question:

Can u be transformed into v in the Thue system $(\Sigma, \mathcal{R})$?

This problem is undecidable. No algorithm can solve it for arbitrary u, v and $\Sigma, \mathcal{R}$.

   Let $(\Sigma, \mathcal{R})$ be a Thue system in which the following holds: for every $a \in \Sigma$ there is a $b \in \Sigma$ such that both $\varepsilon \to ba$ and $ba \to \varepsilon$ are rules of $\mathcal{R}$. Informally, every symbol a has a symbol b that annihilates it. (It follows that every word has an annihilating "inverse" word.) For such Thue systems the following problem arises.

▲ WORD PROBLEM FOR GROUPS
   Let $(\Sigma, \mathcal{R})$ be an arbitrary Thue system where $\forall a \in \Sigma \ \exists b \in \Sigma(\varepsilon \to ba \in \mathcal{R} \wedge ba \to \varepsilon \in \mathcal{R})$, and let $u, v \in \Sigma^*$ be arbitrary words. Question:

"Is $u \xrightarrow{*} v$ in $(\Sigma, \mathcal{R})$?"

The problem in undecidable; no algorithm can solve it for arbitrary $u, v, \Sigma, \mathcal{R}$.

Remark. This is fine, but where are the semi-groups and groups? In a Thue system $(\Sigma, \mathcal{R})$ the relation $\xrightarrow{*}$ is an equivalence relation, so Σ^* is partitioned into equivalence classes. The set $\Sigma^*/\xrightarrow{*}$ of all the classes, together with the operation of concatenation of classes, is a *semi-group* $(\Sigma^*/\xrightarrow{*}, \cdot)$. If, in addition, the rules $\mathcal{R}$ fulfill the above "annihilation" requirement, then $(\Sigma^*/\xrightarrow{*}, \cdot)$ is a *group*.

---

[10] Axel Thue, 1863–1922, Norwegian mathematician.

Let $\mathcal{E}$ be a finite set of equations between words. Sometimes, a new equation can be deduced from $\mathcal{E}$ by a finite number of substitutions and concatenations and by using the transitivity of the relation "=". Here is an example. Let $\mathcal{E}$ contain the equations

1. $bc = cba$
2. $ba = abc$
3. $ca = ac$

We can deduce the equation $abcc = cacacbaa$. How? By concatenating the symbol a to both sides of equation 1 we get the equation $bca = cbaa$. Its left-hand side can be transformed by a series of substitutions as follows: $b\underline{ca} \overset{3.}{=} \underline{ba}c \overset{2.}{=} abcc$. Similarly, we transform its right-hand side: $c\underline{ba}a \overset{2.}{=} ca\underline{bc}a \overset{1.}{=} cac\underline{ba}a \overset{2.}{=} caca\underline{bc}a \overset{1.}{=} cacacbaa$. Since "=" is transitive, we finally obtain the equation $abcc = cacacbaa$. Can we algorithmically check whether or not an equation follows from a set of equations?

▲ EQUALITY OF WORDS

Let $\mathcal{E}$ be an arbitrary finite set of equations between words and let u, v be two arbitrary words. Question:

$$\text{"Does } u = v \text{ follow from } \mathcal{E}\text{?"}$$

The problem is undecidable. No algorithm can solve it for arbitrary $u, v, \mathcal{E}$.

8.3.8 Problems from Analysis

Existence of Zeros of Functions. We say that a function $f(x)$ is *elementary* if it can be constructed from a finite number of exponentials $e^{(\cdot)}$, logarithms $\log(\cdot)$, roots $\sqrt[(\cdot)]{(\cdot)}$, real constants, and the variable x by using function composition and the four basic operations $+, -, \times,$ and $\div$. For example, $\sqrt[2]{\frac{a}{\pi}}e^{-ax^2}$ is an elementary function. If we allow these functions and the constants to be complex, then also trigonometric functions and their inverses become elementary. Now, often we are interested in zeros of functions. Given a function, before we attempt to compute its zeros, it is a good idea to check whether or not they exist. Can we do the checking algorithmically?

▲ EXISTENCE OF ZEROS OF FUNCTIONS

Let $f : \mathbb{R} \to \mathbb{R}$ be an arbitrary elementary function. Question:

$$\text{"Is there a real solution to the equation } f(x) = 0\text{?"}$$

The problem is undecidable. No algorithm can answer this question for arbitrary f. Consequently, the problem is undecidable for general functions f.

8.3.9 Problems from Topology

Classification of Manifolds. Topological *manifolds* play an important role in the fields of mathematics and physics (e.g., topology, general relativity). Intuitively, topological manifolds are like curves and surfaces, except that they can be of higher dimension. (The dimension is the number of independent numbers needed to specify an arbitrary point in the manifold.) An n-dimensional topological manifold is called an *n-manifold* for short.

The crucial property of n-manifolds is, roughly speaking, that they *locally* "look like" Euclidian space $\mathbb{R}^n$. Let us make this more precise. First, two sets $U \subseteq \mathbb{R}^k$ and $V \subseteq \mathbb{R}^n$ are said to be *homeomorphic* if there is a bijection $h : U \to V$ such that both h and h^{-1} are continuous. Such a function h is called the *homeomorphism*. Intuitively, if U and V are homeomorphic, then they can be continuously deformed one into the other without tearing or collapsing. Second, a set $\mathcal{M} \subseteq \mathbb{R}^k$ is said to be *locally Euclidean of dimension n* if every point of $\mathcal{M}$ has a neighborhood in $\mathcal{M}$ that is homeomorphic to a ball in $\mathbb{R}^n$. At last we can define: An n-manifold is a subset of some $\mathbb{R}^k$ that is locally Euclidean of dimension n.

The Euclidean space $\mathbb{R}^3$ is a 3-manifold. But there are many more: 1-manifolds are curves (e.g., line, circle, parabola, other curves), while 2-manifolds are surfaces (e.g., plane, sphere, cylinder, ellipsoid, paraboloid, hyperboloid, torus). For $n \geqslant 3$, n-manifolds cannot be visualized (with the exception of $\mathbb{R}^3$ or parts of it).

There is one more important notion that we will need. A *topological invariant* is a property that is preserved by homeomorphisms. For example, the dimension of a manifold is a topological invariant, but there are others too. If two manifolds have different invariants, they are not homeomorphic.

One of the most important problems of topology is to *classify* manifolds up to topological equivalence. This means that we want to produce, for each dimension n, a *list* of n-manifolds, called *n-representatives*, such that every n-manifold is homeomorphic to exactly one n-representative. Ideally, we would also like to devise an algorithm that would compute, for any given n-manifold, the corresponding n-representative (or, rather, its location in the list). Unfortunately, the problem is incomputable for dimensions $n \geqslant 4$. Specifically, in 1958 Markov proved that for $n \geqslant 4$ the question of whether or not two arbitrary n-manifolds are homeomorphic is undecidable.

▲ CLASSIFICATION OF MANIFOLDS

Let $n \geqslant 4$ and let $\mathcal{M}_1, \mathcal{M}_2$ be arbitrary topological n-manifolds. Question:

> *"Are topological n-manifolds $\mathcal{M}_1$ and $\mathcal{M}_2$ homeomorphic?"*

The problem is undecidable. There is no algorithm capable of distinguishing two arbitrary manifolds with four or more dimensions.

8.3.10 Problems from Mathematical Logic

Decidability of First-Order Theories. In mathematical logic there were also many problems that turned out to be undecidable. After Hilbert's program was set, one of the main problems of mathematical logic was to solve the *Decidability Problem* for **M**, where **M** denotes the theory belonging to the sought-for formal axiomatic system for all mathematics. Then, the *Decidability Problem* for **M** is denoted by $\mathcal{D}_{Dec(\mathbf{M})}$ and defined as $\mathcal{D}_{Dec(\mathbf{M})} \equiv$ "Is F a theorem of **M**?", where F can be an arbitrary formula of **M**. If this problem were decidable, then the corresponding decider D_{Entsch} could today be programmed and run on modern computers. Such programs, called *automatic theorem provers*, would ease the research in mathematics: one would only need to construct a relevant mathematical proposition, submit it to the automatic prover, and wait for the positive or negative answer. However, in 1936, Hilbert's challenge ended with an unexpected result that was simultaneously found by Church and Turing. In short, the problem $\mathcal{D}_{Dec(\mathbf{M})}$ is undecidable. What is more, this holds for any first-order formal axiomatic system **F**.

▲ DECIDABILITY OF FIRST-ORDER THEORIES

Let **F** be an arbitrary first-order theory and F an arbitrary formula of **F**. Question:

$$\mathcal{D}_{Dec(\mathbf{F})} \equiv \text{"Is F a theorem of } \mathbf{F}?"$$

This problem is undecidable; no algorithm can solve it for arbitrary $F \in \mathbf{F}$. Specifically, there is no algorithm D_{Entsch} that could decide whether or not an arbitrary mathematical formula is provable.

Remark. So yet another goal of Hilbert's program, the *Entscheidungsproblem* (goal D, Sect. 4.1.2) has received a negative answer. We will describe the proof on p. 203. Let us add that certain particular theories *can be* decidable; for instance, in 1921, Post proved that the *Propositional Calculus* **P** is a decidable theory. He devised an algorithm that, for an arbitrary proposition of **P**, decides whether or not the proposition is provable in **P**. The algorithm uses truth-tables. An obvious practical consequence of the undecidability of $\mathcal{D}_{Dec(\mathbf{F})}$ is that designers of automatic theorem provers are aware of the hazard that their programs, no matter how improved, may not halt on some inputs.

Satisfiability and Validity of First-Order Formulas. Recall that to interpret a theory one has to choose a particular set $\mathcal{S}$ and particular functions and relations defined on $\mathcal{S}$, and define, for every formula F of **F**, how it is to be understood as a statement about the members, functions, and relations of $\mathcal{S}$ (see Sect. 3.1.2). So let us be given a theory and an interpretation ι of it. If F has no free variable symbols, then its interpretation $\iota(F)$ is either a true or a false statement about the state of affairs in $\mathcal{S}$. If, however, F contains free variable symbols, then, as soon as all the free variable symbols are assigned values, the formula $\iota(F)$ becomes either a true or a false statement about $\mathcal{S}$. Obviously, the assignment of values to free variable symbols can *affect* the truth-value of $\iota(F)$. In general, there are many possible assignments of values to free variable symbols. A formula F is said to be *satisfiable* under the interpretation ι if $\iota(F)$ becomes true for *at least one* assignment of values

to its free variable symbols. And a formula F is said to be *valid* under the interpretation ι if ι(F) becomes true for *every* assignment of values to its free variable
symbols. Finally, if a formula is valid under *every* interpretation, it is said to be *logically valid*. For instance, in a first-order theory the formula $\neg \forall x P(x) \Rightarrow \exists x \neg P(x)$ is
logically valid. Satisfiability and validity are nice properties; it would be beneficial
to be able to algorithmically recognize formulas with such properties. Can we do
this? Unfortunately, we cannot.

▲ SATISFIABILITY OF FIRST-ORDER FORMULAS

Let **F** be a first-order theory, ι an interpretation of **F**, and F an arbitrary formula
of **F**. Question:

$$\mathcal{D}_{Sat(\mathbf{F},\iota)} \equiv \text{``Is F satisfiable under } \iota \text{?''}$$

This problem is undecidable. There is no algorithm capable of solving it for an
arbitrary F. However, if $\mathbf{F} = \mathbf{P}$, the *Propositional Calculus*, the problem *is* decidable.
In that case, the problem is stated as follows:

$$\mathcal{D}_{Sat(\mathbf{P})} \equiv \text{``Can variable symbols of a Boolean expression F be assigned}$$
$$\text{truth-values in such a way that F attains the truth-value 'true'?''}$$

This is the so-called *Satisfiability Problem for Boolean Expressions*, which plays an
important role in *Computational Complexity Theory*.

▲ VALIDITY OF FIRST-ORDER FORMULAS

Let **F** be a first-order theory, ι an interpretation of **F**, and F an arbitrary formula
of **F**. Question:

$$\mathcal{D}_{Val(\mathbf{F},\iota)} \equiv \text{``Is F valid under } \iota \text{?''}$$

The problem is undecidable; no algorithm can solve it for an arbitrary F.

8.3.11 Problems About Games

Tiling Problems. Let $\mathcal{T}$ be a finite set of elements, called *tile templates*. Each tile
template is a 1×1 square object with each of its four sides colored in one of finitely
many colors (see Fig. 8.9). For each tile template there is a stack with a potentially
infinite number of exact copies of the template, called the *tiles*. Now let the plane
be partitioned into 1×1 squares, that is, view the plane as the set $\mathbb{Z}^2$. Any finite
subset of $\mathbb{Z}^2$ can be viewed as a *polygon* with a border consisting only of horizontal
and vertical sides—and vice versa. Such a polygon can be $\mathcal{T}$-*tiled*, which means
that every 1×1 square of the polygon is covered with a tile from an arbitrary stack
associated with $\mathcal{T}$. Without any loss of generality, we assume that the tiles cannot
be rotated or reflected. A $\mathcal{T}$-tiling of a polygon is said to be *regular* if every two
neighboring tiles of the tiling match in the color of their common sides. The question

is: Does the set $\mathcal{T}$ suffice to regularly $\mathcal{T}$-tile an arbitrary polygon? Can we decide this algorithmically for an arbitrary $\mathcal{T}$? The answer is *no*.

▲ DOMINO TILING PROBLEM

Let $\mathcal{T}$ be a finite set of tile templates, each with an unlimited number of copies (tiles). Question:

"Can every finite polygon be regularly $\mathcal{T}$-tiled?"

This problem is undecidable. No algorithm can solve it for an arbitrary set $\mathcal{T}$.

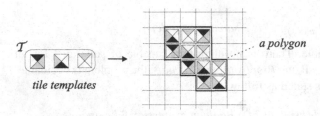

a regular $\mathcal{T}$-tiling of the polygon

Fig. 8.9 A positive instance of the domino tiling problem

A path $p(A,B,X)$ is any finite sequence of 1×1 squares in $\mathbb{Z}^2$ that connects the square A to the square B and does cross the square X (see Fig. 8.10).

▲ DOMINO SNAKE PROBLEM

Let $\mathcal{T}$ be a finite set of tile templates and let A, B and X be arbitrary 1×1 squares in $\mathbb{Z}^2$. Question:

"Is there a path $p(A,B,X)$ that can be regularly $\mathcal{T}$-tiled?"

The problem is undecidable; there is no algorithm capable of solving it for an arbitrary $\mathcal{T}, A, B, X$.

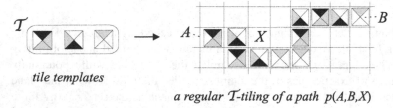

a regular $\mathcal{T}$-tiling of a path $p(A,B,X)$

Fig. 8.10 A positive instance of the domino snake problem

8.4 Can We Outwit Incomputable Problems?

We said that for an incomputable (undecidable) problem $\mathcal{P}$ there exists no *single* algorithm A capable of solving an *arbitrary* instance p of the problem. However, in practice it may happen that we are only confronted with a *particular* instance p_i of $\mathcal{P}$. In this case there exists an algorithm A_i that can solve the particular instance p_i. The algorithm A_i must initially check whether or not the input data actually describe the instance p_i and, if so, start solving p_i. If we succeed in constructing the algorithm A_i, it will necessarily be designed specially for the particular instance p_i. In general, A_i will not be useful for solving any other instance of the problem $\mathcal{P}$.

In the following, we discuss two attempts to get around the harsh statement that A does not exist:

1. It seems that we could construct the sought-for general algorithm A simply by combining all the particular algorithms A_i for all the instances $p_i \in \mathcal{P}$ into one unit. That is,

   > **Algorithm $A(p)$:**
   >   **begin**
   >   **if** $p = p_1$ **then** A_1 **else**
   >   **if** $p = p_2$ **then** A_2 **else**
   > $\vdots$
   >   **if** $p = p_i$ **then** A_i **else**
   > $\vdots$
   >   **end.**

   However, there is a pitfall in this approach. In general, the particular algorithms A_i differ one from another—we say that they are *not uniform*. This is because we lack a single method for constructing the particular algorithms. (If such a method existed, then the construction of A would go along with this method, and the problem $\mathcal{P}$ would be computable.) Consequently, we must be sufficiently ingenious, for each instance $p_i \in \mathcal{P}$ separately, to discover and construct the corresponding particular algorithm A_i.

   But there are infinitely many instances of $\mathcal{P}$ (otherwise, $\mathcal{P}$ would be computable), so the construction of all of the particular algorithms would never end. Even if the construction somehow finished, the encoding of such A would not be finite, and this would violate the fundamental assumption that algorithms are representable by finite programs. (Which Turing program would correspond to A?) Such an A would not be an algorithm in the sense of the *Computability Thesis*.

2. The second attempt avoids the construction of infinitely many particular algorithms A_i. To do this, we first construct a *generator* that is capable of generating, in succession, all of the programs. Specifically, we can easily construct a generator G_{Σ^*} that generates the programs in *shortlex* order (that is, in order of

increasing length, and, in the case of equal length, in lexicographical order; see Sect. 6.3.5 and Appendix A, p. 303). Then we can use G_{Σ^*} as an element in the construction of some other algorithm. In particular, it seems that we can construct a single, finite, and general algorithm B that is capable of solving the problem $\mathcal{P}$:

> **Algorithm** $B(p)$:
> **begin**
>> **repeat**
>>> $P :=$ call G_{Σ^*} and generate the next program;
>>
>> **until** P can solve p;
>> Start P on input p;
>
> **end.**

Unfortunately, there are pitfalls in this attempt too. First, we have tacitly assumed that, for each instance $p \in \mathcal{P}$, there exists a particular program P that solves p. Is this assumption valid? Well, such a program might use a table containing the expected input data (i.e., those that define the instance p), and the answer to p. Upon receiving the actual input the program would check whether or not these define the instance p and, if they do, it would return the associated answer. But now another question arises: who would (pre)compute the answers to various instances and fill them in the tables? Secondly, who will decide (i.e. verify) whether or not the next generated program P can solve the instance $p \in \mathcal{P}$? The tables would enable such a verification, but the assumption that tables are known is unrealistic. So, the verification should be made by a single algorithm. Such a verifier should be capable of deciding, for an *arbitrary* pair P and $p \in \mathcal{P}$, whether or not P correctly solves p. But the existence of such a verifier would imply that B correctly solves $\mathcal{P}$, and hence $\mathcal{P}$ is computable. This would be a contradiction. Moreover, if the verifier could decide, for an *arbitrary* problem $\mathcal{P}$, the question "Does P solve $p \in \mathcal{P}$?", then also the incomputable problem CORRECTNESS OF ALGORITHMS (PROGRAMS) would be computable (see Sect. 8.3.3). Again this would be a contradiction.

Should anyone construct an algorithm and claim that the algorithm can solve an incomputable problem, then we will be able—only by appealing to the *Computability Thesis*—to rebut the claim right away. Indeed, without any analysis of the algorithm, we will be able to predict that the algorithm fails for at least one instance of the problem; when solving such an instance, the algorithm either never halts or returns a wrong result. Moreover, we will know that this holds for any present or future programming language (to code the algorithm), and for any present or future computer (to run the program).

8.5 Chapter Summary

There are several kinds of computational problems: *decision, search, counting,* and *generating* problems. The solution of a decision problem is one of the answers YES or NO. An instance of a decision problem is *positive* or *negative* if the solution of the instance is YES or NO, respectively.

There is a close link between decision problems and *sets.* Because of this we can *reduce* the questions about decision problems to questions about sets. The *language of a decision problem* is the set of codes of all the positive instances of the problem. The question of whether an instance of a decision problem is positive reduces to the question of whether the code of the instance is in the corresponding language.

A decision problem is *decidable, undecidable,* or *semi-decidable* if its language is decidable, undecidable, or semi-decidable, respectively.

Given a Turing machine T and a word w, the *Halting Problem* asks whether or not T eventually halts on w. The *Halting Problem* is undecidable. There are undecidable problems that are semi-decidable; such is the *Halting Problem*. There are also undecidable problems that are not even semi-decidable; such is the *non-Halting Problem*, which asks whether or not T *never* halts on w.

An n-state *stopper* is a TM that has $n \geqslant 1$ non-final states and one final state, and a program which always writes the symbol 1 and moves the window and which, if started on an empty input, eventually halts and leaves a number of 1s on its tape. An n-state *busy beaver* is an n-state stopper that outputs, from among all the n-state stoppers, the maximum number of 1s. The problem of deciding whether or not a TM is an n-state busy beaver for any n is undecidable. The number of 1s output by an n-state busy beaver is a function of n that is *incomputable*.

There are many other and more practical undecidable decision problems (and incomputable computational problems) in various fields of science.

Problems

8.1. Given a *search problem* (see Sect. 8.1.1), we can derive from it the corresponding decision problem. For example, from the search problem $\mathcal{P}(\mathcal{S}) \equiv$ "Find the largest prime in $\mathcal{S}$." we derive a decision problem $\mathcal{D}(\mathcal{S},n) \equiv$ "Does $\mathcal{S}$ contain a prime $> n$?" Discuss the following questions:

(a) Can we construct an algorithm for $\mathcal{P}(\mathcal{S})$, given a decider for $\mathcal{D}(\mathcal{S},n)$? How?

(b) Can a search problem be computable if its derived decision problem is undecidable?

8.2. Similarly to search problems, *counting* and *generating problems* can also be associated in a natural way with the corresponding decision problems.

(a) Give examples of counting and generating problems and the derived decision problems.

(b) Can a counting or generating problem be computable if its decision counterpart is undecidable?

8.3. Let $n \geqslant 1$. Define $\mathcal{T}_n$ to be the set of all the two-way unbounded TMs $T = (Q, \Sigma, \Gamma, \delta, q_1, \sqcup, F)$, where $Q = \{q_1, \ldots, q_{n+1}\}$, $\Sigma = \{0,1\}$, $\Gamma = \{0,1,\sqcup\}$, $\delta : Q \times \Gamma \to Q \times \{1\} \times \{\text{Left}, \text{Right}\}$ and $F = \{q_{n+1}\}$.

(a) Prove: $\mathcal{T}_n$ contains finitely many TMs, for any $n \geqslant 1$.

(b) Compute $|\mathcal{T}_n|$.

Bibliographic Notes

- The undecidability of the *Halting Problem* was proved in Turing [178]. Simultaneously, an explicit undecidable problem was given in Church [25].
- The *Busy Beaver Problem* was introduced in Radó [133]; here, also the incomputability of the busy beaver functions was proved.
- Post proved the undecidability of a certain problem about normal systems in [124]. Then, in [125], he reduced this problem to the *Post Correspondence Problem* and thus proved the undecidability of the latter.
- The reasons for the undecidability of the *Program Verification Problem* are explained in Harel and Feldman [62].
- The missing step in the (negative) solution of *Hilbert's tenth problem* appeared in Matiyasevič [103, 104]. A complete account of Matiyasevič's contribution to the final solution of Hilbert's tenth problem is given in Davis [36] and in the 1982 Dover edition of Davis [34].
- The undecidability of the *Mortal Matrix Problem* as well as the undecidability of several other problems about matrices are discussed in Halava et al. [60].
- The undecidability of the problem of the *Existence of Zeros* of real elementary functions was proved in Wang [185]. Some other undecidable problems involving elementary functions of a real variable appeared in Richardson [137]. For some of the first discovered undecidable problems about systems of polynomials, differential equations, partial differential equations, and functional equations, see Adler [4].
- In describing topological manifolds we leaned on Lee [92]. The undecidability of the problem of *Classifying Manifolds* was proved in Markov [102]. The concepts and ideas used in Markov's proof are described in Stillwell [173].
- The *Domino Tiling Problem* that asks whether the whole plane can be regularly $\mathcal{T}$-tiled was introduced and proved to be undecidable in Wang [184]. The undecidability of a general version of this problem was proved in Berger [13]. The undecidability results of various variants of the *Domino Snake Problem* can be found in Etzion-Petruschka et al. [44]. For a nice discussion of the domino problems, see Harel and Feldman [62].
- The undecidability of the *Word Problem for Semigroups* was proved in Post [126]. See also Kleene [82] for the proof. That the much more difficult *Word Problem for Groups* is also undecidable was proved by Novikov [114] and independently by Boone [17].

Chapter 9
Methods of Proving Incomputability

A method is a particular way of doing something.

Abstract How can we prove the undecidability of the problems listed in the previous chapter? Are there any general methods of proving the undecidability of decision problems? The answer is yes: today we have at our disposal several such methods. These are: 1) proving by *diagonalization*, 2) proving by *reduction*, 3) proving by the *Recursion Theorem*, and 4) proving by *Rices's Theorem*. In addition, we can apply the results discovered by the *relativization* of the notion of computability. In this chapter we will describe the first four methods and postpone the discussion of *relativized computability*—which needs a separate, more extensive treatment—to the following chapters.

9.1 Proving by Diagonalization

We have already encountered diagonalization twice: first, when we proved the existence of a computable numerical function that is not recursive (Sect. 5.3.3), and second, when we proved the undecidability of the *Halting Problem* (Sect. 8.2). In these two cases diagonalization was used in two different ways, directly and indirectly. In this section we will give general descriptions of both.

9.1.1 Direct Diagonalization

This method was first used by Cantor in 1874 in his proof that $2^{\mathbb{N}}$, the power set of $\mathbb{N}$, is uncountable. The generalized idea of the method is as follows.

Let P be a property and $S = \{x \mid P(x)\}$ be the class of *all* the elements with property P. Suppose that we are given a set $T \subseteq S$ such that $T = \{e_0, e_1, e_2, \dots\}$ and each e_i can be uniquely represented by an ordered set, that is, $e_i = (c_{i,0}, c_{i,1}, c_{i,2}, \dots)$, where the components $c_{i,j}$ are members of a set C (say, real numbers).

© Springer-Verlag Berlin Heidelberg 2015   191
B. Robič, *The Foundations of Computability Theory*,
DOI 10.1007/978-3-662-44808-3_9

We may ask whether the elements of S can *all* be exhibited simply by listing the elements of T. In other words, we may ask

$$\text{"Is } T = S \text{?"}$$

If we *doubt* this, we can embark on a proof that $T \subsetneq S$. The method is as follows.

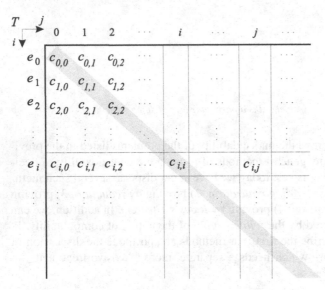

Fig. 9.1 By switching each component of the diagonal $d = (c_{0,0}, c_{1,1}, c_{2,2}, \ldots)$ we obtain $sw(d)$, the switched diagonal. But $sw(d)$ differs from every e_i, because $sw(c_{i,i}) \neq c_{i,i}$ for every $i \in \mathbb{N}$

Imagine a table T (see Fig. 9.1) with the elements $e_0, e_1, e_2, \ldots$ on its vertical axis, the numbers $0, 1, 2, \ldots$ on the horizontal axis, and the entries $T(i, j) = c_{i,j}$. The diagonal components $c_{i,i}$ define an object

$$d = (c_{0,0}, c_{1,1}, c_{2,2}, \ldots)$$

called the *diagonal*. Let $sw : \mathcal{C} \to \mathcal{C}$ be a function such that $sw(c) \neq c$ for every $c \in \mathcal{C}$. We will call sw the *switching function*. Define the *switched diagonal* to be

$$sw(d) = (sw(c_{0,0}), sw(c_{1,1}), sw(c_{2,2}), \ldots).$$

Now observe that, for every i, the switched diagonal $sw(d)$ differs from the element e_i (because $sw(c_{i,i}) \neq c_{i,i}$). Consequently, $sw(d) \notin T = \{e_0, e_1, e_2, \ldots\}$.

This, however, does not yet prove that $T \subsetneq S$ *unless* $sw(d) \in S$. So, *if* $sw(d)$ has the property P, then $sw(d) \in S$, which results in $sw(d) \in S - T$, and hence $T \subsetneq S$.

In summary, if we can find a switching function sw such that the switched diagonal $sw(d)$ has the property P, then $T \neq S$. The steps of the method are then as follows.

Method. (Proof by Diagonalization) To prove that the set of all the elements having a property P is uncountable, proceed as follows:

1. Define $S = \{x \mid P(x)\}$, the class of all elements with property P.

2. Define two sets, $\mathcal{T} = \{e_0, e_1, e_2, \ldots\} \subseteq S$ and C, such that each e_i is uniquely represented by an ordered set $(c_{i,0}, c_{i,1}, c_{i,2}, \ldots)$, where $c_{i,j} \in C$.

3. Consider a table T with the elements $e_0, e_1, e_2, \ldots$ on its vertical axis, the numbers $0, 1, 2, \ldots$ on its horizontal axis, and entries $T(i,j) = c_{i,j}$. Then try to find a function $sw : C \rightarrow C$ with the following properties:

   - $sw(c) \neq c$ for every $c \in C$ (i.e., sw is a switching function);

   - The object which is represented by the switched diagonal $(sw(c_{0,0}), sw(c_{1,1}), sw(c_{2,2}), \ldots)$ has property P.

   If such a function sw is found, then $\mathcal{T} \subsetneq S$, and S is uncountable.

Example 9.1. ($\mathbb{N}$ vs. $\mathbb{R}$) Let us prove that there are more real numbers than natural numbers. To do this, we focus only on real numbers in the interval $[0,1]$.

   Define the property $P(x) \equiv$ "x is a real number in $[0,1]$." So $S = [0,1]$. Define a set $\mathcal{T} = \{e_0, e_1, e_2, \ldots\}$, where $e_i \in S$ for all i, and each e_i is uniquely represented as $e_i = 0.c_{i,0}c_{i,1}c_{i,2}\ldots$, where $c_{i,j} \in C = \{0,1,2,3,4,5,6,7,8,9\}$ are digits. The uniqueness of the representation is achieved if those e_i that would have finitely many non-zero digits (e.g., 0.25) are represented by using infinite sequence of 9s (e.g., $0.24999\ldots$).

   Now define the switching function by $sw : c \mapsto c + 1 \pmod{10}$. The switched diagonal $sw(d)$ defines a number $0.c'_{0,0}c'_{1,1}c'_{2,2}\ldots$, where $c'_{i,i} = sw(c_{i,i})$. Obviously, this number is an element of $[0,1] = S$. So, $sw(d) \in S - \mathcal{T}$. It follows that $S = [0,1]$ is uncountable. As $[0,1] \subseteq \mathbb{R}$, the same holds for $\mathbb{R}$.   □

Example 9.2. (Non-recursive Computable Function) Let us prove that there are functions that are computable but not recursive.

   Define the property $P(f) \equiv$ "f is a numerical function that is computable." So S is the set of all the numerical functions whose values can be algorithmically computed whenever they are defined. Recall (Box 5.1, p. 74) that every recursive function is uniquely defined by its construction. Since every construction can be encoded by a finite sequence of symbols 0 and 1 and then interpreted as a (binary representation of a) natural number, it follows that each recursive function can be viewed, for some natural i, as the ith element in the sequence of all the recursive functions. So let us define $\mathcal{T}$ to be the set $\mathcal{T} = \{f_0, f_1, f_2, \ldots\}$, where f_i is the ith recursive function. Because recursive functions are total by definition, each $f_i \in \mathcal{T}$ is perfectly defined by the values it attains on the elements of the set $\mathbb{N}$. So we can represent f_i by $f_i = (c_{i,0}, c_{i,1}, c_{i,2}, \ldots)$, where $c_{i,j} = f_i(j)$ and $c_{i,j} \in C = \mathbb{N}$.

   Now define the switching function by $sw : c \mapsto c + 1$. Observe that the switched diagonal $sw(d)$ *represents a total and computable function*. (To see this note that $sw(c_{i,i})$ is defined for every i and that the algorithm for computing the value of $sw(d)$ at an arbitrary n constructs f_n, computes $f_n(n)$ and adds 1 to it.) However, $sw(d)$ *differs from every recursive function* $f_i \in \mathcal{T}$, because $sw(d)$ and f_i attain different values at i. Consequently, $sw(d) \in S - \mathcal{T}$. This completes the proof.   □

9.1.2 Indirect Diagonalization

First, recall that we encoded Turing machines T by encoding their programs by words $\langle T \rangle \in \{0,1\}^*$ and, consequently, by natural numbers that we called indexes (see Sect. 6.2.1). We then proved that every natural number is the index of exactly one Turing program. Hence, we can speak of the first, second, third, and so on Turing program, or, due to the *Computability Thesis*, of the first, second, third, and so on algorithm. So there is a sequence $A_0, A_1, A_2, \ldots$ such that *every* algorithm that we can conceive of is an A_i for some $i \in \mathbb{N}$. Remember that A_i is encoded (described) by its index i.

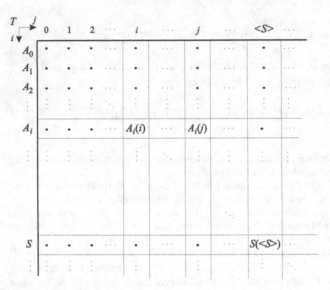

Fig. 9.2 Table T contains the results $A_i(j)$ of applying every algorithm A_i on every input j. The diagonal values are $A_i(i)$, $i = 0,1,2,\ldots$. The shrewd algorithm is S. If applied on its own code, S uncovers the inability of the alleged decider D_P to decide whether or not S has the property P

Second, imagine an infinite table T (see Fig. 9.2) with $A_0, A_1, A_2, \ldots$ on the vertical axis, and $0, 1, 2, \ldots$ on the horizontal axis. Define the components of T as follows: for each pair $i, j \in \mathbb{N}$, let the component $T(i, j) = A_i(j)$, the result of algorithm A_i when given the input j.

Now let P be a property that is sensible for algorithms. Consider the question

> "Is there an algorithm D_P capable of deciding, for an arbitrary algorithm A,
> whether or not A has property P?"

If it exists, the algorithm D_P is the *decider* of the class of the algorithms that have property P. If we doubt that D_P exists, we can try to prove this with the following method, which is a generalization of the method used to prove Lemma 8.1 (p. 168).

Method. (Proof by Diagonalization) To prove that there is no algorithm D_P capable of deciding whether or not an arbitrary algorithm has a given property P, proceed as follows:

1. Suppose that the decider D_P exists.

2. Try to construct a *shrewd* algorithm S with the following properties:

   - S uses D_P;

   - if S is given as an input its own code $\langle S \rangle$, it uncovers the inability of D_P to decide whether or not S has the property P.

   If such an algorithm S is constructed, then D_P does not exist, and the property P is undecidable.

The second step requires S to expose (probably in some shrewd way) the inability of the alleged D_P to output the correct answer on the input $\langle S \rangle$. But how do we construct such an S? Initially, S must call D_P and hand over to it the input $\langle S \rangle$. By supposition, D_P will answer either YES (i.e., S has property P) or NO (i.e., S does not have property P). Then—and this is the hard part of the proof—we must construct the rest of S in such a way that the resulting S will have the property P *if and only if D_P has decided the contrary. If* we succeed in constructing such an S, then the alleged D_P errs on at least one input, namely $\langle S \rangle$, contradicting our supposition in the first step. Consequently, the decider D_P does not exist and the decision problem "Does A have the property P?" is undecidable.

Of course, the existence of S and its construction both depend on property P. If S exists, it may take considerable ingenuity to construct it.

Example 9.3. (Shrewd Algorithm for the *Halting Problem*) Is there a decider D_P capable of deciding, for an arbitrary TM, whether TM halts on its own code?

We construct the shrewd TM S as follows. When S reads $\langle S \rangle$, it sends it to D_P to check whether S halts on $\langle S \rangle$. If D_P has answered YES (i.e., saying that S halts on $\langle S \rangle$), we must make sure that the following execution of S will *never* halt. This is why we loop the execution back to D_P. Otherwise, if D_P has answered NO (i.e., indicated that S does not halt on $\langle S \rangle$), we force S to do just the contrary: so S must halt immediately. The algorithm S is depicted in Fig. 8.3 on p. 168.   □

Remark. We can apply this method even if we use some other equivalent model of computation. In that case, A_i still denotes an algorithm, but the notion of the algorithm is now defined according to the chosen model of computation. For example, if the model of computation is λ-calculus, then A_i is a λ-term.

9.2 Proving by Reduction

Until the undecidability of the first decision problem was proved, diagonalization was the only method capable of producing such a proof. Indeed, we have seen that the undecidability of the *Halting Problem* was proven by diagonalization. However, after the undecidability of the first problem was proven, one more method of proving the undecidability became applicable. The method is called *reduction*. In this section we will first describe reduction in general, and then focus on two special kinds of this method, the so-called *m*-reduction and 1-reduction.

9.2.1 Reductions in General

Suppose we are given a problem $\mathcal{P}$. Instead of solving $\mathcal{P}$ directly, we might try to solve it *indirectly* by executing the following scenario (Fig. 9.3):

1. *express* $\mathcal{P}$ in terms of some other problem, say $\mathcal{Q}$;
2. solve $\mathcal{Q}$;
3. *construct* the solution to $\mathcal{P}$ by using the solution to $\mathcal{Q}$ only.

If 1 and 3 have actually been implemented, we say that $\mathcal{P}$ has been *reduced* to $\mathcal{Q}$.

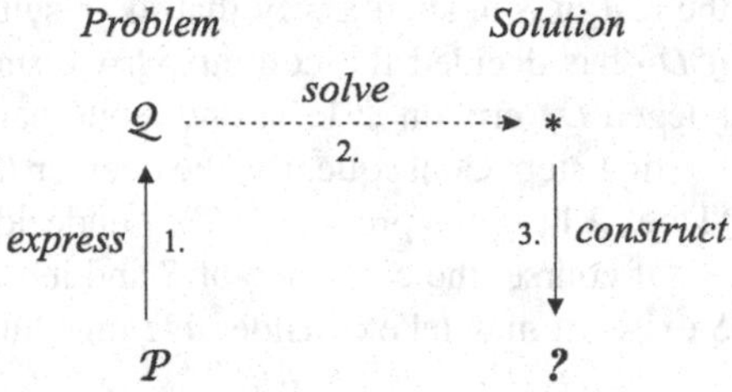

Fig. 9.3 Solving the problem $\mathcal{P}$ is reduced to (or substituted by) solving the problem $\mathcal{Q}$

How can we implement steps 1 and 3? To express $\mathcal{P}$ in terms of $\mathcal{Q}$ we need a function—call it r—that maps every instance $p \in \mathcal{P}$ into an instance $q \in \mathcal{Q}$. Such an r must meet two *basic conditions*:

a. Since problem instances are encoded by words in Σ^* (see Sect. 8.1.2), r must be a function $r : \Sigma^* \to \Sigma^*$ that assigns to every code $\langle p \rangle$ a code $\langle q \rangle$, where $p \in \mathcal{P}$ and $q \in \mathcal{Q}$.
b. Since we want to be able to compute $\langle q \rangle = r(\langle p \rangle)$ for arbitrary $\langle p \rangle$, the function r must be computable on Σ^*.

To be able to construct the solution to $\mathcal{P}$ from the solution to $\mathcal{Q}$ we add one more basic condition:

c. The function r must be such that, for arbitrary $p \in \mathcal{P}$, the solution to p can be computed from the solution to $q \in \mathcal{Q}$, where $\langle q \rangle = r(\langle p \rangle)$.

If such a function r is found, it is called the *reduction* of problem $\mathcal{P}$ to problem $\mathcal{Q}$. In this case we say that $\mathcal{P}$ is *reducible* to $\mathcal{Q}$, and denote the fact by

$$\mathcal{P} \leq \mathcal{Q}.$$

The informal interpretation of the above relation is as follows.

> If $\mathcal{P} \leq \mathcal{Q}$, then the problem $\mathcal{P}$ is *not harder to solve* than the problem $\mathcal{Q}$.

The reduction r may still be too general to serve a particular, intended purpose. In this case we define a set $\mathcal{C}$ of additional conditions to be fulfilled by r. If such a function r is found, we call it the *$\mathcal{C}$-reduction* of problem $\mathcal{P}$ to problem $\mathcal{Q}$, and say that $\mathcal{P}$ is *$\mathcal{C}$-reducible* to $\mathcal{Q}$. We denote this by

$$\mathcal{P} \leq_C \mathcal{Q}.$$

What sort of conditions should be included in $\mathcal{C}$? The answer depends on the *kind* of problems $\mathcal{P}, \mathcal{Q}$ and on our *intentions*. For example, $\mathcal{P}$ and $\mathcal{Q}$ can be decision, search, counting, or generating problems. Our intention may be to employ reduction in order to see whether some of the problems $\mathcal{P}$ and $\mathcal{Q}$ have a certain property. For instance, some properties of interest are (i) the decidability of a problem, (ii) the computability of a problem, (iii) the computability of a problem in polynomial time, and (iv) the approximability of a problem. The $\mathcal{C}$-reductions that correspond to these properties are the *m-reduction* ($\leq_m$), *Turing* reduction ($\leq_T$), *polynomial-time bounded* Turing reduction ($\leq_T^P$), and *L-reduction* ($\leq_L$), respectively.

In the following we will focus on the *m*-reduction ($\leq_m$) and postpone the discussion of the Turing reduction ($\leq_T$) and some of its *stronger* cases to Chaps. 11 and 14. The *resource-bounded* reductions, such as $\leq_m^P$, $\leq_T^P$ and $\leq_L$, are used in *Computational Complexity Theory*, so we will not discuss them.

9.2.2 The m-Reduction

If $\mathcal{P}$ and $\mathcal{Q}$ are decision problems, then they are represented by the languages $L(\mathcal{P}) \subseteq \Sigma^*$ and $L(\mathcal{Q}) \subseteq \Sigma^*$, respectively. We use this in the next definition.

> **Definition 9.1.** (*m*-Reduction) Let $\mathcal{P}$ and $\mathcal{Q}$ be decision problems. A reduction $r : \Sigma^* \to \Sigma^*$ is said to be the *m*-**reduction** of $\mathcal{P}$ to $\mathcal{Q}$ if the following additional condition is met:
>
> $$\mathcal{C}: \quad \langle p \rangle \in L(\mathcal{P}) \Longleftrightarrow r(\langle p \rangle) \in L(\mathcal{Q}), \quad \text{for every } p \in \mathcal{P}.$$
>
> In this case we say that $\mathcal{P}$ is *m*-**reducible** to $\mathcal{Q}$ and denote this by $\mathcal{P} \leq_m \mathcal{Q}$.

The condition $\mathcal{C}$ enforces that r transforms the codes of the *positive* instances $p \in \mathcal{P}$ into the codes of the *positive* instances $q \in \mathcal{Q}$, and the codes of the *negative* instances $p \in \mathcal{P}$ into the codes of the *negative* instances $q \in \mathcal{Q}$. This is depicted in Fig. 9.4.

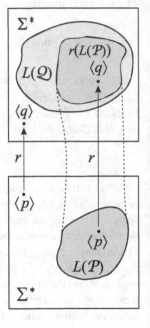

Fig. 9.4 *m*-reduction of a decision problem $\mathcal{P}$ to a decision problem $\mathcal{Q}$. Instead of solving $\mathcal{P}$ directly, we show how to express arbitrary instance p of $\mathcal{P}$ with an instance q of $\mathcal{Q}$, so that the answer to q will also be the answer to p. The mapping $r : \langle p \rangle \mapsto \langle q \rangle$ must be computable, but not necessarily injective.

Obviously, *m*-reduction r maps the set $L(\mathcal{P})$ into the set $L(\mathcal{Q})$, i.e., $r(L(\mathcal{P})) \subseteq L(\mathcal{Q})$. Here, we distinguish between two possible situations:

1. $r(L(\mathcal{P})) \subsetneq L(\mathcal{Q})$. In this case r transforms $\mathcal{P}$ into a (proper) *subproblem* of $\mathcal{Q}$. We say that problem $\mathcal{P}$ is properly *contained* in problem $\mathcal{Q}$.

2. $r(L(\mathcal{P})) = L(\mathcal{Q})$. In this case r merely restates $\mathcal{P}$ into $\mathcal{Q}$. We say that problem $\mathcal{P}$ is *equal* to problem $\mathcal{Q}$.

But Fig. 9.4 reveals even more:

a) *Suppose that $L(\mathcal{Q})$ is a decidable set.* Then also $L(\mathcal{P})$ is a decidable set.

Proof. Given an arbitrary $\langle p \rangle \in \Sigma^*$, we compute the answer to the question $\langle p \rangle \in ? L(\mathcal{P})$ as follows: 1) compute $r(\langle p \rangle)$ and 2) ask $r(\langle p \rangle) \in ? L(\mathcal{Q})$. Since $L(\mathcal{Q})$ is decidable, the answer to the question can be computed. If the answer is YES, then $r(\langle p \rangle) \in L(\mathcal{Q})$ and, because of the condition $\mathcal{C}$ in the definition of the *m*-reduction, also $\langle p \rangle \in L(\mathcal{P})$. If, however, the answer is NO, then $r(\langle p \rangle) \notin L(\mathcal{Q})$ and hence $\langle p \rangle \notin L(\mathcal{P})$. Thus, the answer to the question $\langle p \rangle \in ? L(\mathcal{P})$ can be computed for an arbitrary $\langle p \rangle \in \Sigma^*$.   □

Summary: Let $\mathcal{P} \leq_m \mathcal{Q}$. Then: $L(\mathcal{Q})$ is decidable $\Longrightarrow L(\mathcal{P})$ is decidable.

b) Suppose that $L(Q)$ is a semi-decidable set. Then $L(P)$ is a semi-decidable set.

> *Proof.* Given an arbitrary $\langle p \rangle \in \Sigma^*$, let us ask $\langle p \rangle \in ? L(P)$. If in truth $\langle p \rangle \in L(P)$, then $r(\langle p \rangle) \in r(L(P)) \subseteq L(Q)$, and since $L(Q)$ is semi-decidable, we obtain the answer YES to the question $r(\langle p \rangle) \in ? L(Q)$. But this also answers the question $\langle p \rangle \in ? L(P)$ (due to the condition C). ☐

So: Let $P \leq_m Q$. Then: $L(Q)$ is semi-decidable $\Longrightarrow L(P)$ is semi-decidable.

We have proved the following theorem.

Theorem 9.1. *Let P and Q be decision problems. Then:*

a) $P \leq_m Q \;\wedge\; Q$ *decidable problem* $\Longrightarrow P$ *decidable problem*
b) $P \leq_m Q \;\wedge\; Q$ *semi-decidable problem* $\Longrightarrow P$ *semi-decidable problem*

Recall that a set which is not semi-decidable must be undecidable (see Fig. 8.4 on p. 170). Using this in the contraposition of the case b) of Theorem 9.1, substituting U for P, and assuming that $U \leq_m Q$, we obtain the following important corollary.

Corollary 9.1. U *undecidable problem* $\wedge\; U \leq_m Q \Longrightarrow Q$ *undecidable problem*

9.2.3 Undecidability and m-Reduction

The above corollary is the backbone of the following method of proving that a decision problem Q is undecidable. Informally, the method proves that a known undecidable problem U is a subproblem of, or equal to, the problem Q.

Method. The undecidability of a decision problem Q can be proved as follows:

1. Select: an undecidable problem U;
2. Prove: $U \leq_m Q$;
3. Conclude: Q is undecidable problem.

Example 9.4. (*Halting Problem* Revisited) Using this method we proved the undecidability of the *Halting Problem* $\mathcal{D}_{Halt}$ (see Sect. 8.4). To see this, select $U = \mathcal{D}_H$ and observe that U is a subproblem of $\mathcal{D}_{Halt}$. Note also that after we proved the undecidability of the problem $\mathcal{D}_H$ (by diagonalization), $\mathcal{D}_H$ was the only choice for U. ☐

Another method of proving the undecidability of a decision problem Q combines the proof by contradiction and the case a) of Theorem 9.1 with U substituted for P.

Method. The undecidability of a decision problem Q can be proved as follows:

1. Suppose: Q is a decidable problem;   // Supposition.
2. Select: an undecidable problem U;
3. Prove: $U \leq_m Q$;
4. Conclude: U is a decidable problem;   // 1 and 3 and Theorem 9.1a.
5. Contradiction between 2 and 4!
6. Conclude: Q is an undecidable problem.

Let us comment on steps 2 and 3 (the other steps are trivial).

- *Step 2:* For U we select an undecidable problem that seems to be the most promising one to accomplish step 3. This demands good knowledge of undecidable problems and of problem Q, and often a mix of inspiration, ingenuity, and luck. Namely, until step 3 is accomplished, it is not clear whether the task of m-reducing U to Q is impossible or possible (but we are not equal to the task of constructing this m-reduction). If we give up, we may want to try with another U.

- *Step 3:* Here, the reasoning is as follows: Since Q is decidable (as supposed in step 1), there is a decider $D_{L(Q)}$ that solves Q (i.e., answers the question $x \in? L(Q)$ for an arbitrary x). We can use $D_{L(Q)}$ as a building block and try to construct a decider $D_{L(U)}$; this would be capable of solving U (by answering the question $w \in? L(U)$ for an arbitrary w). If we succeed in this construction, we will be able to conclude (in step 4) that U is decidable.

We now see how to tackle step 3: Using the supposed decider $D_{L(Q)}$, construct the decider $D_{L(U)}$. A possible construction of $D_{L(U)}$ is depicted in Fig. 9.5.

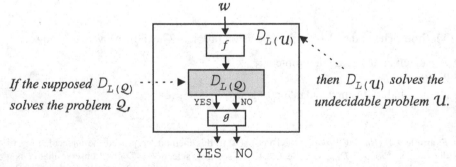

Fig. 9.5 Suppose that $D_{L(Q)}$ solves the decision problem Q. Using $D_{L(Q)}$, construct $D_{L(U)}$ that solves an undecidable problem U. Then $D_{L(Q)}$ cannot exist and Q is undecidable

The machine $D_{L(\mathcal{U})}$ operates as follows: It reads an arbitrary input $w \in \Sigma^*$, transforms w into $f(w)$ and hands this over to the decider $D_{L(\mathcal{Q})}$. The latter understands this as the question $f(w) \in? L(\mathcal{Q})$ and always answers with YES or NO. The answer $D_{L(\mathcal{Q})}(f(w))$ is then transformed by a function g and output to the environment as the answer of the machine $D_{L(\mathcal{U})}$ to the question $w \in? L(\mathcal{U})$.

What do the functions f and g look like? Of course, we expect f and g to depend on the problems $\mathcal{Q}$ and $\mathcal{U}$. However, because we want the constructed $D_{L(\mathcal{U})}$ to *always* halt, we must require the following:

- f and g are *computable* functions;

- $g(D_{L(\mathcal{Q})}(f(w))) = \begin{cases} \text{YES} & \text{if } w \in L(\mathcal{U}), \\ \text{NO} & \text{if } w \notin L(\mathcal{U}), \end{cases}$ for every $w \in \Sigma^*$.

The next (demanding) example highlights the main steps in our reasoning when we use this method. To keep it as clear as possible, we will omit certain details and refer the reader to the Problems to this chapter to fill in the missing gaps.

Example 9.5. (Computable Proper Set) Let $\mathcal{Q} \equiv$ "Is $L(T)$ computable?" We suspect that $\mathcal{Q}$ is undecidable. Let us prove this by reduction. *Suppose $\mathcal{Q}$ were decidable.* Then there would exist a decider $D_{L(\mathcal{Q})}$ that could tell, for any $\langle T \rangle$, whether or not $L(T)$ is computable. Now, what would be the *undecidable* problem $\mathcal{U}$ for which a *contradictory* TM could be constructed by using $D_{L(\mathcal{Q})}$? This is the tricky part of the proof, and to answer the question, a great deal of inspiration is needed.

Here it is. First, we show that there is a TM A which on input $\langle T, w \rangle$ constructs (the code of) a TM B such that $L(B) = \mathcal{K}_0$ (if T accepts w) and $L(B) = \emptyset$ (if T does not accept w). (Problem 9.5a.) Then, using A and the alleged $D_{L(\mathcal{Q})}$, we can construct a *recognizer* R for $\overline{\mathcal{K}_0}$. (Problem 9.5b.)

But this is a contradiction! The existence of R would mean that $\overline{\mathcal{K}_0}$ is c.e. (which is not). So, our supposition is wrong and $\mathcal{Q}$ is undecidable. □

9.2.4 The 1-Reduction

Until now we did not discuss whether or not the reduction function r is injective. We do this now. If r is not injective, then several instances, say $m \geqslant 1$, of the problem $\mathcal{P}$ can transform into the same instance $q \in \mathcal{Q}$. If, however, r is injective, then different instances of $\mathcal{P}$ transform into different instances of $\mathcal{Q}$. This is a special case of the previous situation, where $m = 1$. In this case we say that $\mathcal{P}$ is 1-*reducible* to $\mathcal{Q}$ and denote this by

$$\mathcal{P} \leq_1 \mathcal{Q}.$$

Since it holds that

$$\mathcal{P} \leq_1 \mathcal{Q} \Longrightarrow \mathcal{P} \leq_m \mathcal{Q},$$

we can try to prove the undecidability of a problem by using 1-reduction. Notice that the general properties of m-reductions hold for 1-reductions too.

Example 9.6. (*Halting Problem* Revisited) Let us revisit the proof that the *Halting Problem* $\mathcal{D}_{Halt}(= \mathcal{Q})$ is undecidable (see Sect. 8.2). After we proved that $\mathcal{D}_H$ is undecidable (see Lemma 8.1, p. 168), we took $\mathcal{U} = \mathcal{D}_H$. Then, the reasoning was: *if* $\mathcal{D}_{Halt}$ were decidable, then the associated decider $D_{L(\mathcal{D}_{Halt})}$ could be used to construct a decider for the problem $\mathcal{D}_H$. This decider is depicted in Fig. 9.6 a). The function f is total and computable since it only doubles the input $\langle T \rangle$. So is g, because it is the identity function. □

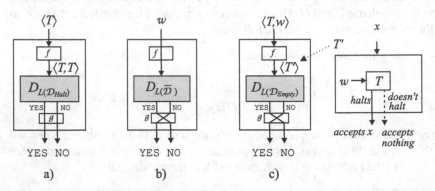

Fig. 9.6 Undecidability of problems: a) $\mathcal{D}_{Halt}$; b) $\overline{\mathcal{D}}$, if $\mathcal{D}$ undecidable; c) $\mathcal{D}_{Empty}$

Example 9.7. (Complementary Problem) If a problem $\mathcal{D}$ is undecidable, then the complementary problem $\overline{\mathcal{D}}$ is undecidable. To prove this, take $\mathcal{U} = \mathcal{D}$, $\mathcal{Q} = \overline{\mathcal{D}}$ and suppose that $\overline{\mathcal{D}}$ were decidable. We could then use the supposed decider $D_{L(\overline{\mathcal{D}})}$, which is the same as $D_{\overline{L(\mathcal{D})}}$, to construct the decider $D_{L(\mathcal{D})}$. This is depicted in Fig. 9.6 b). Now f is the identity function and g transforms an arbitrary answer into the opposite one. □

Example 9.8. (Empty Proper Set) Recall from Sect. 6.3.3 that the proper set of a Turing machine is the set of all the words accepted by the machine. So, given a Turing machine T, does the proper set $L(T)$ contain any words? This is the decision problem $\mathcal{D}_{Empty} \equiv$ "Is $L(T) = \emptyset$?". Let us prove that it is undecidable. For the undecidable problem $\mathcal{U}$ we pick the *Halting Problem* $\mathcal{D}_{Halt} \equiv$ "Does T halt on w?" and suppose that $\mathcal{D}_{Empty}(= \mathcal{Q})$ is decidable. Then we can use the associated decider $D_{L(\mathcal{D}_{Empty})}$ to construct a decider for the language $L(\mathcal{D}_{Halt})$! The construction is in Fig. 9.6 c). When the constructed machine reads an arbitrary input $\langle T, w \rangle$, it uses the function f to construct the code $\langle T' \rangle$ of a new machine T'. This machine, if started, would operate as follows: given an arbitrary input x, the machine T' would simulate T on w and, if T halted, then T' would accept x; otherwise, T' would not recognize x. Consequently, we have $L(T') \neq \emptyset \Longleftrightarrow T$ halting on w. But the validity of $L(T') \neq \emptyset$ can be decided by the supposed decider $D_{L(\mathcal{D}_{Empty})}$. After swapping the answers (using the function g) we obtain the answer to the question "Does T halt on w?". So the constructed machine is capable of deciding the *Halting Problem*. Contradiction. □

Example 9.9. (Entscheidungproblem) Recall that the *Decidability Problem* for a theory **F** is the question *"Is a theory **F** decidable?"*. In other words, the problem asks whether there exists a decider that, for arbitrary formula $F \in \mathbf{F}$, finds out whether or not F is provable in **F**. Hilbert was particularly interested in the *Decidability Problem* for **M**, the sought-for theory that would formalize all mathematics, and, of course, in the associated decider D_{Entsch}. But Church and Turing independently proved the following theorem.

Theorem 9.2. *The Entscheidungsproblem $\mathcal{D}_{Entsch}$ is undecidable.*

Proof. (Reduction $\mathcal{D}_H \leq_m \mathcal{D}_{Entsch}$.) Suppose the problem $\mathcal{D}_{Entsch}$ were decidable. Then there would exist a decider D_{Entsch} capable of deciding, for an arbitrary formula of the *Formal Arithmetic* **A**, whether or not the formula is provable in **A** (see Sects. 4.2.1, 4.2.2). Now consider the statement

$$\text{``Turing machine } T \text{ halts on input } \langle T \rangle \text{.''} \qquad (*)$$

The breakthrough was made by Turing when he showed that this statement can be represented by a formula in **A**. In the construction of the formula, he used the concept of *internal configuration* of the Turing machine T (see Definition 6.2 on p. 104). Let us designate this formula by

$$K(T).$$

Then, using the supposed decider D_{Entsch} we could easily construct a decider D capable of deciding, for arbitrary T, whether or not T halts on $\langle T \rangle$. The decider D is depicted in Fig. 9.7. Here, the function f maps the code $\langle T \rangle$ into the formula $K(T)$, while the function g is the identity function and maps the unchanged answer of D_{Entsch} into the answer of D.

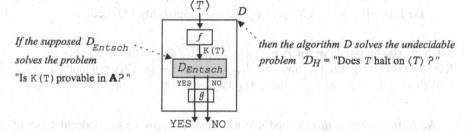

If the supposed D_{Entsch} solves the problem "Is $K(T)$ provable in **A**?" *then the algorithm D solves the undecidable problem* $\mathcal{D}_H$ = "Does T halt on $\langle T \rangle$?"

Fig. 9.7 Solving the *Halting Problem* with the alleged decider for the *Entscheidungsproblem*

Suppose that the *Formal Arithmetic* **A** were such as Hilbert expected, that is, consistent and complete. Then, D's answer would tell us, for arbitrary T, whether or not the statement (*) is true. But this would in turn mean that the halting problem $\mathcal{D}_H$ is decidable—which it is provably not! We must conclude that the problem $\mathcal{D}_{Entsch}$ is undecidable. (A fortiori, thanks to Gödel we know that *Formal Arithmetic* **A** is not complete.) This completes the proof. □

Thus, the decider D_{Entsch} does not exist. What about semi-decidability? Is it perhaps that $\mathcal{D}_{Entsch}$ is a semi-decidable problem? The answer is yes.

Theorem 9.3. *The Entscheidungsproblem $\mathcal{D}_{Entsch}$ is semi-decidable.*

Proof. Let $F \in M$ be an arbitrary formula. We have seen in Sect. 4.2.2 that we can systematically generate all the sequences of symbols of **M** and, for each generated sequence, algorithmically check whether or not it is a proof of F in **M**. If the formula F is in truth a theorem of **M**, then a proof of F in **M** exists and will, therefore, be eventually generated and recognized. So the described algorithm is a recognizer of the set of all the mathematical theorems. If, however, F is in truth *not* a theorem of **M**, then the recognizer will never halt. Nevertheless, this suffices for the problem $\mathcal{D}_{Entsch}$ to be semi-decidable. This completes the proof. □

Remark. After 16 years, yet another goal of *Hilbert's Program* received a negative response: *Formal Arithmetic* **A** *is not a decidable theory.* Thus, the recognition of arithmetical theorems cannot be fully automated, regardless of the kind of algorithms we might use to achieve this goal. Since **A** is a subtheory of **M**, the same holds for mathematics: No matter what kind of algorithms we use, only a proper subclass of all mathematical theorems can be algorithmically recognized. So there will always be mathematical theorems whose theoremship is algorithmically undecidable.

After the researchers learned this, some focused on formal axiomatic systems and theories that are weaker than *Formal Arithmetic* yet algorithmically decidable. An example of such a theory is *Presburger Arithmetic*,[1] introduced in 1929. In this arithmetic the operation of multiplication is omitted, thus allowing on ℕ only the the operation of addition and the equality relation. □

9.3 Proving by the Recursion Theorem

Recall from Sect. 7.4 that the *Recursion Theorem* can be restated as the *Fixed-Point Theorem*, which tells us that *every computable function has a fixed point.* This reveals the following method for proving the incomputability of functions.

Method. (Incomputability of Functions) Suppose that a function g has no fixed point. Then, g is not computable (i.e., it is not total, or it is incomputable, or both). If we somehow prove that g *is* total, then g must be incomputable.

We further develop this method into a method for proving the undecidability of problems as follows. Let $\mathcal{D}$ be a decision problem which we want to prove is undecidable. *Suppose* that $\mathcal{D}$ is decidable. Then the characteristic function $\chi_{L(\mathcal{D})}$ is computable. We can use $\chi_{L(\mathcal{D})}$ as a (necessary) component and construct a function $g : \mathbb{N} \to \mathbb{N}$ in such a way that g is computable. We then try to prove that g has no fixed point. If we succeed in this, we have a contradiction with the *Fixed-Point Theorem.* Thus, g cannot be computable and, consequently, $\chi_{L(\mathcal{D})}$ cannot be computable either. But characteristic functions are total by definition. Hence, $\chi_{L(\mathcal{D})}$ must be incomputable. So $\mathcal{D}$ is undecidable. We summarize this in the following method.

Method. (Undecidability of Problems) Undecidability of a decision problem $\mathcal{D}$ can be proved as follows:

1. Suppose: $\mathcal{D}$ is a decidable problem;
2. Construct: a computable function g using the characteristic function $\chi_{L(\mathcal{D})}$;
3. Prove: g has no fixed point;
4. Contradiction with the *Fixed-Point Theorem*!
5. Conclude: $\mathcal{D}$ is an undecidable problem.

---

[1] Mojżesz Presburger, 1904–1943, Polish mathematician, logician, and philosopher.

Example 9.10. We will use this method to prove *Rice's Theorem* in the next subsection.   □

Example 9.11. See the proof of the incomputability of a *search problem* in Sect. 9.5.   □

9.4 Proving by Rice's Theorem

All the methods of proving the undecidability of problems that we have described until now may require a considerable degree of inspiration, ingenuity, and even luck. In contrast, the method which we will consider in this section is far less demanding. It is based on a theorem that was discovered and proved in 1951 by Rice.[2] There are three versions of the theorem: for partial computable functions, for index sets, and for computably enumerable sets. They state:

> for p.c. functions: *Every non-trivial property of p.c. functions is undecidable.*
> for index sets: *Every index set different from $\emptyset$ and $\mathbb{N}$ is undecidable.*
> for c.e. sets: *Every non-trivial property of c.e. sets is undecidable.*

The theorem reveals that the undecidability of certain kinds of decision problems is more of a rule than an exception. Let us see the details.

9.4.1 Rice's Theorem for Functions

Let P be a property that is sensible for functions, and φ an arbitrary partial computable function. We define the following decision problem:

$$\mathcal{D}_P \equiv \text{``Does a p.c. function } \varphi \text{ have the property } P?\text{''}$$

We will say that the property P is *decidable* if the problem $\mathcal{D}_P$ is decidable. Thus, if P is a decidable property of p.c. functions, then there exists a Turing machine that decides, for an arbitrary p.c. function φ, whether or not φ has the property P.

What are the properties P that we will be interested in? We will not be interested in a property if it is such that a function has it in some situations and does not have it in other situations. This could happen when the property depends on the way in which function values are computed. For example, whether or not a function φ has the property defined by $P(\varphi) \equiv \text{``}\varphi(x)$ is computed in x^3 steps'' certainly depends on the algorithm used to compute $\varphi(x)$. It may also depend on the program (i.e., algorithm encoding) as well as on the machine where the function values are computed. (The machine can be an actual computer or an abstract computing machine.) For this reason, we will only be interested in the properties that are *intrinsic* to functions, i.e., the properties of functions where the functions are viewed only as mappings from one set to another. Such properties are because of their basic nature insensitive to

---

[2] Henry Gordon Rice, 1920–2003, American logician and mathematician.

the machine, algorithm, and program, that are used to compute function values. For instance, being total is an intrinsic property of functions, because every function φ is either or not total, irrespective of the way the φ values are computed.

So let P be an arbitrary intrinsic property of p.c. functions. It may happen that *every* p.c. function has this property P; and it also may happen that *no* p.c. function has the property P. In each of the two cases we will say that the property P is *trivial*.

This is where *Rice's Theorem* enters. Rice discovered the following surprising connection between the decidability and the triviality of the intrinsic properties of partial computable functions.

Theorem 9.4. (Rice's Theorem for p.c. Functions) *Let P be an arbitrary intrinsic property of p.c. functions. Then:*

$$P \text{ is decidable} \iff P \text{ is trivial.}$$

Proof. See below.   □

9.4.2 Rice's Theorem for Index Sets

Before we embark on the proof of Theorem 9.4, we restate the theorem in an equivalent form that will be more convenient to prove. The new form will refer to *index sets* instead of p.c. functions.

So let P be an arbitrary intrinsic property of p.c. functions. Define $\mathcal{F}$ to be the class of all the p.c. functions having the property P,

$$\mathcal{F} = \{\psi \mid \psi \text{ has the property } P\}.$$

Then the decision problem $\mathcal{D}_P$ can be written as

$$\mathcal{D}_P \equiv \varphi \in? \mathcal{F}.$$

Let $\mathrm{ind}(\mathcal{F})$ be the set of all of the indexes of all the Turing machines that compute any of the functions in $\mathcal{F}$. Obviously $\mathrm{ind}(\mathcal{F}) = \bigcup_{\psi \in \mathcal{F}} \mathrm{ind}(\psi)$, where $\mathrm{ind}(\psi)$ is the index set of the function ψ (see Sect. 7.2). Now the critical observation is: Because P is insensitive to the program used to compute φ's values, it must be that $\mathrm{ind}(\mathcal{F})$ contains *either all* of the indexes of the Turing machines computing φ, *or none* of these indexes—depending on whether or not φ has property P. So the question $\varphi \in? \mathcal{F}$ is equivalent to the question $x \in? \mathrm{ind}(\mathcal{F})$, where x is the index of an arbitrary Turing machine computing φ. Hence,

$$\mathcal{D}_P \equiv x \in? \mathrm{ind}(\mathcal{F}).$$

From the last formulation of $\mathcal{D}_P$ it follows that

$\mathcal{D}_P$ is a decidable problem $\iff$ ind($\mathcal{F}$) is a decidable set.

But when is the set ind($\mathcal{F}$) decidable? The answer gives the following version of *Rice's Theorem*.

Theorem 9.5. (Rice's Theorem for Index Sets) *Let $\mathcal{F}$ be an arbitrary set of p.c. functions. Then:*

$$\text{ind}(\mathcal{F}) \text{ is decidable} \iff \text{ind}(\mathcal{F}) \text{ is either } \emptyset \text{ or } \mathbb{N}.$$

This form of *Rice's Theorem* can now easily be proved by the *Fixed-Point Theorem* (see Box 9.1).

Box 9.1 (Proof of Theorem 9.5).

Both $\emptyset$ and $\mathbb{N}$ are decidable sets. Consequently, ind($\mathcal{F}$) is decidable if it is any of the two. Now, take an arbitrary ind($\mathcal{F}$) that is neither $\emptyset$ nor $\mathbb{N}$. *Suppose* that ind($\mathcal{F}$) is decidable. Then we can deduce a contradiction as follows. Because ind($\mathcal{F}$) is a proper and non-empty subset of $\mathbb{N}$, there must be two different natural numbers a and b such that $a \in$ ind($\mathcal{F}$) and $b \in \overline{\text{ind}(\mathcal{F})}$. Define a function $f : \mathbb{N} \to \mathbb{N}$ that maps every $x \in$ ind($\mathcal{F}$) into b and every $x \in \overline{\text{ind}(\mathcal{F})}$ into a.

Clearly, f is total. But it is also computable. (Why? By supposition ind($\mathcal{F}$) is decidable, so in order to compute $f(x)$ for an arbitrary $x \in \mathbb{N}$, we first decide whether $x \in$ ind($\mathcal{F}$) or $x \in \overline{\text{ind}(\mathcal{F})}$ and, depending on the answer, assign $f(x) = b$ or $f(x) = a$, respectively.)

Consequently, f has a fixed point (by the *Fixed-Point Theorem*). But it is obvious that there can be *no* fixed point: namely, for an arbitrary $x \in \mathbb{N}$, the numbers x and $f(x)$ are in different sets ind($\mathcal{F}$) and $\overline{\text{ind}(\mathcal{F})}$, so $x \neq f(x)$. This is a contradiction. Hence, ind($\mathcal{F}$) cannot be decidable. □

There is some bad news brought by *Rice's Theorem*. We have just proved that the question $\mathcal{D}_P \equiv$ "Does φ have the property P?" is only decidable for trivial properties P. But trivial properties are not very interesting. A property is more interesting if it is non-trivial, i.e., if it is shared by some functions and not by others. Non-trivial properties are less predictable, so getting the answer to the above question is usually more "dramatic," particularly when the consequences of different answers are very different. In this respect *Rice's Theorem* brings to us a disillusion because it tells us that any attempt to algorithmically fully recognize an interesting property of functions is in vain.

But there is good news too. *Rice's Theorem* tells us that determining whether the problem $\mathcal{D}_P$ is decidable or not can be reduced to determining whether or not the property P of functions is trivial. But the latter is usually easy to do. So we can set up the following method of proving the undecidability of decision problems of the kind $\mathcal{D}_P$. In this way we can prove the undecidability of the decision problems listed in Section 8.3.5.

Method. (Undecidability of Problems) Given a property P, the undecidability of the decision problem $\mathcal{D}_P \equiv$ "Does a p.c. function φ have the property P?" can be proved as follows:

1. Try to show that P meets the following conditions:
   a. P is a property sensible for functions;
   b. P is insensitive to the machine, algorithm, or program used to compute φ.
2. If P fulfills the above conditions, then try to show that P is non-trivial. To do this,
   a. find a p.c. function that has property P;
   b. find a p.c. function that does not have property P.

If all the steps are successful, then the problem $\mathcal{D}_P$ is undecidable.

9.4.3 Rice's Theorem for Sets

Let R be a property that is sensible for sets and let $\mathcal{X}$ be an arbitrary c.e. set. We define the following decision problem:

$$\mathcal{D}_R \equiv \text{"Does a c.e. set } \mathcal{X} \text{ have the property } R\text{?"}$$

As with functions, we say that the property R of c.e. sets is *decidable* if the problem $\mathcal{D}_R$ is decidable. So, if R is decidable, then there is a Turing machine capable of deciding, for an arbitrary c.e. set $\mathcal{X}$, whether or not R holds for $\mathcal{X}$. For the same reasons as above, we are only interested in the *intrinsic* properties of sets. These are the properties R that are independent of the way of recognizing the set $\mathcal{X}$. Hence, the answer to the question of whether or not $\mathcal{X}$ has the property R is insensitive to the *local* behavior of the machine, algorithm, and program that are used to recognize $\mathcal{X}$. For example, being finite is an intrinsic property of sets, because every set $\mathcal{X}$ either is or is not finite, irrespective of the way in which $\mathcal{X}$ is recognized. Finally, we say that R is *trivial* if it holds for all c.e. sets or for none.

Theorem 9.6. (Rice's Theorem for c.e. Sets) *Let R be an arbitrary intrinsic property of c.e. sets. Then: R is decidable $\Longleftrightarrow$ R is trivial.*

Box 9.2 (Proof of Theorem 9.6).

Let $\mathscr{C}$ be the class of all c.e. sets, $\mathcal{X}$ an arbitrary c.e. set, and R an arbitrary intrinsic property of sets. Define $\mathscr{R} = \{\mathcal{S} \in \mathscr{C} \,|\, R(\mathcal{S})\}$, the set of c.e. sets with the property R. Then we have

$$\mathcal{D}_R \equiv \mathcal{X} \in? \mathscr{R}. \tag{1}$$

Since a c.e. set is the domain of a computable function (see Theorem 7.4 on p. 144), we have $\mathcal{X} = \text{dom}(\varphi_x)$ for some $x \in \mathbb{N}$, and $\mathcal{R} = \{\text{dom}(\varphi_i) \mid \varphi_i \text{ is computable} \wedge R(\text{dom}(\varphi_i))\}$. Now let $\mathcal{F}$ be the set of all of the computable functions whose domains have the property R. Then question (1) transforms to the equivalent question

$$\varphi_x \in ? \mathcal{F}. \tag{2}$$

We now proceed in the same fashion as before. First, we introduce the set $\text{ind}(\mathcal{F})$ of all of the indexes of all of the p.c. functions that are in $\mathcal{F}$. This allows us to reduce question (2) to the equivalent question

$$x \in ? \text{ind}(\mathcal{F}). \tag{3}$$

We see that $\mathcal{D}_R$ is decidable *iff* $x \in ? \text{ind}(\mathcal{F})$ is decidable. But we have already proved that the question $x \in ? \text{ind}(\mathcal{F})$ is decidable *iff* $\text{ind}(\mathcal{F})$ is either $\emptyset$ or $\mathbb{N}$. So, $\mathcal{D}_R$ is decidable *iff* R is trivial.□

Therefore, determining whether the problem $\mathcal{D}_R$ is decidable or not can be reduced to determining whether or not R is a trivial property of c.e. sets. Again the latter is much easier to do.

Based on this we can set up a method of proving undecidability of problems $\mathcal{D}_R$ for different properties R. The method can easily be obtained from the previous one (just substitute R for P and c.e. sets for p.c. functions). In this way we can prove the undecidability of many decision problems, such as: *"Is a in $\mathcal{X}$? Is $\mathcal{X}$ equal to $\mathcal{A}$? Is $\mathcal{X}$ a regular set?"*, where a and $\mathcal{A}$ are parameters.

Remarks. (1) Since c.e. sets are semi-decidable and vice versa, *Rice's Theorem* for c.e. sets can be stated in terms of semi-decidable sets: Any nontrivial property of semi-decidable sets is undecidable. (2) What about the properties of decidable (i.e. computable) sets? Is it also true that a nontrivial property of decidable sets is undecidable? Observe that this does *not* automatically follow from the *Rice's Theorem* (for c.e. sets), because the class of decidable (computable) sets is properly contained in the class of semi-decidable (c.e.) sets.

9.4.4 Consequences: Behavior of Abstract Computing Machines

Let us return to *Rice's Theorem* for index sets. Since indexes represent Turing programs (and Turing machines), this version of *Rice's Theorem* refers to TPs (and TMs). In essence, it tells us that only trivial properties of TPs (and TMs) are decidable. For example, whether or not an arbitrary TM halts is an undecidable question, because the property of halting is not trivial.

This version of *Rice's Theorem* also discloses an unexpected relation between the local and global behavior of TMs and, consequently, of abstract computing machines in general. Recall from p. 146 that we identified the *global behavior* of a TM T with its proper function (i.e., the function ψ_T the machine T computes) or with its proper set (i.e., the set $L(T)$ the machine accepts and hence recognizes). On the other hand, the *local behavior* of T refers to the way in which T's program δ executes. *Rice's Theorem* tells us that, given an arbitrary T, we perfectly know how T behaves locally, but we are unable to algorithmically determine its global behavior—unless this behavior is trivial. Since the models of computation are equivalent, we conclude: In general, we cannot algorithmically predict the global behavior of an abstract computing machine from the machine's local behavior.

9.5 Incomputability of Other Kinds of Problems

In the previous sections we were mainly interested in *decision* problems. Now that we have developed a good deal of the theory and methods for dealing with decision problems, it is time to take a look at other kinds of computational problems. As mentioned in Sect. 8.1.1, there are also search problems, counting problems, and generating problems. For each of these kinds there exist incomputable problems.

In this section we will prove the incomputability of a certain search problem. Search problems are of particular interest because they frequently arise in practice. For example, the *sorting problem* is a search problem: to sort a sequence $a_1, a_2, \ldots, a_n$ of numbers is the same as to find a permutation $i_1, i_2, \ldots, i_n$ of indexes $1, 2, \ldots, n$ so that $a_{i_1} \leqslant a_{i_1} \leqslant \ldots \leqslant a_{i_n}$. Actually, every *optimization problem* is a search problem because it involves the search for a feasible solution that best fulfills a given set of conditions.

We now focus on the following search problem.

▲ SHORTEST EQUIVALENT PROGRAM

Let code(A) be an arbitrary program describing an algorithm A. The aim is:

"Given a program code(A), *find the shortest equivalent program."*

So the problem asks for the shortest description of an arbitrary given algorithm A.

Before we continue we must clear up several things. First, we consider two programs to be *equivalent* if, for every input, they return the same results (although they compute the results in different ways). Secondly, in order to speak about the shortest programs, we must define the *length* of a program. Since there are several ways to do this, we ask, which of them make sense? Here, the following must hold if a definition of the program length is to be considered reasonable:

If programs get longer, then, eventually,
their shortest equivalent programs get longer too.

To see this, observe that if the lengths of the shortest equivalent programs were bounded above by a constant, then there would be only a finite number of shortest programs. This would imply that all of the programs could be shortened into only a finite number of shortest equivalent programs. But this would mean that only a finite number of computational problems can be algorithmically solved.

Now we are prepared to state the following proposition.

Proposition 9.1. *The problem* SHORTEST EQUIVALENT PROGRAM *is incomputable.*

Box 9.3 (Proof of Proposition 9.1).

(Contradiction with the *Recursion Theorem*) We focus on Turing programs and leave the generalization to arbitrary programming languages to the reader. For the sake of contradiction we make the following supposition.

Supposition (∗): There exists a Turing program S that transforms every Turing program into the shortest equivalent Turing program.

Three questions immediately arise: (1) What is the transformation of a Turing program? (2) What is the length of a Turing program? (3) When are two Turing programs equivalent? The answers are:

1. The *transformation* of a Turing program: Every Turing program can be represented by a natural number, the index of this Turing program (see Sect. 6.2.1). Considering this, we can view the supposed Turing program S as computing a function $s : \mathbb{N} \to \mathbb{N}$ that *transforms* (i.e., maps) an arbitrary index into another index. The above supposition then states (among other things) that s is a *total computable* function.

2. The *length* of a Turing program: Indexes of Turing machines have two important properties. First, *every* Turing program is encoded by an index. Secondly, indexes increase (decrease) simultaneously with the increasing (decreasing) of (a) the number of instructions in Turing programs; (b) the number of different symbols used in Turing programs; and (c) the number of states needed in Turing programs. Therefore, to measure the *length* of a Turing program by its *index* seems to be a natural choice.

3. The *equivalence* of Turing programs: Recall (Sect. 6.3.1) that a Turing program P having index n computes the values of the computable function ψ_n (i.e., the proper function of the Turing machine T_n). We say that a Turing program P' is *equivalent* to P if P' computes, for every input, exactly the same values as P. Remember also (Sect. 7.2) that the indexes of all the Turing programs equivalent to P constitute the index set $\mathrm{ind}(\psi_n)$. Now it is obvious that $s(n)$ must be the *smallest* element in the set $\mathrm{ind}(\psi_n)$.

Since $s(n)$ is an index (i.e., the code of a Turing program), we see that, by the same argument as above, the function s cannot be bounded above by any constant. In other words, if the lengths of Turing programs increase, so eventually do the lengths of their shortest equivalent programs; that is,

$$\text{For every } n \text{ there exists an } n' > n \text{ such that } s(n) < s(n'). \tag{1}$$

Using the function s we can restate our supposition (∗) more precisely:

Supposition (∗∗): There is a (total) computable function $s : \mathbb{N} \to \mathbb{N}$ such that

$$s(n) \overset{\text{def}}{=} \text{the smallest element in } \mathrm{ind}(\varphi_n). \tag{2}$$

We now draw a contradiction from the supposition (∗∗). First, let $f : \mathbb{N} \to \mathbb{N}$ be the function whose values $f(x)$ are computed by the following Turing machine S. Informally, given an arbitrary input $x \in \mathbb{N}$, the machine S computes and generates the values $s(k)$, $k = 0, 1, \ldots$, until an $s(m)$, for which $s(m) > x$, has been generated. At that point S returns the result $f(x) := s(m)$ and halts. Here is a more detailed description. The machine S has one input tape, one output tape, and two work tapes, W_1 and W_2 (see Fig. 9.8). The Turing program of S operates as follows:

1. Let $x \in \mathbb{N}$ be an arbitrary number written on the input tape of S.
2. S generates the numbers $k = 0,1,2,\ldots$ to W_1 and after each generated k executes steps 3–6:
3. S computes the value $s(k)$;
4.   if $s(k)$ has not yet been written to W_2,
5.   then S generates $s(k)$ to W_2;
6.   if $s(k) > x$, then S copies $s(k)$ to the output tape (so $f(x) = s(k)$) and halts.

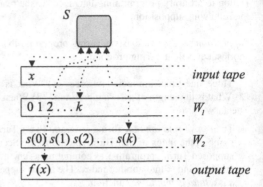

Fig. 9.8 The configuration of
the Turing machine S

In step 3, the machine S computes the function s. Because s is supposed to be a computable function (2), and (1) holds, the machine S halts for arbitrary x. Consequently, f is a computable function.

We now show that such an f *cannot* exist. Let x be an arbitrary natural number. We saw that there exists an $m \in \mathbb{N}$ such that $s(m) > x$. But $s(m)$ is by definition the *smallest* index in the set $\mathrm{ind}(\varphi_m)$, so the relation $s(m) > x$ tells us that

$$x \notin \mathrm{ind}(\varphi_m).$$

At the same time, $f(x) = s(m)$, so

$$f(x) \in \mathrm{ind}(\varphi_m).$$

Since x and $f(x)$ are not both in the same index set, the functions φ_x and $\varphi_{f(x)}$ cannot be equal; that is,

$$\varphi_x \neq \varphi_{f(x)}, \text{ for every } x \in \mathbb{N}. \tag{3}$$

The inequality (3) reveals that f has *no* fixed point. But we have seen that f is a computable function, so—according to the *Fixed-Point Theorem*—f *should* have a fixed point. This is a contradiction. Consequently, f is not a computable function. This means that the machine S does not halt for arbitrary x. Since (1) holds, it must be that (2) does not hold. Our supposition must be rejected.   □

9.6 Chapter Summary

General methods of proving the undecidability of decision problems use diagonalization, reduction, the *Recursion Theorem*, or *Rices's Theorem*.

  Diagonalization can be used to prove that a set S of all elements having a given property P is uncountable. It uses a table whose vertical axis is labelled with the members of S and the horizontal axis by all natural numbers. The supposition is that each member of S can be uniquely represented by a sequence of symbols, written in the corresponding row, over an alphabet. If we change each diagonal symbol, and the obtained sequence of diagonal symbols represents an element of S, then the set is uncountable. Another use of diagonalization is to interpret S as the set of all indexes of all algorithms having a given property P. We want to know if the property is decidable. The labels on the horizontal axis of the table are interpreted as inputs to algorithms. Each entry in the table contains the result of applying the corresponding algorithm on the corresponding input. Of special interest is the diagonal of the table. The idea is to design a shrewd algorithm that will use the alleged decider of the set S, and uncover the inability of the decider to decide whether or not the shrewd algorithm has the property P. Then the property P is undecidable.

  Reduction is a method where a given problem is expressed in terms of another problem in such way that the solution to the second problem would give rise to the solution of the first problem. For decision problems, we can use the m-reduction or 1-reduction. The existence of the m-reduction can be used to prove the undecidability of a decision problem. More powerful is the Turing reduction (see Part III).

  The *Recursion Theorem* can be used to prove the undecidability of a decision problem. The method relies on the fact that every computable function has a fixed point. If we prove that a given function has no fixed point, then the function cannot be computable. Since the characteristic function of a set is a function, this method can be used to prove the undecidability of a set.

  Rices's Theorem is used to prove that certain properties of partial computable (p.c.) functions or computably enumerable (c.e.) sets are undecidable. The properties must be intrinsic to functions or sets, and thus independent of the way the functions are computed or the sets recognized. *Rice's Theorem* states that every nontrivial intrinsic property is undecidable.

Problems

9.1. Prove:

(a) $\leq_m$ is a reflexive and transitive relation;

(b) $\mathcal{A} \leq_m \mathcal{B}$ if $r(\mathcal{A}) \subseteq \mathcal{B}$ and $r(\overline{\mathcal{A}}) \subseteq \overline{\mathcal{B}}$, for some computable function r.

Definition 9.2. A set $\mathcal{A}$ is said to be m-**complete** if $\mathcal{A}$ is c.e. and $\mathcal{B} \leq_m \mathcal{A}$ for every c.e. set $\mathcal{B}$. Similarly, a set $\mathcal{A}$ is 1-**complete** if $\mathcal{A}$ is c.e. and $\mathcal{B} \leq_1 \mathcal{A}$ for every c.e. set $\mathcal{B}$.

9.2. Prove:

(a) $\mathcal{K}$ is m-complete;

(b) $\mathcal{K}_0$ is m-complete.

9.3. Prove:

(a) $\mathcal{K}$ is 1-complete;

(b) $\mathcal{K}_0$ is 1-complete.

9.4. Prove: $\mathcal{A}$ is 1-complete $\Longrightarrow$ $\mathcal{A}$ is m-complete.

9.5. Construct the following Turing machines (which we used in the Example 9.5 on p. 201):

(a) a TM A that takes $\langle T, w \rangle$ as input and outputs a TM B such that

$$L(B) = \begin{cases} \mathcal{K}_0 & \text{if } T \text{ accepts } w; \\ \emptyset & \text{if } T \text{ does not accept } w. \end{cases}$$

[*Hint.* The actions of B on input x are as follows: B starts T on w; if T accepts w, then B starts U (the recognizer of $\mathcal{K}_0$) on x and outputs U's answer (if it is a YES) as its own; if, however, T doesn't accept w, then U is not started, so B outputs nothing. It remains to be shown that there exists a Turing machine A that is capable of computing the code $\langle B \rangle$.]

(b) a TM R that recognizes $\overline{\mathcal{K}_0}$ if allowed to call the above A and the alleged decider $D_{L(\mathcal{Q})}$ for the problem $\mathcal{Q} \equiv$ "Is $L(T)$ computable?"

[*Hint.* The actions of R on input $\langle T, w \rangle$ are as follows: R calls A on the input $\langle T, w \rangle$; hands over the result $\langle B \rangle$ of A to $D_{L(\mathcal{Q})}$; and outputs $D_{L(\mathcal{Q})}$'s answer (if it is a YES) as its own answer.]

9.6. Prove that the following decision problems are undecidable (use any method of this chapter):

(a) "Does T halt on empty input?"

(b) "Does T halt on every input?"

(c) "Does M halt on w?" (M is a fixed TM)

(d) "Do T and T' halt on the same inputs?"

(e) "Is the language $L(T)$ empty?" ($\equiv \mathcal{D}_{Emp}$)

　　　[*Hint.* Reduce $\mathcal{D}_{Halt}$ to $\mathcal{D}_{Emp}$.]

(f) "Does an algorithm A terminate on every input data?" ($\equiv \mathcal{D}_{Term}$)

(g) "Does an algorithm A terminate on input data d?"

(h) "Is dom(φ) empty?" ($\equiv \mathcal{D}_{\mathcal{K}_1}$)

(i) "Is dom(φ) finite?" ($\equiv \mathcal{D}_{\mathcal{F}in}$)

(j) "Is dom(φ) infinite?" ($\equiv \mathcal{D}_{\mathcal{I}nf}$)

(k) "Is $\mathcal{A} -$ dom(φ) finite?" ($\equiv \mathcal{D}_{Cof}$)

(l) "Is φ total?" ($\equiv \mathcal{D}_{Tot}$)

(m) "Can φ be extended to total computable function?" ($\equiv \mathcal{D}_{\mathcal{E}xt}$)

(n) "Is φ surjective?" ($\equiv \mathcal{D}_{Sur}$)

(o) "Is φ defined at x?"

(p) "Is $\varphi(x) = y$ for at least one x?"

(q) "Is dom(φ) = dom(ψ)?"

(r) "Is $\varphi \simeq \psi$?"

Bibliographic notes

- The undecidability of the *Entscheidungsproblem* was proved in Church [24] and Turing [178].
- The method of *diagonalization* first appeared in Cantor [20], where he proved that the set of real numbers is uncountable and that there are other uncountable sets. For additional applications of the diagonalization method and its variants, see Odifreddi [116].
- *Reduction* as a method of proving the undecidability of problems was extensively analyzed in Post [123] and Post [124]. Here, in addition to the 1-reduction and *m*-reduction, other so-called *strong* reductions were defined (e.g., the *bounded truth-table* and *truth-table* reductions). The paper also describes the *Turing* reduction as the most general among them.
- For the *resource-bounded* reductions, such as the *polynomial-time bounded* Turing or the *m*-reduction, see Ambos-Spies [7].
- *Rice's Theorem* was proved in Rice [135].
- A number of examples of using the above methods in incomputability proofs can be found in the general monographs on *Computability Theory* cited in the Bibliographic Notes to Chap. 6.
- *Presburger Arithmetic* was introduced in Presburger [129]. For the proof that this theory is decidable, see Boolos et al. [15].

Part III
RELATIVE COMPUTABILITY

In the previous part we described how the world of incomputable problems was discovered. This resulted in an awareness that there exist computational problems that are unsolvable by any reasonable means of computing, e.g., the Turing machine. In Part III, we will focus on decision problems only. We will raise the questions, "What if an unsolvable decision problem had been somehow made solvable? Would this have turned all the other unsolvable decision problems into solvable problems?" We suspect that this might be possible *if* all the unsolvable decision problems were somehow reducible one to another. However, it will turn out that this is not so; some of them would indeed become solvable, but there would still remain others that are unsolvable. We might speculate even further and suppose that one of the remaining unsolvable decision problems was somehow made solvable. As before, this would turn many unsolvable decision problems into solvable; yet, again, there would remain unsolvable decision problems. We could continue in this way, but we would never exhaust the class of unsolvable problems.

Questions of the kind "Had the problem Q been solvable, would this have made the problem P solvable too?" are characteristic of the *relativized computability*, a large part of the *Computability Theory*. This theory analyzes the solvability of problems *relative to* (or in view of) the solvability of other problems. Although such questions seem to be overly speculative and the answers to be of questionable practical value, they nevertheless reveal a surprising *fact:* Unsolvable decision problems can differ in the degree of their unsolvability. We will show that the class of all decision problems partitions into infinitely many subclasses, called *degrees of unsolvability*, each of which consists of all equally difficult decision problems. It will turn out that, after defining an appropriate relation on the class of all degrees, the degrees of unsolvability are intricately connected into a lattice-like structure. In addition to having many interesting properties per se, the structure will reveal many surprising facts about the unsolvability of decision problems.

We will show that there are several approaches to partitioning the class of decision problems into degrees of unsolvability. Each of them will establish a particular *hierarchy* of degrees of unsolvability, such as the *jump* and *arithmetical* hierarchies, and thus offer yet another view of the solvability of computational problems.

Chapter 10
Computation with External Help

An oracle was an ancient priest who made statements about
future events or about the truth.

Abstract According to the *Computability Thesis, all* models of computation, including those yet to be discovered, are equivalent to the Turing machine and formalize the intuitive notions of computation. In other words, what cannot be solved on a Turing machine cannot be solved in nature. But what if Turing machines could get external help from a *supernatural* assistant? In this chapter we will describe the birth of this idea, its development, and the formalization of it in the concept of the *oracle Turing machine*. We will then briefly describe how external help can be added to other models of computation, in particular to partial recursive functions. We will conclude with the *Relative Computability Thesis*, which asserts that all such models are equivalent, one to the other, thus formalizing the intuitive notion of the "computation with external help." Based on this, we will adopt the oracle Turing machine as *the* model of computation with external help.

10.1 Turing Machines with Oracles

"What if an unsolvable decision problem were *somehow* made solvable?" For instance, what if we somehow managed to construct a procedure that is capable of mechanically solving an arbitrary instance of the *Halting Problem*? But in making such a supposition, we must be cautious: since we have already proved that *no* algorithm can solve the *Halting Problem*, such a supposition would be in contradiction with this fact *if* we meant by the hypothetical procedure an algorithm as formalized by the *Computability Thesis*. Making such a supposition would render our theory inconsistent and entail all the consequences described in Box 4.1 (p. 50). So, to avoid such an inconsistency, we must assume that the hypothetical procedure would *not* be the ordinary Turing machine.

Well, if the hypothetical procedure is not the ordinary Turing machine, then what is it? The answer was suggested by Turing himself.

© Springer-Verlag Berlin Heidelberg 2015
B. Robič, *The Foundations of Computability Theory*,
DOI 10.1007/978-3-662-44808-3_10

219

10.1.1 Turing's Idea of Oracular Help

In 1939, Turing published yet another seminal idea, this time of a machine he called the *o-machine*. Here is the citation from his doctoral thesis:

> Let us suppose that we are supplied with some unspecified means of solving number-theoretic problems; a kind of oracle as it were. We shall not go any further into the nature of this oracle apart from saying that it cannot be a machine. With the help of the oracle we could form a new kind of machine (call them *o*-machines), having as one of its fundamental processes that of solving a given number-theoretic problem. More definitely these machines are to behave in this way. The moves of the machine are determined as usual by a table except in the case of moves from a certain internal configuration ο. If the machine is in the internal configuration ο and if the sequence of symbols marked with *l* is then the well-formed formula *A*, then the machine goes into the internal configuration þ or t according as it is or is not true that *A* is dual. The decision as to which is the case is referred to the oracle.
>
> These machines may be described by tables of the same kind as those used for the description of *a*-machines, there being no entries, however, for the internal configuration ο. We obtain description numbers from these tables in the same way as before. If we make the convention that, in assigning numbers to internal configurations, ο, þ, t are always to be q_2, q_3, q_4, then the description numbers determine the behavior of the machines uniquely.

Remarks. Let us add some remarks about the terminology and notation used in the above passage. The "internal configuration" is what we now call the state of the Turing machine. The "sequence of symbols marked with *l*" is a sequence of symbols on the tape that are currently marked in a certain way. We will see shortly that this marking can be implicit. The "duality" is a property that a formula has or does not have; we will not need this notion in the subsequent text. The "table" is the table Δ representing the Turing program δ as decribed in Fig. 6.3 (see p. 103). To comply with our notation, we will from now on substitute Gothic symbols as follows: $ο \mapsto q_?$, $þ \mapsto q_+$ and $t \mapsto q_-$.

Now we see that the set Q of the states of the *o*-machine would contain, in addition to the usual states, three distinguished states $q_?, q_+, q_-$. The operation of the *o*-machine would be directed by an ordinary Turing program, i.e., a transition function $\delta : Q \times \Gamma \to Q \times \Gamma \times \{\text{Left}, \text{Right}, \text{Stay}\}$. The function would be undefined in the state $q_?$, for every $z \in \Gamma$, i.e., $\forall z \in \Gamma \, \delta(q_?, z) \uparrow$. However, upon entering the state $q_?$, the *o*-machine would *not* halt; instead, an *oracle* would miraculously supply, in the very next step of the computation, a cue telling the *o*-machine which of the states q_+ and q_- to enter. In this way the oracle would answer the *o*-machine's question asking whether or not the currently marked word on the tape is in a certain, predefined set of words, a set which can be decided by the oracle. The *o*-machine would immediately take up the cue, enter the suggested state, and continue executing its program δ.

Further Development of Turing's Idea

Turing did not further develop his *o*-machine. The idea remained dormant until 1944, when Post awakened it and brought it into use. Let us describe how the idea evolved. We will focus on the interplay between the o-machine and its oracle. In doing so, we will lean heavily on Definition 6.1 of the ordinary TM (see Sect. 6.1.1).

A Turing machine *with an oracle*, i.e., the o-machine, consists of the following components: (1) a control unit with a program; (2) a potentially infinite tape divided into cells, each of which contains a symbol from a tape alphabet $\Gamma = \{z_1, z_2, \ldots, z_t\}$, where $z_1 = 0, z_2 = 1$ and $z_t = \sqcup$; (3) a window that can move to the neighboring (and, eventually, to any) cell, thus making the cell accessible to the control unit for reading or writing; and (4) an *oracle*. (See Fig. 10.1.)

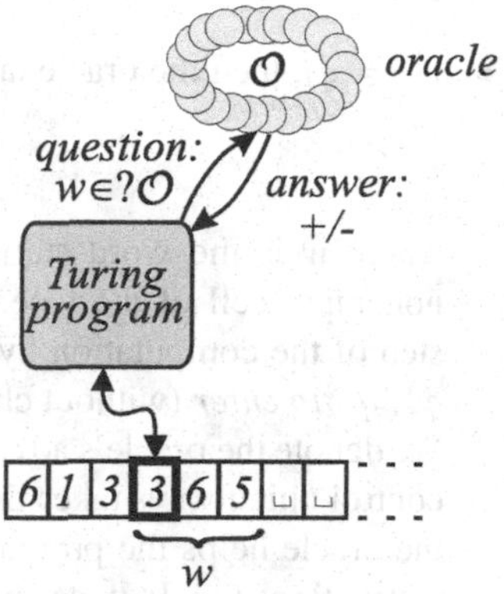

Fig. 10.1 The o-machine with the oracle for the set $\mathcal{O}$ can ask the oracle whether or not the word w is in $\mathcal{O}$. The oracle answers in the next step and the o-machine continues its execution appropriately

The control unit is always in some state from a set $Q = \{q_1, \ldots, q_s, q_?, q_+, q_-\}$. As usual, the state q_1 is initial and some of $q_1, \ldots, q_s$ are final states. But now there are also three additional, *special states*, denoted by $q_?$, q_+, and q_-.

The program is characteristic of the particular o-machine; as usual, it is a partial function $\delta : Q \times \Gamma \to Q \times \Gamma \times \{\text{Left}, \text{Right}, \text{Stay}\}$. But, in addition to the usual instructions of the form $\delta(q_i, z_r) = (q_j, z_w, D)$, for $1 \leqslant i, j \leqslant s$ and $1 \leqslant r, w \leqslant t$, there are *four special instructions for each* $z_r \in \Gamma$:

$$\delta(q_?, z_r) = (q_+, z_r, \text{Stay}) \tag{$*$}$$
$$\delta(q_?, z_r) = (q_-, z_r, \text{Stay})$$

$$\delta(q_+, z_r) = (q_{j_1}, z_{w_1}, D_1) \tag{$**$}$$
$$\delta(q_-, z_r) = (q_{j_2}, z_{w_2}, D_2)$$

where $q_{j_1}, q_{j_2} \in Q - \{q_+, q_-\}$, $z_{w_1}, z_{w_2} \in \Gamma$, and $D_1, D_2 \in \{\text{Left}, \text{Right}, \text{Stay}\}$.

Before the o-machine is started the following preparative arrangement is made: (1) an input word belonging to the set Σ^*, such that $\{0, 1\} \subseteq \Sigma \subseteq \Gamma - \{\sqcup\}$ is written on the tape; (2) the window is shifted over the leftmost symbol of the input word; (3) the control unit is set to the initial state q_1; and (4) an arbitrary set $\mathcal{O} \subseteq \Sigma^*$ is fixed. We call $\mathcal{O}$ the *oracle set*.

From now on the o-machine operates in a mechanical stepwise fashion as directed *either by the program δ or by the oracle*. The details are as follows. Suppose that the control unit is in a state $q \in Q$ and the cell under the window contains a symbol $z_r \in \Gamma$. (See Fig. 10.1.) Then:

- If $q \neq q_?$, then the o-machine reads z_r into its control unit. If $\delta(q,z_r)\downarrow = (q',z_w,D)$, the control unit executes this instruction as usual (i.e., writes z_w through the window, moves the window in direction D, and enters the state q'); otherwise, the o-machine halts. In short, when the o-machine is not in state $q_?$, it operates as an ordinary Turing machine.

- If $q = q_?$, then the oracle takes this as the question

$$w \in ?\mathcal{O}, \qquad\qquad (***)$$

where w is the word starting under the window and ending in the rightmost nonempty cell of the tape. The oracle answers the question $(***)$ in the *next* step of the computation by *advising the control unit which of the special states q_+, q_- to enter* (without changing z_r under the window or moving the window). We denote the oracle's advice by "+" when $w \in \mathcal{O}$, and by "−" when $w \notin \mathcal{O}$. The control unit always takes the advice and enters the state either q_+ or q_-. In short, the oracle helps the program δ to choose the "right" one of the two alternative instructions $(*)$. In this way the oracle miraculously resolves the nondeterminism arisen from the state $q_?$.

After that, the o-machine continues executing its program. In particular, if the previous state has been $q_?$, the program will execute the instruction either $\delta(q_+,z_r) = (q_{j_1},z_{w_1},D_1)$ or $\delta(q_-,z_r) = (q_{j_2},z_{w_2},D_2)$, depending on the oracle's advice. Since the execution continues in one or the other way, this allows the program δ to react differently to different advice. In short, the oracle's advice will be taken into account by the Turing program δ.

The o-machine can ask its oracle, in succession, many questions. Since between two consecutive questions the o-machine can move the window and change the tape's contents, the questions $w \in ?\mathcal{O}$, in general, differ in words w. But note that they all refer to the set $\mathcal{O}$, which was fixed before the computation started.

The o-machine halts if the control unit has entered a final state, or the program specifies no next instruction (the machine reads z_r in a state $q(\neq q_?)$ and $\delta(q,z_r)\uparrow$).

Remark. Can the o-machine ask the oracle a question $u \in ?\mathcal{O}$, where u is *not* the word currently starting under the window and ending at the rightmost nonempty symbol of the tape? The answer is yes. The machine must do the following: it remembers the position of the window, moves the window past the rightmost nonempty symbol, writes u to the tape, moves the window back to the beginning of u, and enters the state $q_?$. Upon receiving the answer to the question $u \in ?\mathcal{O}$, the machine deletes the word u, returns the window to the remembered position, enters the state q_+ or q_- depending on the oracle's answer, and continues the "interrupted" computation.

10.1.2 The Oracle Turing Machine (o-TM)

To obtain a modern definition of the Turing machine with an oracle, we make the following two adaptations to the above o-machine:

1. *We model the oracle with an oracle tape.* The *oracle tape* is a one-way unbounded, read-only tape that contains all the values of the characteristic function χ_O. (See Fig. 10.2.) We assume that it takes, for arbitrary $w \in \Sigma^*$, only one step to search the tape and return the value $\chi_O(w)$.

2. *We eliminate the special states $q_?, q_+, q_-$ and redefine the instructions' semantics.* Since the special states behave differently, we will get rid of them. How can we do that? The idea is to 1) adapt the o-machine so that it will interrogate the oracle at *each* instruction of the program, and 2) leave it to the machine to decide whether or not the oracle's advice will be taken into account. To achieve goal 2 we must only use what we learned in the previous subsection: the oracle's advice to the question $\delta(q_?, z_r)$ will have no impact on the computation *iff* $\delta(q_+, z_r) = \delta(q_-, z_r)$. To achieve goal 1, however, we must redefine what actions of the machine will be triggered by the instructions.[1] In particular, we must redefine the instructions in such a way that each instruction will make the machine consult the oracle. Clearly, this will call for a change in the definition of the transition function.

Remark. We do not concern ourselves with how the values $\chi_O(w)$ emerged on the oracle tape and how it is that any of them can be found and read in just one step. A realistic implementation of such an access to the contents of the oracle tape would be a tremendous challenge when the set O is infinite. Actually, had we known how to do this for any set O, according to the *Computability Thesis* there would have existed an ordinary Turing machine capable of providing, in finite time, any data from the oracle tape. Hence, the oracle Turing machines could have been simulated by the ordinary TMs. This would have dissolved the supernatural flavor of the oracle Turing machine and brought back the threat of inconsistency (as we explained on p. 219.)

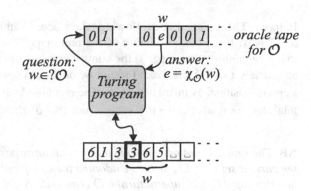

Fig. 10.2 The oracle Turing machine has an oracle tape. Given an oracle set $O \subseteq \Sigma^*$, the oracle tape contains all the values of the characteristic function χ_O. Here, $e = \chi_O(w)$, for arbitrary $w \in \Sigma^*$

---

[1] In other words, we must redefine the *operational semantics* of the instructions.

After making these adaptations we obtain a modern definition of the Turing machine that uses external help. We will call it the *oracle Turing machine*, or in short, *o*-TM.

Definition 10.1. (Oracle Turing Machine) The **oracle Turing machine with oracle set** $\mathcal{O}$ (in short, *o*-TM *with oracle set* $\mathcal{O}$, or $\mathcal{O}$-TM) consists of a control unit, an input tape, an oracle Turing program, an oracle tape, and a set $\mathcal{O}$. Formally, $\mathcal{O}$-TM is an eight-tuple $T^{\mathcal{O}} = (Q, \Sigma, \Gamma, \widetilde{\delta}, q_1, \sqcup, F, \mathcal{O})$.

The *control unit* is in a state from the set of states $Q = \{q_1, \dots, q_s\}$, $s \geqslant 1$. We call q_1 the initial state; some states are final and belong to the set $F \subseteq Q$.

The *input tape* is a usual one-way unbounded tape with a window. The tape alphabet is $\Gamma = \{z_1, \dots, z_t\}$, where $t \geqslant 3$. For convenience we fix $z_1 = 0$, $z_2 = 1$, and $z_t = \sqcup$. The input alphabet is a set Σ, where $\{0, 1\} \subseteq \Sigma \subseteq \Gamma - \{\sqcup\}$.

The *oracle set* $\mathcal{O}$ is an arbitrary subset of Σ^*.

The *oracle tape* contains, for each $w \in \Sigma^*$, the value $\chi_{\mathcal{O}}(w)$ of the characteristic function $\chi_{\mathcal{O}} : \mathcal{O} \to \{0, 1\}$. It is a read-only tape; although lacking a window, it can immediately find and return the value $\chi_{\mathcal{O}}(w)$, for any $w \in \Sigma^*$.

The *oracle Turing program* (in short, *o*-TP) resides in the control unit; it is a partial function

$$\widetilde{\delta} : Q \times \Gamma \times \{0, 1\} \to Q \times \Gamma \times \{\text{Left, Right, Stay}\}.$$

Thus, any instruction is of the form

$$\widetilde{\delta}(q, z, e) = (q', z', D),$$

which is interpreted as follows: If the control unit is in the state q, and reads z and e from the input and oracle tape, respectively, then it changes to the state q', writes z' to the input tape, and moves the window in the direction D. Here, e denotes the value $\chi_{\mathcal{O}}(w)$, where w is the word starting under the window and ending in the rightmost nonempty cell of the input tape.

Before *o*-TM is started, the following take place: 1) an input word belonging to Σ^* is written to the beginning of the input tape; 2) the window is shifted to the beginning of the tape; 3) the control unit is set to the initial state q_1; 4) an oracle set $\mathcal{O} \subseteq \Sigma^*$ is fixed. From now on $\mathcal{O}$-TM operates in a mechanical stepwise fashion, as instructed by $\widetilde{\delta}$. The machine halts when either enters a final state, or reads z and e in a state q such that $\widetilde{\delta}(q, z, e) \uparrow$.

NB *The oracle is not omnipotent; it is just an unsurpassable expert at recognizing the current set $\mathcal{O} \subseteq \Sigma^*$. Since we adopted the Computability Thesis, we must admit that this capability is supernatural if $\mathcal{O}$ is an undecidable set (in the ordinary sense). In such a case we will not ask where the oracle's expertise comes from.*

10.1.3 Some Basic Properties of o-TMs

Let us list a few basic properties of o-TMs that follow directly from Definition 10.1.

First, looking at Definition 10.1 we find that if we change the oracle set, say from $\mathcal{O}$ to $\mathcal{O}' \subseteq \Sigma^*$, *no* change is needed in the o-TP $\widetilde{\delta}$, i.e., no instruction $\widetilde{\delta}(q,z,e) = (q',z',D)$ must be changed, added, or deleted. The program $\widetilde{\delta}$ remains capable of running with the new oracle set $\mathcal{O}'$. In other words, the o-TP $\widetilde{\delta}$ is *insensitive* to the oracle set $\mathcal{O}$. (By the way, this is why the symbol $\widetilde{\delta}$ has no symbol $\mathcal{O}$ attached.)

Consequence 10.1. *Oracle Turing programs $\widetilde{\delta}$ are not affected by changes in $\mathcal{O}$.*

Second, although changing $\mathcal{O}$ does not affect the o-TP $\widetilde{\delta}$, it does affect, in general, the *execution* of $\widetilde{\delta}$. Namely, if $\mathcal{O}$ changes, $\chi_{\mathcal{O}}$ also changes, so the values e read from the oracle tape change too. But these values are used to select the next instruction to be executed. Nevertheless, we *can* construct o-TPs whose executions are insensitive to changes in the oracle set. Such an o-TP must just strictly ignore the values e read from the oracle tape. To achieve this, the o-TP must be such that

$$o\text{-TP has instr. } \widetilde{\delta}(q,z,0) = (q',z',D) \iff o\text{-TP has instr. } \widetilde{\delta}(q,z,1) = (q',z',D),$$

for every pair $(q,z) \in Q \times \Gamma$. So, if o-TM reads z in state q, it performs (q',z',D) regardless of the value e. If the above only holds for certain pairs (q,z), the o-TP $\widetilde{\delta}$ ignores the oracle for those pairs only. We sum up this in the following statement.

Consequence 10.2. *During the execution, oracle Turing programs $\widetilde{\delta}$ can ignore $\mathcal{O}$.*

Third, we ask: What is the relation between the computations performed by the o-TMs and the computations performed by the ordinary TMs? Let T be an arbitrary ordinary TM and δ its TP. We can easily construct an o-TM that simulates TM T. Informally, the o-TP $\widetilde{\delta}$ does, while ignoring its oracle set, what the TP δ would do. To achieve this, we must construct the program $\widetilde{\delta}$ in the following way:

    let $\widetilde{\delta}$ have no instructions;
    for each instruction $\delta(q,z) = (q',z',D)$ in δ **do**
      add instructions $\widetilde{\delta}(q,z,0) = (q',z',D)$ and $\widetilde{\delta}(q,z,1) = (q',z',D)$ to $\widetilde{\delta}$.

So we can state the following conclusion.

Theorem 10.1. *Oracle computation is a generalization of ordinary computation.*

Fourth, we now see that each $\mathcal{O}$-TM, say $T^{\mathcal{O}} = (Q,\Sigma,\Gamma,\widetilde{\delta},q_1,\sqcup,F,\mathcal{O})$, is characterized by two independent components:

1. *a particular o*-TM, denoted by $T^* = (Q,\Sigma,\Gamma,\widetilde{\delta},q_1,\sqcup,F,*)$, that is capable of consulting *any* particular oracle set when it is selected and substituted for $*$;
2. *a particular* oracle set $\mathcal{O}$ which is to be "plugged into" T^* in place of $*$.

Formally,

$$T^{\mathcal{O}} = (T^*,\mathcal{O}).$$

10.1.4 Coding and Enumeration of o-TMs

Like ordinary Turing programs, oracle Turing programs can also be encoded and enumerated. The coding proceeds in a similar fashion to that in the ordinary case (see Sect. 6.2.1). But there are differences too: firstly, we must take into account that the oracle Turing programs $\widetilde{\delta}$ are defined differently from the ordinary ones δ, and, secondly, oracle sets must be taken into account somehow. Let us see the details.

Coding of o-TMs

To encode an $\mathcal{O}$-TM $T^{\mathcal{O}} = (T^*,\mathcal{O})$ we will only encode the component T^*. The idea is that we only encode *o*-TP $\widetilde{\delta}$ of T^*, but in such a way that the other components Q,Σ,Γ,F, which determine the particular T^*, can be restored from the code. An appropriate coding alphabet is $\{0,1\}$. This is because $\{0,1\}$ is contained in the input alphabet of every *o*-TM, so every *o*-TM will be able to read codes of other *o*-TMs as input data. Here are the details.

Let $T^{\mathcal{O}} = (Q,\Sigma,\Gamma,\widetilde{\delta},q_1,\sqcup,F,\mathcal{O})$ be an arbitrary $\mathcal{O}$-TM. If

$$\widetilde{\delta}(q_i,z_j,e) = (q_k,z_\ell,D_m)$$

is an instruction of the program $\widetilde{\delta}$, we encode the instruction by the word

$$K = 0^i 10^j 10^e 10^k 10^\ell 10^m, \qquad (*)$$

where $D_1 = $ Left, $D_2 = $ Right, and $D_3 = $ Stay.

In this way, we encode each instruction of the program $\widetilde{\delta}$. From the obtained codes $K_1,K_2,\ldots,K_r$ we construct the *code* of the *o*-TP $\widetilde{\delta}$ as follows:

$$\langle\widetilde{\delta}\rangle = 111K_1 11K_2 11 \ldots 11K_r 111. \qquad (**)$$

The restoration of Q, Σ, Γ, F would proceed in a similar way as in the ordinary case (see Sect. 7.2). We can therefore identify the code $\langle T^* \rangle$ with the code $\langle \tilde{\delta} \rangle$:

$$\langle T^* \rangle \stackrel{\text{def}}{=} \langle \tilde{\delta} \rangle.$$

Enumeration of o-TMs

Since $\langle T^* \rangle$ is a word in $\{0,1\}^*$, we can interpret it as the binary representation of a natural number. We call this number the *index* of the o-TM T^* (and of its o-TP $\tilde{\delta}$).

Clearly, some natural numbers have binary representations that are not of the form (∗∗). Such numbers are not indexes of any o-TM T^*. This is a weakness of the above coding, because it prevents us from establishing a *surjective* mapping from the set of all o-TMs on the set $\mathbb{N}$ of natural numbers. However, as in the ordinary case, we can easily patch this by introducing a special o-TM, called the *empty o-TM*, and making the following **convention**: any natural number whose binary representation is not of the form (∗∗) is an index of the empty o-TM. (The o-TP $\tilde{\delta}$ of the empty o-TM is everywhere undefined, so it immediately halts on any input word.)

We are now able to state the following proposition.

Proposition 10.1. *Every natural number is the index of exactly one o-TM.*

Given an arbitrary index $\langle T^* \rangle$ we can now restore from it the components Q, Σ, Γ, F of the corresponding o-TM $T^* = (Q, \Sigma, \Gamma, \tilde{\delta}, q_1, \sqcup, F, *)$. In other words, we have implicitly defined a *total* function $g : \mathbb{N} \to \mathcal{T}^*$ from the set $\mathbb{N}$ of natural numbers to the set $\mathcal{T}^*$ of all o-TMs. Given an arbitrary $n \in \mathbb{N}$, we view $g(n)$ as the nth o-TM and therefore denote this o-TM by T_n^*. By letting $n = 0, 1, 2, \ldots$ we can *generate* the sequence

$$T_0^*, T_1^*, T_2^*, \ldots$$

of all o-TMs. Thus, we have *enumerated* oracle Turing machines. Clearly, the function g enumerates oracle Turing programs too, so we also obtain the sequence

$$\tilde{\delta}_0, \tilde{\delta}_1, \tilde{\delta}_2, \ldots$$

Remarks. (1) Given an arbitrary index n, we can reconstruct from n both the ordinary Turing machine T_n and the oracle Turing machine T_n^*. Of course, the two machines are different objects. So are different the corresponding sequences $T_1, T_2, T_3, \ldots$ and $T_1^*, T_2^*, T_3^*, \ldots$. The same holds for the sequences $\delta_0, \delta_1, \delta_2, \ldots$ and $\tilde{\delta}_0, \tilde{\delta}_1, \tilde{\delta}_2, \ldots$. (2) Plugging different oracle sets into an o-TM T_n^* affects neither the o-TM nor its index. For this reason we will relax our pedantry and from now on also say that a natural number n is the *index of the* $\mathcal{O}$-TM $T_n^{\mathcal{O}}$. Conversely, given an $\mathcal{O}$-TM $T^{\mathcal{O}}$, we will denote its index by $\langle T^{\mathcal{O}} \rangle$, while keeping in mind that the index is independent of $\mathcal{O}$. Furthermore, we will say that the sequence $T_1^{\mathcal{O}}, T_2^{\mathcal{O}}, T_3^{\mathcal{O}}, \ldots$ is the *enumeration of* $\mathcal{O}$-TMs, and remember that the ordering of the elements of the sequence is independent of the particular $\mathcal{O}$.

10.2 Computation with Oracles

In this section we will continue with setting the stage for the theory of oracular computation. Specifically, we will define the basic notions about the *computability* of functions and the *decidability* of sets by oracle Turing machines. Since we will build on Definition 10.1, we will define the new notions using the universe Σ^*. Then we will switch, w.l.g., to the equivalent universe $\mathbb{N}$. This will enable us to develop and present the theory in a simpler, unified, and more standard way.

10.2.1 Generalization of Classical Definitions

The new definitions will be straightforward generalizations of definitions of the proper function, computable function, decidable set, and index set for the ordinary TM. We will therefore move at a somewhat faster pace.

Proper Functionals

First, we generalize Definition 6.4 (p. 125) of the proper function. Let T_n^* be an arbitrary o-TM and k a natural number. Then we can define a mapping $\Psi_n^{(k+1)}$ with $k+1$ arguments in the following way:

> Given an arbitrary set $\mathcal{O} \subseteq \Sigma^*$ and arbitrary words $u_1,\ldots,u_k \in \Sigma^*$, write the words to the input tape of the $\mathcal{O}$-TM $T_n^{\mathcal{O}}$ and start it. *If* the machine halts leaving on its tape a single word $v \in \Sigma^*$, *then* let $\Psi_n^{(k+1)}(\mathcal{O},u_1,\ldots,u_k) \overset{\text{def}}{=} v$; otherwise, let $\Psi_n^{(k+1)}(\mathcal{O},u_1,\ldots,u_k) \overset{\text{def}}{=} \uparrow$.

In the above definition, $\mathcal{O}$ is variable and can be instantiated to any subset of Σ^*. Identifying subsets of Σ^* with their characteristic functions we see that $\chi_{\mathcal{O}}$, the characteristic function of the set $\mathcal{O}$, can be any member of $\{0,1\}^{\Sigma^*}$, the set of all the functions from Σ^* to $\{0,1\}$. Consequently, $\Psi_n^{(k+1)}$ can be viewed as a partial function from the set $\{0,1\}^{\Sigma^*} \times (\Sigma^*)^k$ into the set Σ^*. Informally, $\Psi_n^{(k+1)}$ maps a function and k words into a word. Now, the functions that map other functions we usually call *functionals*. So, $\Psi_n^{(k+1)}$ is a functional. Since it is associated with a particular o-TM T_n^*, we will call $\Psi_n^{(k+1)}$ the $k+1$-ary *proper functional* of the o-TM T_n^*.

If we fix $\mathcal{O}$ to a particular subset of Σ^*, the argument $\mathcal{O}$ in $\Psi_n^{(k+1)}(\mathcal{O},x_1,\ldots,x_k)$ turns into a parameter, so the functional $\Psi_n^{(k+1)}(\mathcal{O},x_1,\ldots,x_k)$ becomes dependent of k arguments $x_1,\ldots,x_k$ only, and hence becomes a function from $(\Sigma^*)^k$ to Σ^*. We will denote this function(al) by $\Psi_n^{\mathcal{O},(k)}(x_1,\ldots,x_k)$, in short $\Psi_n^{\mathcal{O},(k)}$, and call it the *proper functional* of the $\mathcal{O}$-TM $T_n^{\mathcal{O}}$.

In short, each o-TM is associated with a proper functional $\Psi^{(k+1)}$, and each $\mathcal{O}$-TM is associated with a proper functional $\Psi^{\mathcal{O},(k)}$, for arbitrary natural k.

Remarks. (1) It is a common convention to use capital Greek letters in order to distinguish proper functionals of *o*-TMs from proper functions ψ of ordinary TMs. The distinction is needed because the Ψs are computed by the $\tilde{\delta}$s, while the ψs are computed by ordinary δs. (2) When the number k is understood, we will omit the corresponding superscripts in $\Psi_n^{(k+1)}$ and $\Psi_n^{\mathcal{O},(k)}$; for example, we will simply say "Ψ_n is a proper functional of T_n^*," or "$\Psi_n^{\mathcal{O}}$ is a proper functional of $T_n^{\mathcal{O}}$."

Corresponding to the enumeration of *o*-TMs $T_1^*, T_2^*, T_3^*, \ldots$ is, for each natural k, the enumeration of proper functionals $\Psi_1^{(k+1)}, \Psi_2^{(k+1)}, \Psi_3^{(k+1)}, \ldots$. Similarly, corresponding to the enumeration of $\mathcal{O}$-TMs $T_1^{\mathcal{O}}, T_2^{\mathcal{O}}, T_3^{\mathcal{O}}, \ldots$ is, for each natural k, the enumeration of proper functionals $\Psi_1^{\mathcal{O},(k)}, \Psi_2^{\mathcal{O},(k)}, \Psi_3^{\mathcal{O},(k)}, \ldots$.

Function Computation

Next, we generalize Definition 6.5 (p. 126) of the computable function.

Definition 10.2. Let $\mathcal{O} \subseteq \Sigma^*$, $k \geqslant 1$, and $\varphi : (\Sigma^*)^k \to \Sigma^*$ a function. We say:[2]

φ is $\mathcal{O}$-**computable** *if* there is an $\mathcal{O}$-TM that can compute φ anywhere on dom(φ) *and* dom(φ) = $(\Sigma^*)^k$;

φ is **partial $\mathcal{O}$-computable** (or $\mathcal{O}$-**p.c.**) *if* there is an $\mathcal{O}$-TM that can compute φ anywhere on dom(φ);

φ is $\mathcal{O}$-**incomputable** *if* there is *no* $\mathcal{O}$-TM that can compute φ anywhere on dom(φ).

So, a function $\varphi : (\Sigma^*)^k \to \Sigma^*$ is $\mathcal{O}$-p.c. if it is the k-ary proper functional of some $\mathcal{O}$-TM; that is, if there is a natural n such that $\varphi \simeq \Psi_n^{\mathcal{O},(k)}$.

Let $\mathcal{S} \subseteq (\Sigma^*)^k$ be an arbitrary set. In accordance with Definition 6.6 (p. 126), we will say that

- φ is $\mathcal{O}$-computable *on* $\mathcal{S}$ if φ can be computed by an $\mathcal{O}$-TM for any $x \in \mathcal{S}$;
- φ is $\mathcal{O}$-p.c. *on* $\mathcal{S}$ if φ can be computed by an $\mathcal{O}$-TM for any $x \in \mathcal{S}$ such that $\varphi(x)\downarrow$;
- φ is $\mathcal{O}$-incomputable *on* $\mathcal{S}$ if there is no $\mathcal{O}$-TM capable of computing φ for any $x \in \mathcal{S}$ such that $\varphi(x)\downarrow$.

---

[2] Alternatively, we can say that φ is computable (p.c., incomputable) *relative to* the set $\mathcal{O}$.

Set Recognition

Finally, we can now generalize Definition 6.10 (p. 132) of the decidable set.

Definition 10.3. Let $\mathcal{O} \subseteq \Sigma^*$ be an oracle set. For an arbitrary set $\mathcal{S} \subseteq \Sigma^*$ we say:[3]

$\mathcal{S}$ is $\mathcal{O}$-**decidable** (or $\mathcal{O}$-**computable**) in Σ^* *if* $\chi_{\mathcal{S}}$ is $\mathcal{O}$-computable on Σ^*;

$\mathcal{S}$ is $\mathcal{O}$-**semi-decidable** (or $\mathcal{O}$-**c.e.**) in Σ^* *if* $\chi_{\mathcal{S}}$ is $\mathcal{O}$-computable on $\mathcal{S}$.

$\mathcal{S}$ is $\mathcal{O}$-**undecidable** (or $\mathcal{O}$-**incomputable**) in Σ^* *if* $\chi_{\mathcal{S}}$ is $\mathcal{O}$-incomputable on Σ^*.

(Remember that $\chi_{\mathcal{S}} : \Sigma^* \to \{0,1\}$ is total.)

Index Sets

We have seen that each natural number is the index of exactly one o-TM (p. 227). What about the other way round? Is each o-TM represented by exactly one index? The answer is no; an o-TM has countably infinitely many indexes. We prove this in the same fashion as we proved the *Padding Lemma* (see Sect. 7.2): other indexes of a given o-TM are constructed by padding its code with codes of redundant instructions, and permuting the codes of instructions. So, if e is the index of T^* and x is constructed from e in the described way, then $\Psi_x^{(k+1)} \simeq \Psi_e^{(k+1)}$. This, combined with Definition 10.2, gives us the following generalization of the *Padding Lemma*.

Lemma 10.1. (Generalized Padding Lemma) *An $\mathcal{O}$-p.c. function has countably infinitely many indexes. Given one of them, the others can be generated.*

Similar to the ordinary case, the *index set of an $\mathcal{O}$-p.c. function* φ contains all the indexes of all $\mathcal{O}$-TMs that compute φ.

Definition 10.4. (Index Set of $\mathcal{O}$-p.c. Function) The **index set** of an $\mathcal{O}$-p.c. function φ is the set $\mathrm{ind}^{\mathcal{O}}(\varphi) \stackrel{\text{def}}{=} \{x \in \mathbb{N} \mid \Psi_x^{\mathcal{O}} \simeq \varphi\}$.

Let $\mathcal{S}$ be an arbitrary $\mathcal{O}$-c.e. set, and $\chi_{\mathcal{S}}$ its characteristic function. The index set $\mathrm{ind}^{\mathcal{O}}(\chi_{\mathcal{S}})$ we will also denote by $\mathrm{ind}^{\mathcal{O}}(\mathcal{S})$ and call the *index set of the $\mathcal{O}$-c.e. set $\mathcal{S}$*.

---

[3] We can also say that $\mathcal{S}$ is decidable (semi-decidable, undecidable) *relative to* the set $\mathcal{O}$.

10.2.2 Convention: The Universe ℕ and Single-Argument Functions

So far we have been using Σ^* as the universe. But we have shown in Sect. 6.3.6 that there is a bijection from Σ^* onto ℕ, which allows us to arbitrarily choose whether to study the decidability of *subsets of Σ^** or of *subsets of* ℕ—the findings apply to the alternative too. It is now the right time to make use of this and switch to the universe ℕ. The reasons for this are that, first, ℕ is often used in the study of relative computability, and second, the presentation will be simpler.

NB *From now on* ℕ *will be the universe.*

Some definitions will tacitly (and trivially) adapt to the universe ℕ. For instance:

- an oracle set $\mathcal{O}$ is an arbitrary subset of ℕ (Definition 10.1);
- a function $\varphi : \mathbb{N}^k \to \mathbb{N}$ is $\mathcal{O}$-computable if there is an $\mathcal{O}$-TM that can compute φ anywhere on $\mathrm{dom}(\varphi) = \mathbb{N}$ (Definition 10.2);
- a set $\mathcal{S} \subseteq \mathbb{N}$ is $\mathcal{O}$-decidable in ℕ if $\chi_{\mathcal{S}}$ is $\mathcal{O}$-computable on ℕ (Definition 10.3).

The next two adapted definitions come into play when the impact of different oracle sets on oracular computability is studied:

- a proper functional of the o-TM T_n^* is $\Psi_n^{(k+1)} : \{0,1\}^{\mathbb{N}} \times \mathbb{N}^k \to \mathbb{N}$ (Sect. 10.2.1).
- a proper functional of the $\mathcal{O}$-TM $T_n^{\mathcal{O}}$ is $\Psi_n^{\mathcal{O},(k)} : \mathbb{N}^k \to \mathbb{N}$ (Sect. 10.2.1).

Next, w.l.g., we will simplify our discussion by focusing on the case $k = 1$.

NB *From now on we will focus on single-argument functions (the case $k = 1$).*

10.3 Other Ways to Make External Help Available

External help can be made available to other models of computation too.

Kleene introduced external help to partial recursive (p.r.) functions. The idea is simple: given an arbitrary set $\mathcal{O} \subseteq \mathbb{N}$, add the characteristic function $\chi_{\mathcal{O}}$ to the set $\{\zeta, \sigma, \pi_i^k\}$ of initial functions (see Sect. 5.2.1). Any function that can be constructed from $\{\zeta, \sigma, \pi_i^k, \chi_{\mathcal{O}}\}$ by finitely many applications of the rules of construction (i.e., composition, primitive recursion, and μ-operation) uses external help *if* $\chi_{\mathcal{O}}$ appears in the function's construction. Kleene called such a function *p.r. relative to $\mathcal{O}$*.

Post introduced external help into his canonical systems by hypothetically adding primitive assertions expressing the (non)membership in $\mathcal{O}$.

Davis proved the equivalence of Kleene's and Post's approaches. The equivalences of each of the two and Turing's approach were proved by Kleene and Post, respectively. Consequently, Turing's, Kleene's, and Post's definitions of functions computable with external help are equivalent, in the sense that if a function is computable according to one definition it is also computable according to the other.

10.4 Relative Computability Thesis

Based on these equivalences and following the example of the *Computability Thesis*, a thesis was proposed stating that the intuitive notion of the "algorithm with external help" is formalized by the concept of the oracle Turing machine.

Relative Computability Thesis.

"algorithm with external help" $\longleftrightarrow$ *oracle Turing program (or equivalent model)*

Also, the intuitive notion of being "computable with external help" is formalized by the notion of being $\mathcal{O}$-computable.

10.5 Practical Consequences: o-TM with a Database or Network

Since oracles are supernatural entities, the following question arises: Can we replace a given oracle for a set $\mathcal{O}$ with a more realistic concept that will simulate the oracle? Clearly, if $\mathcal{O}$ is a decidable set, the replacement is trivial—we only have to replace the oracle by the ordinary TM, which is the decider of the set $\mathcal{O}$ (or, alternatively, the computer of the function $\chi_{\mathcal{O}}$).

A different situation occurs when $\mathcal{O}$ is undecidable (and, hence, $\chi_{\mathcal{O}}$ incomputable). Such an $\mathcal{O}$ cannot be finite, because finite sets are decidable. But in practice, we do not really need the *whole* set $\mathcal{O}$, i.e., the values $\chi_{\mathcal{O}}(x)$ for *every* $x \in \mathbb{N}$. This is because the o-TM may only issue the question "$x \in ?\mathcal{O}$" for finitely many different numbers $x \in \mathbb{N}$. (Otherwise, the machine certainly would not halt.) Consequently, there is an $m \in \mathbb{N}$, which is the largest of all xs for which the question "$x \in ?\mathcal{O}$" is issued during the computation. Of course, m depends on the input word that was submitted to the o-TM.

Now we can make use of the ideas discussed in Sect. 8.4. We could compute the values $\chi_{\mathcal{O}}(i)$, $i = 0, 1, \ldots, m$ in advance, where each $\chi_{\mathcal{O}}(i)$ would be computed by an algorithm A_i specially designed to answer only the particular question "$i \in ?\mathcal{O}$." The computed values would then be stored in an external *database* and accessed during the oracular computation. Alternatively, we might compute the values $\chi_{\mathcal{O}}(i)$ on the fly during the oracular computation: upon issuing a question "$i \in ?\mathcal{O}$," an external *network of computers* would be engaged in the computation of the answer to the question. Of course, the o-TM would idle until the answer arrived.

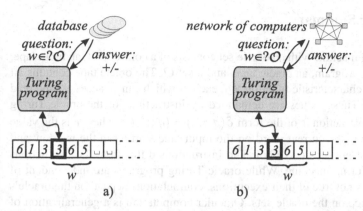

Fig. 10.3 The oracle tape is replaced by a) a database with a finite number of precomputed values $\chi_O(i), i = 0, 1, \ldots, m$; b) a network of computers that compute each value $\chi_O(i)$ separately upon the o-TM's request

10.6 Practical Consequences: Online and Offline Computation

In the real world of computer technology and telecommunications, often a device A is associated with some other device B. This association can be dynamic, changing between the states of connectedness and disconnectedness. We say that A is *online* when it is connected to B; that is, either A is under the direct control of B, or B is available for immediate use on A's demand. Otherwise, we say that A is *offline* (disconnected).

Let A be a computer and B any device capable of providing data. We say that A executes an *online algorithm* if the algorithm processes its input data in the order in which the data are fed to the algorithm by B. Thus, an online algorithm cannot know the entire input that will be fed to it, so it may be forced to correct its past actions when new input data arrive. In contrast, we say that A executes an *offline algorithm* if the algorithm is given the whole input data at the beginning of the computation. This allows it to first inspect the whole input and then choose the appropriate computation strategy.

We can now see the parallels between online/offline computing and oracular/ordinary computing. Ordinary Turing machines are offline; they are disconnected from any oracle, or any other device, such as external database or computer network. Ordinary Turing programs are offline algorithms, because they are given the entire input before they are started. In contrast, oracle Turing machines are online; they are connected to an oracle or a device such as an external database or a computer network. Oracle Turing programs are online algorithms: although a part of their input is written on the input tape at the beginning of the computation, the other part—the oracle's advices—is processed piece by piece in a serial fashion, without knowing the advices that will be given to future questions.

10.7 Chapter Summary

An oracle Turing machine with an oracle set consists of a control unit, an input tape, an oracle Turing program, an oracle tape, and a set $\mathcal{O}$. The oracle tape contains all the values of the characteristic function $\chi_{\mathcal{O}}$, each of which can be accessed and read in a single step. These values are demanded by instructions of the oracle Turing program. Each instruction is of the form $\widetilde{\delta}(q, z, e) = (q', z', D)$, where e is the value of $\chi_{\mathcal{O}}(w)$ and w is the current word on the input tape starting in the window and ending at the last non-space symbol. The instructions differ from the instructions of an ordinary Turing machine. While oracle Turing programs are independent of oracle sets, this is not true of their executions: computations depend on the oracle's answers and hence on the oracle sets. Oracular computation is a generalization of ordinary computation. Oracle TMs and their programs can be coded and enumerated. With each oracle TM is associated its proper functional. Given an oracle set $\mathcal{O}$, a partial function can be $\mathcal{O}$-computable, $\mathcal{O}$-p.c., or $\mathcal{O}$-incomputable; and a set can be $\mathcal{O}$-decidable, $\mathcal{O}$-c.e., or $\mathcal{O}$-undecidable. The *Relative Computability Thesis* states that the oracle Turing machine formalizes the intuitive notion of the "algorithm with external help." In practice, the contents of the unrealistic oracle tape could be approximated 1) by a finite sequence of precomputed values of $\chi_{\mathcal{O}}$ (stored in an external database) or 2) by an on-the-fly calculation of each demanded value of $\chi_{\mathcal{O}}$ separately (computed by a network of computers).

Bibliographic Notes

- The idea of an *oracle* for a possibly incomputable set $\mathcal{O}$ that is attached to a Turing machine was introduced in Turing [180, §4]. The citation on p. 220 is from there.
- Post developed Turing's idea above in the influential paper [124, §11]. See also Post [127].
- *External help* was also added to other models of computation. Kleene added external help to partial recursive functions in [81] and Post to his canonical systems in [127].
- The *equivalence* of Kleene's and Post's approaches to external help was proved in Davis [33]. A proof of the equivalence of Turing's and Kleene's approaches was published in Kleene's book [82, Sects. 68,69]. Later, other models and proofs of their equivalence appeared. For example, the register machine (with external help) is described and proved to be equivalent to partial computable functions (with external help) in Enderton [42].
- The *Relative Computability Thesis* stating that Turing's approach is the formalization of the intuitive notion of "computation with external help" appeared (in terms of Turing and "effective" reducibility) in Post [124, §11].
- The basic notions and concepts of relativized computability obtained their mature form in Kleene and Post [84]. For other important results that appeared in this paper, see the Bibliographic Notes to the following chapters.
- An account of the development of oracle computing is given in Soare [167]. The reader can find out more about Post's research and life in Urquhart [181].
- In defining the modern *oracle Turing machine*, we have leaned on Soare [164, 168].
- The idea of substituting the oracle tape with a *database* or some other kind of *external interaction* was described in Soare [167, 168].

Chapter 11
Degrees of Unsolvability

Degree indicates the extent to which something happens or the amount something is felt.

Abstract In Part II, we proved that besides computable problems there are also incomputable ones. So, given a computational problem, it makes sense to talk about its *degree of unsolvability*. Of course, at this point we only know of *two* such degrees: one is shared by all computable problems, and the other is shared by all incomputable ones. (This will change, however, in the next chapter.) Nevertheless, the main aim of this chapter is to formalize the intuitive notion of the degree of unsolvability. Building on the concept of the oracle Turing machine, we will first define the concept of the *Turing reduction*, the most general reduction between computational problems. We will then proceed in a natural way to the definition of *Turing degree*—the formal counterpart of the intuitive notion of the degree of unsolvability.

11.1 Turing Reduction

Until now we have been using the generic symbol $\mathcal{O}$ to denote an oracle set. From now on we will be talking about particular oracle sets and denote them by the usual symbols, e.g., $\mathcal{A}, \mathcal{B}$.

What does it mean when we say, for two sets $\mathcal{A}, \mathcal{B} \subseteq \mathbb{N}$, that "$\mathcal{A}$ is $\mathcal{B}$-decidable in $\mathbb{N}$"? By Definition 10.3 (p. 230), it means that the characteristic function $\chi_\mathcal{A}$ is $\mathcal{B}$-computable on $\mathbb{N}$. So, by Definition 10.2 (p. 229), there is a $\mathcal{B}$-TM $T^\mathcal{B}$ that can compute $\chi_\mathcal{A}(x)$ for any $x \in \mathbb{N}$. (Recall that $\chi_\mathcal{A}$ is a total function by definition.) Since the oracle can answer the question $n \in ?\mathcal{B}$ for any $n \in \mathbb{N}$, it makes the set $\mathcal{B}$ *appear decidable* in $\mathbb{N}$ in the ordinary sense (see Definition 6.10 on p. 132).

We conclude:

If $\mathcal{A}$ is $\mathcal{B}$-decidable, then the decidability of $\mathcal{B}$ would imply the decidability of $\mathcal{A}$.

This relation between sets deserves a special name. Hence, the following definition.

© Springer-Verlag Berlin Heidelberg 2015
B. Robič, *The Foundations of Computability Theory*,
DOI 10.1007/978-3-662-44808-3_11

Definition 11.1. (Turing Reduction) Let $\mathcal{A}, \mathcal{B} \subseteq \mathbb{N}$ be arbitrary sets. We say that $\mathcal{A}$ is **Turing reducible** (in short *T-reducible*) *to* $\mathcal{B}$, if $\mathcal{A}$ is $\mathcal{B}$-decidable. We denote this by

$$\mathcal{A} \leq_T \mathcal{B}, \qquad\qquad\qquad (*)$$

which reads as follows: *If $\mathcal{B}$ is decidable, then also $\mathcal{A}$ is decidable.* The relation $\leq_T$ is called the **Turing reduction** (in short *T-reduction*).

If $\mathcal{B}$ is decidable (in the ordinary sense), then the oracle for $\mathcal{B}$ is no mystery. Such an oracle can be replaced by an ordinary decider $D_{\mathcal{B}}$ of the set $\mathcal{B}$ (see Sect. 6.3.3). Consequently, the $\mathcal{B}$-TM $T^{\mathcal{B}}$ is equivalent to an ordinary TM, in the sense that what can compute one can also compute the other. The construction of this TM is straightforward: the TM must simulate the program $\widetilde{\delta}$ of $T^{\mathcal{B}}$, with the exception that whenever $T^{\mathcal{B}}$ asks $x \in ?\mathcal{B}$, the TM must call $D_{\mathcal{B}}$, submit x to it, and wait for its decision.

The situation is quite different when $\mathcal{B}$ is undecidable (in the ordinary sense). In this case, the oracle for $\mathcal{B}$ is *more powerful* than any ordinary TM (because no ordinary TM can decide $\mathcal{B}$). This makes $T^{\mathcal{B}}$ more powerful than any ordinary TM (because $T^{\mathcal{B}}$ can decide $\mathcal{B}$, simply by asking whether or not the input is in $\mathcal{B}$). In particular, if we replace a decidable oracle set with an undecidable c.e. (semi-decidable) set, this may have "big" implications (see Problem 11.3).

11.1.1 Turing Reduction of a Computational Problem

Why have we named $\leq_T$ a *reduction*? What is reduced here, and what does the reducing? Recall that each subset of $\mathbb{N}$ represents a decision problem (see Sect. 8.1.2). Thus, $\mathcal{A}$ is associated with the decision problem $\mathcal{P} \equiv$ "$x \in ?\mathcal{A}$," and $\mathcal{B}$ with the decision problem $\mathcal{Q} \equiv$ "$x \in ?\mathcal{B}$." The relation $(*)$ can now be interpreted as follows:

If $\mathcal{Q} \equiv$ "$x \in ?\mathcal{B}$" were decidable, then also $\mathcal{P} \equiv$ "$x \in ?\mathcal{A}$" would be decidable.

Because of this we also use the sign $\leq_T$ to relate the associated decision problems:

$$\mathcal{P} \leq_T \mathcal{Q} \qquad\qquad\qquad (**)$$

As regards the oracle for $\mathcal{B}$, we know *what* the oracle does: since it can answer any question $w \in ?\mathcal{B}$, it *solves the problem* $\mathcal{Q}$. We can view it as a procedure—call it B—for solving the problem $\mathcal{Q}$. But when $\mathcal{B}$ is undecidable, we cannot know *how* the oracle finds the answers. In this case, B is a supernatural "algorithm," whose operation cannot be described and understood by a human, a black box, as it were.

Let us now look at $T^{\mathcal{B}}$, the machine that computes the values of $\chi_{\mathcal{A}}$, and let $\widetilde{\delta}$ be its *o*-TP. For $\widetilde{\delta}$ we know *what* it does and *how* it does it, *when* it asks the oracle and *how* it uses its answers. Hence, we can view $\widetilde{\delta}$ as an ordinary algorithm A for solving $\mathcal{P}$, which can call the mysterious "algorithm" B. There is no limit on the number of calls as long as this number is finite (otherwise, A would not halt).

Considering all this, we can interpret the relation (∗∗) as follows:

If there were an algorithm B for solving Q,
then there would be an algorithm A for solving P,
where A could make finitely many calls to B.

Since P would in principle be solved *if* Q were solved, we can focus on the problem Q and on the design of the algorithm B. We say that we have *reduced* P *to* Q. If and when Q is solved (i.e., B designed), also the problem P is, in principle, solved (by A). Note that this means that the problem P is *not more difficult to solve* than the problem Q. This is an alternative interpretation of the relation (∗∗).

Observe that the situation described in the above interpretation is not bound to decision problems only; it may as well occur between computational problems of other kinds. Take, for instance, $P \equiv$ SHORTEST EQUIVALENT PROGRAM (Sect. 9.5) and $Q \equiv$ EXISTENCE OF SHORTER EQUIVALENT PROGRAMS (Sect. 8.3.3). In this case, there is an A that calls B finitely often to find the shortest equivalent program.

NB *Nevertheless, from now on we will limit our discussion to decision problems. In doing so, we will develop the theory on sets of natural numbers. (Decision problems will be mentioned only to give another view of notions, concepts, and theorems.)*

11.1.2 Some Basic Properties of the Turing Reduction

We now list some of the properties of the T-reduction. First, we check that there indeed exist two sets $A, B \subseteq \mathbb{N}$ that are related by $\leq_T$. To see this, consider the situation where A is an arbitrary decidable set. From Definition 11.1 it immediately follows that $A \leq_T B$ for *arbitrary* set B. Hence the following theorem.

Theorem 11.1. *Let A be a decidable set. Then $A \leq_T B$ for arbitrary set B.*

But a set A need not be decidable to be T-reducible to some other set B. This will follow from the next simple theorem.

Theorem 11.2. *For every set S it holds that $\overline{S} \leq_T S$.*

Proof. If S were decidable, then (by Theorem 7.1, p. 144) also $\overline{S} = \mathbb{N} - S$ would be decidable. □

Let S be undecidable in the ordinary sense. Then $\overline{S}$ is undecidable too (Sect. 8.2.2). Now Theorem 11.2 tells us that $\overline{S} \leq_T S$, i.e., making S appear decidable (by using the oracle for S), also makes $\overline{S}$ appear decidable. Thus, a set A *need not* be decidable to be T-reducible to some other set B.

Is the T-reduction related to the m-reduction, which we defined in Sect. 9.2.2? If so, is one of them more "powerful" than the other? The answer is yes. The next two theorems tell us that the T-reduction is a nontrivial generalization of the m-reduction.

Theorem 11.3. *If two sets are related by $\leq_m$, then they are also related by $\leq_T$.*

Proof. Let $\mathcal{A} \leq_m \mathcal{B}$. Then there is a computable function r such that $x \in \mathcal{A} \Longleftrightarrow r(x) \in \mathcal{B}$ (see Sects. 9.2.1 and 9.2.2). Consequently, the characteristic functions $\chi_\mathcal{A}$ and $\chi_\mathcal{B}$ are related by the equation $\chi_\mathcal{A} = \chi_\mathcal{B} \circ r$, where $\circ$ denotes the function composition. We now see: if the function $\chi_\mathcal{B}$ were computable, then the composition $\chi_\mathcal{B} \circ r$ would also be computable (because r is computable, and a composition of computable functions is a computable function)—hence, $\mathcal{A} \leq_T \mathcal{B}$.   □

However, the converse is not true, as we explain in the next theorem.

Theorem 11.4. *If two sets are related by $\leq_T$, they may not be related by $\leq_m$.*

Proof. Theorem 11.2 states that $\overline{\mathcal{S}} \leq_T \mathcal{S}$ for any set $\mathcal{S}$. Is the same true of the relation $\leq_m$? The answer is no; there exist sets $\mathcal{S}$ such that $\overline{\mathcal{S}} \leq_m \mathcal{S}$ does *not* hold. To see this, let $\mathcal{S}$ be an arbitrary undecidable c.e. set. (Such is the set $\mathcal{K}$; see Sect. 8.2.1.) The set $\overline{\mathcal{S}}$ is not c.e. (otherwise $\mathcal{S}$ would be decidable by Theorem 7.3, p. 144). Now, if $\overline{\mathcal{S}} \leq_m \mathcal{S}$ held, then $\overline{\mathcal{S}}$ would be c.e. (by Theorem 9.1, p. 197), which would be a contradiction.

In particular, $\overline{\mathcal{K}} \leq_T \mathcal{K}$, but $\overline{\mathcal{K}} \nleq_m \mathcal{K}$. The same is true of $\mathcal{K}_0$: $\overline{\mathcal{K}_0} \leq_T \mathcal{K}_0$, but $\overline{\mathcal{K}_0} \nleq_m \mathcal{K}_0$.   □

We can use the T-reduction for proving the undecidability of sets as we use the m-reduction. But the T-reduction must satisfy fewer conditions than the m-reduction (see Definition 9.1, p. 197). This makes T-reductions easier to construct than m-reductions. Indeed, Theorem 11.4 indicates that there are situations where T-reduction is possible while m-reduction is not. So, let us develop a method of proving the undecidability of sets that will use T-reductions. We will closely follow the development of the method for m-reductions (see Sect. 9.2.3). First, from Definition 11.1 we obtain the following theorem.

Theorem 11.5. *For arbitrary sets $\mathcal{A}$ and $\mathcal{B}$ it holds:*

$$\mathcal{A} \leq_T \mathcal{B} \wedge \mathcal{B} \text{ is decidable} \Longrightarrow \mathcal{A} \text{ is decidable}$$

The contraposition is: $\mathcal{A}$ *is undecidable* $\Longrightarrow \mathcal{A} \nleq_T \mathcal{B} \vee \mathcal{B}$ *is undecidable*. Assuming that $\mathcal{A} \leq_T \mathcal{B}$, and using this in the contraposition, we obtain the next corollary.

Corollary 11.1. *For arbitrary sets $\mathcal{A}$ and $\mathcal{B}$ it holds:*

$$\mathcal{A} \text{ is undecidable} \wedge \mathcal{A} \leq_T \mathcal{B} \Longrightarrow \mathcal{B} \text{ is undecidable}$$

This reveals the following method for proving the undecidability of sets.

Method. The undecidability of a set $\mathcal{B}$ can be proved as follows:

1. Suppose: $\mathcal{B}$ is decidable;   // Supposition.
2. Select: an undecidable set $\mathcal{A}$;
3. Prove: $\mathcal{A} \leq_T \mathcal{B}$;
4. Conclude: $\mathcal{A}$ is decidable;   // 1 and 3 and Theorem 11.5.
5. Contradiction between 2 and 4!
6. Conclude: $\mathcal{B}$ is undecidable.

Remark. The method can easily be adapted to prove that a decision problem $\mathcal{Q}$ is undecidable. First, *suppose* that there is a decider B for $\mathcal{Q}$. Then choose a known undecidable decision problem $\mathcal{P}$ and try to construct a decider A for $\mathcal{P}$, where A can make calls to B. If we succeed, $\mathcal{P}$ is decidable. Since this is a contradiction, we drop the supposition. So, $\mathcal{Q}$ is undecidable.

Turing reduction is a relation that has another two important properties.

Theorem 11.6. *Turing reduction $\leq_T$ is a reflexive and transitive relation.*

Proof. (Reflexivity) This is trivial. Let $\mathcal{S}$ be an arbitrary set. If $\mathcal{S}$ were decidable, then, of course, the same $\mathcal{S}$ would be decidable. Hence, $\mathcal{S} \leq_T \mathcal{S}$ by Definition 11.1. (Transitivity) Let $\mathcal{A}, \mathcal{B}, \mathcal{C}$ be arbitrary sets and suppose that $\mathcal{A} \leq_T \mathcal{B} \wedge \mathcal{B} \leq_T \mathcal{C}$. So, if $\mathcal{C}$ were decidable, $\mathcal{B}$ would also be decidable (because $\mathcal{B} \leq_T \mathcal{C}$), but the latter would then imply the decidability of $\mathcal{A}$ (because $\mathcal{A} \leq_T \mathcal{B}$). Hence, the decidability of $\mathcal{C}$ would imply the decidability of $\mathcal{A}$, i.e., $\mathcal{A} \leq_T \mathcal{C}$. □

Generally, a reflexive and transitive binary relation is called a *preorder*, and a set equipped with such a relation is said to be preordered by this relation.

Given a preordered set, one of the first things to do is to check whether its preorder qualifies for any of the more interesting orders (see Appendix A). Because these orders have additional properties, they reveal much more about their domains. Two such orders are the equivalence relation and the partial order.

The above theorem tells us that $\leq_T$ is a preorder on $2^{\mathbb{N}}$. Is it, perhaps, even an equivalence relation? To check this we must see whether $\leq_T$ is a symmetric relation, i.e., whether $\mathcal{A} \leq_T \mathcal{B}$ implies $\mathcal{B} \leq_T \mathcal{A}$, for arbitrary $\mathcal{A}, \mathcal{B}$. But we are already able to point at two sets for which the implication *does not* hold: these are the empty set $\emptyset$ and the diagonal set $\mathcal{K}$ (see Definition 8.6, p. 167). Namely, $\emptyset \leq_T \mathcal{K}$ (due to Theorem 11.1) while $\mathcal{K} \not\leq_T \emptyset$ (because $\emptyset$ is decidable and $\mathcal{K}$ undecidable). Thus we conclude:

Turing reduction is not symmetric and, consequently, not an equivalence relation.

Although this result is negative, we will use it in the next subsection.

11.2 Turing Degrees

We are now ready to formalize the intuitive notion of *degree of unsolvability*. Its formal counterpart will be called *Turing degree*. The path to the definition of Turing degree will be a short one: first, we will use the relation $\leq_T$ to define a new relation $\equiv_T$; then we will prove that $\equiv_T$ is an equivalence relation; and finally, we will define Turing degrees to be the equivalence classes of $\equiv_T$.

We have proved in the previous subsection that $\leq_T$ is not symmetric, because there exist sets A and B such that $A \leq_T B$ and $B \not\leq_T A$. However, neither is $\leq_T$ asymmetric, because there do exist sets A and B for which $A \leq_T B$ and $B \leq_T A$. For example, this is the case when both A and B are decidable. (This follows from Theorem 11.1.) Thus, for some pairs of sets the relation $\leq_T$ is symmetric, while for others it is not. In this situation, we can define a new binary relation that will tell, for any two sets, whether or not $\leq_T$ is symmetric for them. Here is the definition of the new relation.

Definition 11.2. (Turing Equivalence) Let $A, B \subseteq \mathbb{N}$ be arbitrary sets. We say that A is **Turing-equivalent** (in short *T-equivalent*) to B, if $A \leq_T B \wedge B \leq_T A$. We denote this by

$$A \equiv_T B$$

and read: *If one of A, B were decidable, also the other would be decidable.* The relation $\equiv_T$ is called the **Turing equivalence** (in short, *T-equivalence*).

From the above definition it follows that $A \equiv_T B$ for any *decidable* sets A, B. What about undecidable sets? Can two such sets be *T*-equivalent? The answer is yes; it will follow from the next theorem.

Theorem 11.7. *For every set S it holds that $S \equiv_T \overline{S}$.*

Proof. Let S be an arbitrary set. Then $\overline{S} \leq_T S$ (by Theorem 11.2). Now focus on the set $\overline{S}$. The same theorem tells us that $\overline{\overline{S}} \leq_T \overline{S}$, i.e., $S \leq_T \overline{S}$. So $S \equiv_T \overline{S}$. Note that S can be undecidable. *Alternatively:* If one of the characteristic functions $\chi_S, \chi_{\overline{S}}$ were computable, the other would also be computable (because $\chi_{\overline{S}} = 1 - \chi_S$.) $\square$

Calling $\equiv_T$ "equivalence relation" is justified. The reader should have no trouble in proving that the relation $\equiv_T$ is reflexive, transitive, and symmetric. Therefore, $\equiv_T$ is an equivalence relation on $2^{\mathbb{N}}$, the power set of the set $\mathbb{N}$.

Now, being an equivalence relation, the relation $\equiv_T$ partitions the set $2^{\mathbb{N}}$ into $\equiv_T$-*equivalence classes*. Each $\equiv_T$-equivalence class contains as its elements all the subsets of $\mathbb{N}$ that are *T*-equivalent one to another. It will soon turn out that $\equiv_T$-equivalence classes are one of the central notions of the relativized computability. The next definition introduces their naming.

Definition 11.3. (Turing Degree) A **Turing degree** (in short *T-degree*) of a set $\mathcal{S}$, denoted by $\deg(\mathcal{S})$, is the equivalence class $\{\mathcal{X} \in 2^{\mathbb{N}} \mid \mathcal{X} \equiv_T \mathcal{S}\}$.

Given a set $\mathcal{S}$, the T-degree $\deg(\mathcal{S})$ contains, by definition, as its elements all the sets that are T-equivalent to $\mathcal{S}$. Since $\equiv_T$ is symmetric and transitive, these sets are also T-equivalent to one another. Thus, if one of them were decidable, all would be decidable. In other words, the question $x \in ?\mathcal{X}$ (i.e., the membership problem) is equally (un)decidable for each of the sets $\mathcal{X} \in \deg(\mathcal{S})$. Informally, this means that the information about what is and what is not in one of them is equal to the corresponding information of any other set. In short, the sets $\mathcal{X} \in \deg(\mathcal{S})$ bear the same *information* about their contents.

Remark. We now put $\deg(\mathcal{S})$ in the light of decision problems. Let $\mathcal{X}, \mathcal{Y} \in \deg(\mathcal{S})$ be any sets. Associated with $\mathcal{X}$ and $\mathcal{Y}$ are decision problems $\mathcal{P} \equiv$ "$x \in ?\mathcal{X}$" and $\mathcal{Q} \equiv$ "$y \in ?\mathcal{Y}$," respectively. Since $\mathcal{X} \equiv_T \mathcal{Y}$, we have $\mathcal{P} \leq_T \mathcal{Q}$ and $\mathcal{Q} \leq_T \mathcal{P}$. This means that if any of the problems $\mathcal{P}, \mathcal{Q}$ were decidable, the other one would also be decidable. Thus, $\mathcal{P}$ and $\mathcal{Q}$ are equally (un)decidable. Now we see that the class of all decision problems whose languages are elements of the T-degree $\deg(\mathcal{S})$ represents a certain *degree of unsolvability*, which we are faced with when we try to solve any of these problems. Problems associated with $\deg(\mathcal{S})$ are equally (un)solvable, i.e., equally difficult.

Based on this we declare the following formalization of the intuitive notion of the degree of unsolvability of decision problems.

Formalization. *The intuitive notion of the degree of unsolvability is formalized by*

$$\text{``degree of unsolvability''} \longleftrightarrow \textit{Turing degree}$$

Remark. Since the concept of "degree of unsolvability" is formalized by the T-degree, we will no longer distinguish between the two. We will no longer use quotation marks to distinguish between its intuitive and formal meaning.

NB *This formalization opens the door to a mathematical treatment of our intuitive, vague awareness that solvable and unsolvable problems differ in something that we intuitively called the degree of unsolvability.*

Intuitively, we expect that the degree of unsolvability of decidable sets differs from the degree of unsolvability of undecidable sets. So let us prove that indeed there are two different corresponding T-degrees.

First, let $\mathcal{S}$ be an arbitrary *decidable* set. Then $\deg(\mathcal{S})$ contains exactly all decidable sets. As the empty set $\emptyset$ is decidable, we have $\emptyset \in \deg(\mathcal{S})$, so $\deg(\mathcal{S}) = \deg(\emptyset)$. This is why we usually denote the class of all decidable sets by $\deg(\emptyset)$.

Second, we have seen (p. 239) that $\emptyset \leq_T \mathcal{K} \wedge \mathcal{K} \not\leq_T \emptyset$. Therefore, $\emptyset \not\equiv_T \mathcal{K}$ and hence $\deg(\emptyset) \neq \deg(\mathcal{K})$.

But $deg(\emptyset)$ and $deg(\mathcal{K})$ are $\equiv_T$-equivalence classes, so they share no elements. We can now conclude as we expected to: $deg(\emptyset)$ and $deg(\mathcal{K})$ represent two different degrees of unsolvability. We have proved the following theorem.

Theorem 11.8. *There exist at least two T-degrees:*

$$deg(\emptyset) = \{\mathcal{X} \mid \mathcal{X} \equiv_T \emptyset\},$$
$$deg(\mathcal{K}) = \{\mathcal{X} \mid \mathcal{X} \equiv_T \mathcal{K}\}.$$

The Relation <

It is natural to say that a decidable decision problem is "less difficult to solve" than an undecidable one. We will now formalize the intuitively understood relation of "being less difficult to solve." To do this, we will introduce a new binary relation, denoted by $<$, which will be capable of expressing formally that $deg(\emptyset)$ represents a degree of unsolvability that is "lower" than the degree of unsolvability represented by $deg(\mathcal{K})$. The definition of $<$ will be straightforward. Let us denote the irreflexive reduction of the relation $\leq_T$ as usual by $<_T$, i.e., $A <_T B \overset{\text{def}}{\Longleftrightarrow} A \leq_T B \wedge A \not\equiv_T B$. (Thus $A <_T B \Longleftrightarrow A \leq_T B \wedge B \not\leq_T A$.) Then the sought-for relation $<$ is *induced* by the relation $<_T$, as the following definition describes.

Definition 11.4. (Relation <) Let $deg(\mathcal{A})$ and $deg(\mathcal{B})$ be arbitrary T-degrees. Then $deg(\mathcal{A})$ is **lower** than $deg(\mathcal{B})$, denoted by $deg(\mathcal{A}) < deg(\mathcal{B})$, if $\mathcal{A} <_T \mathcal{B}$.

When $deg(\mathcal{A}) < deg(\mathcal{B})$, we also say that $deg(\mathcal{B})$ is *higher* than $deg(\mathcal{A})$.

Fig. 11.1 Turing degree $deg(\mathcal{A})$ is lower than $deg(\mathcal{B})$, i.e., $deg(\mathcal{A}) < deg(\mathcal{B})$. The relation $<$ between the degrees $deg(\mathcal{A})$ and $deg(\mathcal{B})$ is induced by the relation $<_T$ between the representatives $\mathcal{A}$ and $\mathcal{B}$.

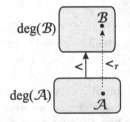

Intuitively, a T-degree should not be lower than itself, because this would not agree with the intended meaning of the relation $<$. So we ask: Is the relation $<$ irreflexive? The answer is yes, as is assured by the next, simple theorem.

Theorem 11.9. *The relation $<$ is irreflexive.*

Proof. Let S be an arbitrary set. Suppose that $\deg(S) < \deg(S)$. Then, $S <_T S$ (by Definition 11.4), i.e., $S \leq_T S \wedge S \not\equiv_T S$ (by definition of $<_T$). Hence $S \not\equiv_T S$. But this is a contradiction, because $\equiv_T$ is reflexive. $\qquad\square$

Since $\emptyset <_T \mathcal{K}$, we can now write the statement

The degree of unsolvability of decidable decision problems is lower than the degree of unsolvability of undecidable decision problems T-equivalent to the Halting Problem

in a formal way as

$$\deg(\emptyset) < \deg(\mathcal{K}).$$

But, we have intuitively anticipated this! What is so special about the above formal statement? Isn't this much ado about nothing? The answer is that the formalization that we have just made will enable us to discover in the next chapter a surprising fact that there are many more other degrees of unsolvability. We will show this by the construction of T-degrees that differ from $\deg(\emptyset)$ and $\deg(\mathcal{K})$.

11.3 Chapter Summary

A set $\mathcal{A}$ is Turing reducible to a set $\mathcal{B}$, written $\mathcal{A} \leq_T \mathcal{B}$, if the decidability of $\mathcal{B}$ implies the decidability of $\mathcal{A}$. The T-reduction $\leq_T$ is a generalization of $\leq_m$, the m-reduction; if two sets are related by $\leq_T$, they may not be related by $\leq_m$. A decidable set is T-reducible to any set. The complement of a set is T-reducible to the set. If a set $\mathcal{A}$ is T-reducible to a decidable set, then $\mathcal{A}$ is decidable. If an undecidable set is T-reducible to a set $\mathcal{B}$, then $\mathcal{B}$ is undecidable.

Correspondingly, a decision problem $\mathcal{P}$ is T-reducible to a decision problem $\mathcal{Q}$, written $\mathcal{P} \leq_T \mathcal{Q}$, if the decidability of $\mathcal{Q}$ implies the decidability of $\mathcal{P}$. If a decision problem $\mathcal{P}$ is T-reducible to a decidable decision problem, then $\mathcal{P}$ is decidable. If an undecidable decision problem is T-reducible to a decision problem $\mathcal{Q}$, then $\mathcal{Q}$ is undecidable.

Generally, a computational problem $\mathcal{P}$ is T-reducible to a computational problem $\mathcal{Q}$, written $\mathcal{P} \leq_T \mathcal{Q}$, if the computability of $\mathcal{Q}$ implies the computability of $\mathcal{P}$. If a computational problem $\mathcal{P}$ is T-reducible to a computable computational problem, then $\mathcal{P}$ is computable. If an incomputable computational problem is T-reducible to a computational problem $\mathcal{Q}$, then $\mathcal{Q}$ is incomputable.

The T-reduction is reflexive and transitive, but not symmetric. Two sets are Turing equivalent if each of them is T-reducible to the other. T-equivalence is an equivalence relation. A Turing degree of a set $\mathcal{A}$ is the set of all sets that are T-equivalent to $\mathcal{A}$. T-degree is a formalization of the notion of "degree of unsolvability." There are (at least) two T-degrees: $\deg(\emptyset)$ and $\deg(\mathcal{K})$. The first corresponds to the degree of unsolvability of all decidable sets (i.e., decidable decision problems), and the second corresponds to the degree of unsolvability of all undecidable sets that are T-equivalent to the set $\mathcal{K}$ (i.e., undecidable decision problems T-equivalent to the *Halting Problem* $\mathcal{D}_H$). A T-degree $\deg(\mathcal{A})$ is lower than a T-degree $\deg(\mathcal{B})$ if $\mathcal{A}$ is T-reducible to $\mathcal{B}$, but not vice versa.

Problems

11.1. The relations $\leq_T$ and $\subseteq$ are, in general, independent of each other. That is, we can have $\mathcal{A} \leq_T \mathcal{B}$ simultaneously with any of the relations $\mathcal{A} \subseteq \mathcal{B}$ and $\mathcal{A} \supseteq \mathcal{B}$, or neither of the two. Can you give examples of such sets $\mathcal{A}$ and $\mathcal{B}$?

11.2. Prove:

(a) $\mathcal{A}$ is a c.e. set $\Longrightarrow \mathcal{A} \leq_T \mathcal{K}$.
(b) $\mathcal{A}, \mathcal{B}$ are disjoint c.e. sets $\Longrightarrow \mathcal{A} \leq_T \mathcal{A} \cup \mathcal{B}$ and $\mathcal{B} \leq_T \mathcal{A} \cup \mathcal{B}$.

11.3. Prove: There exist sets $\mathcal{A}$ and $\mathcal{B}$ such that $\mathcal{A}$ is *not* c.e., $\mathcal{B}$ is (undecidable) c.e., and $\mathcal{A} \leq_T \mathcal{B}$.

Remark. Therefore, a c.e. oracle set $\mathcal{B}$ can make decidable even a non-c.e. set $\mathcal{A}$.
[*Hint.* Consider the *Halting Problem* and the *non-Halting Problem.*]

11.4. Prove: $\equiv_T$ is an equivalence relation.

Bibliographic Notes

- The concept of *Turing reducibility* was first described intuitively in Post [124]. There, he explained how it can be that some decision problem is more difficult than some other undecidable decision problem. He also explored other less general reducibilities, called strong reducibilities, such as 1-reducibility, m-reducibility, btt-reducibility, and tt-reducibility. Although the informal concept of Turing reducibility was used in the following years over and over, it was not formally defined until Kleene [82, §61]. For an exposition on Turing reducibility, see Davis [37].
- The intuitive concept of *degree of unsolvability* was introduced via Turing reducibility in Post [124], and, based on Post [123], used in the abstract of (unpublished paper) Post [127].
- The program to determine the *relative computability* of undecidable decision problems, based on the notions of Turing reducibility, was set forth in Post [124].
- The above concepts were formally defined and became fully understood in Kleene and Post [84].
- Degrees of undecidability are also covered in Ambos-Spies and Fejer [8], Cooper [29], Cutland [32], Enderton [42], Lerman [93], Odifreddi [116], and Soare [164, 169].

Chapter 12
The Turing Hierarchy of Unsolvability

A hierarchy is a system of organizing things into different ranks,
levels, or positions, depending on how important they are.

Abstract At this point we only know of two degrees of unsolvability: the T-degree shared by all the decidable decision problems, and the T-degree shared by all undecidable decision problems that are T-equivalent to the *Halting Problem*. In this chapter we will prove that, surprisingly, for every undecidable decision problem there exists a more difficult decision problem. This will in effect mean that there is an infinite hierarchy of degrees of unsolvability and that there is no most difficult decision problem.

12.1 The Perplexities of Unsolvability

Turing degrees $\deg(\emptyset)$ and $\deg(\mathcal{K})$ are the only T-degrees whose existence we have intuitively anticipated and formally proved. Are $\deg(\emptyset)$ and $\deg(\mathcal{K})$ the only existing T-degrees? Put differently: Is every decision problem *either* decidable *or* undecidable as is the *Halting Problem* $\mathcal{D}_H$? If the answer were *no*, then there would be a decision problem that would be undecidable because of some reason essentially different from the reason for which the $\mathcal{D}_H$ is undecidable. If so, this would immediately raise the following questions:

1. Are there *undecidable* decision problems that are *more difficult* than $\mathcal{D}_H$?
2. Are there *undecidable* decision problems that are *less difficult* than $\mathcal{D}_H$?
3. Are there *undecidable* decision problems of difficulty *incomparable* to $\mathcal{D}_H$'s?

But, would all this make any sense? We could not compare the difficulties of undecidable problems just by comparing the times elapsed to obtain their solutions, as the computations could run indefinitely and return no solutions at all. For the same reason, neither could we use any other measure of the quality of the solutions. The situation we would face is illustrated in Fig. 12.1 (for computational problems).

A way out of this situation is to use the Turing reduction. The idea is that for two undecidable decision problems $\mathcal{P}$ and $\mathcal{Q}$, we consider $\mathcal{Q}$ to be more difficult than $\mathcal{P}$ if $\mathcal{P} <_T \mathcal{Q}$. Equivalently, $\mathcal{Q}$ is considered more difficult than $\mathcal{P}$ if $\deg(\mathcal{P}) < \deg(\mathcal{Q})$.

© Springer-Verlag Berlin Heidelberg 2015   245
B. Robič, *The Foundations of Computability Theory*,
DOI 10.1007/978-3-662-44808-3_12

Fig. 12.1 How much later than never can we obtain a solution to a more difficult incomputable problem? How much less than no solution can we get when solving a more difficult incomputable problem? Is there any sensible definition of the property "to be a more difficult incomputable problem" at all? The answer is yes; the way out of these perplexities is to use the Turing reduction

We will now focus on question 1: Are there *undecidable* decision problems that are *more difficult* than $\mathcal{D}_H$?[1] So, is there a problem $\mathcal{Q}$ such that $\mathcal{D}_H <_T \mathcal{Q}$? If so, can we construct it? The answer is yes; to construct such a $\mathcal{Q}$ we must use a mapping, called the Turing jump operator. This is the subject of the next section.

12.2 The Turing Jump

Let $\mathcal{S}$ be an arbitrary set. The *Turing jump* operator is a mapping $' : 2^{\mathbb{N}} \to 2^{\mathbb{N}}$ that assigns (i.e., constructs) to the set $\mathcal{S}$ another set, denoted by $\mathcal{S}'$, whose T-degree is higher than $\deg(\mathcal{S})$. How does the mapping $'$ do this?

First, recall the *Halting Problem* $\mathcal{D}_H$:

"Does T halt on input $\langle T \rangle$?"

The language of this problem is the set $\mathcal{K}$; it contains the codes of all ordinary Turing machines T that halt on their own codes $\langle T \rangle$. Since $\langle T \rangle$ is just a binary represented index x, and ψ_x is the proper function of the ordinary Turing machine T_x, we can rewrite $\mathcal{K}$ as follows:

$$\mathcal{K} \overset{\text{def}}{=} \{\langle T \rangle \,|\, T \text{ halts on input } \langle T \rangle\}$$
$$= \{x \,|\, T_x \text{ halts on input } x\}$$
$$= \{x \,|\, \psi_x(x)\!\downarrow\}.$$

Secondly, let $\mathcal{S}$ be an arbitrary set. Let us wake up the oracle for $\mathcal{S}$, make it available to each o-TM T^*, and consider the obtained $\mathcal{S}$-TM $T^{\mathcal{S}}$. Recall from Sect. 10.1.4 that we can encode each $T^{\mathcal{S}}$ with $\langle T^{\mathcal{S}} \rangle$ and interpret this as a natural number, the index of $T^{\mathcal{S}}$. Following the above definition of the *Halting Problem*, we now define the halting problem for oracle Turing machines $T^{\mathcal{S}}$:

---

[1] We will answer question 2 in Chapter 14, and question 3 in Chapter 13.

"Does T^S halt on input $\langle T^S \rangle$?"

Let us denote the language of this problem by $\mathcal{K}^S$. The set $\mathcal{K}^S$ contains the codes of all S-TMs T^S that halt on their own codes $\langle T^S \rangle$. Again, each $\langle T^S \rangle$ is a binary represented index x of T^S. Recalling from Sect. 10.2.1 that Ψ_x^S is the proper functional of T_x^S, we rewrite the set $\mathcal{K}^S$ as follows:

$$
\begin{aligned}
\mathcal{K}^S &\overset{\text{def}}{=} \{ \langle T^S \rangle \mid T^S \text{ halts on input } \langle T^S \rangle \} \\
&= \{ x \mid T_x^S \text{ halts on input } x \} \\
&= \{ x \mid \Psi_x^S(x)\!\downarrow \}.
\end{aligned}
$$

So, given an arbitrary set $S \subseteq \mathbb{N}$, we have constructed a new set $\mathcal{K}^S \subseteq \mathbb{N}$.

Thirdly, we define $'$ to be the mapping that sends a set S to the set $\mathcal{K}^S$. In plain words, the mapping $'$ operates so that it "elevates" its argument S to an oracle set, i.e., makes S "jump" on the set $\mathcal{K}$. This is why $'$ is called the *Turing jump* operator, and the set $\mathcal{K}^S$, the *Turing jump of the set S*. We also denote $\mathcal{K}^S$ by S'. Here is the official definition.

Definition 12.1. (Turing Jump of a Set) The **Turing jump of a set** S is the set S', defined by

$$
S' = \mathcal{K}^S \overset{\text{def}}{=} \{ x \mid \Psi_x^S(x)\!\downarrow \}.
$$

12.2.1 Properties of the Turing Jump of a Set

The main result of this subsection will be Corollary 12.1, which states that S and S' are of different and comparable T-degrees.

First, of course, both S and S' are sets. But the oracle for S' is more powerful than the oracle for S. Indeed, the following lemma tells us that if the oracle for S makes a set appear S-c.e., then the oracle for S' makes the same set appear S'-decidable.

Lemma 12.1. A *is* S-c.e $\implies A$ *is* S'-decidable.

Proof. Let A be an arbitrary S-c.e. set. Define a binary functional Φ_x^S as follows: $\Phi_x^S(y,z) = 1$ if $y \in A$, and $\Phi_x^S(y,z)\!\uparrow$ if $y \notin A$. We will not need the actual value of x, but note that x is fixed. The argument z has no impact on Φ_x^S; it is there only because we want it to remain the only argument after the application of the *Parametrization Theorem*. The functional Φ_x^S is S-p.c. (as it is S-computable on A). Let us apply the *Parametrization Theorem* and move y from Φ_x^S to the index; hence $\Phi_x^S(y,z) = \Psi_{s(x,y)}^S(z)$, for an injective computable function s (see Sect. 7.3). Now observe that the following equivalences hold: $y \in A \iff \Psi_{s(x,y)}^S(s(x,y))\!\downarrow \iff s(x,y) \in \mathcal{K}^S$. So we have $y \in A \iff s(x,y) \in S'$, where s is a computable function and x fixed. This means that A is m-reducible to S', i.e., $A \leq_m S'$. Then, $A \leq_T S'$ (by Theorem 11.3), and A is S'-decidable. $\quad\square$

The next theorem states that every set S is S'-decidable.

Theorem 12.1. *Let S be an arbitrary set. Then $S \leq_T S'$.*

Proof. Since $S \leq_T S$ holds for every S, the set S is S-decidable and, a fortiori, S-c.e. Then Lemma 12.1 (with $\mathcal{A} := S$) tells us that S is S'-decidable; that is, $S \leq_T S'$.   □

The converse is not true. The next theorem states that S' is S-undecidable. However, the theorem guarantees that the S-undecidability of S' is not "excessive"; specifically, it tells us that S' is S-c.e. In short, although there is no S-TM capable of deciding the set S', there is an S-TM capable of recognizing S'.

Theorem 12.2. *Let S be an arbitrary set. Then:*

a) S' is S-undecidable   (i.e., $S' \not\leq_T S$).
b) S' is S-c.e.

Proof. The proof runs along the same lines as the proof of Lemma 8.1 (see Sect. 8.2) that $\mathcal{K}$ is undecidable. The main difference is that now we will be talking of S-TMs (instead of ordinary TMs) and of the set $\mathcal{K}^S$ (instead of $\mathcal{K}$). We therefore move at a somewhat faster pace.

a) Suppose, that $S' \leq_T S$. Then the set $S' \stackrel{\text{def}}{=} \{\langle T^S \rangle \mid T^S$ halts on input $\langle T^S \rangle \}$ is S-decidable, and there is an S-TM, capable of deciding the question "Does T^S halt on input $\langle T^S \rangle$?" for any T^S. Let us denote this hypothetical decider by $D_{\mathcal{K}}^S$.

Now we construct a new S-TM that will use $D_{\mathcal{K}}^S$. Since it will call an S-TM, it will itself be an S-TM. So let us denote it by N^S. The input to N^S will be the code $\langle T^S \rangle$ of an arbitrary T^S. The machine N^S must operate as follows. First, it doubles the input $\langle T^S \rangle$ into $\langle T^S, T^S \rangle$, and then sends this to $D_{\mathcal{K}}^S$. The decider takes this as the question $\langle T^S, T^S \rangle \in ?\mathcal{K}^S$, eventually halts, and answers either YES or NO. If the answer is YES, then N^S calls $D_{\mathcal{K}}^S$ again with the same question; otherwise, N^S outputs its own answer YES and halts.

But there is a catch: if N^S is given as input its own code $\langle N^S \rangle$, it puts the $D_{\mathcal{K}}^S$ in trouble. Namely, if $D_{\mathcal{K}}^S$ has answered the first question $\langle N^S, N^S \rangle \in ?\mathcal{K}^S$ with YES, then N^S starts endless cycling, during which $D_{\mathcal{K}}^S$ stubbornly repeats that N^S *will* halt. If, however, $D_{\mathcal{K}}^S$ has answered to $\langle N^S, N^S \rangle \in ?\mathcal{K}^S$ with NO, and so predicted that N^S would *not* halt, N^S halts in the very next step.

In short, it is *not* true that $D_{\mathcal{K}}^S$ correctly answers *any* question $\langle T^S, T^S \rangle \in ?\mathcal{K}^S$. Actually, it fails when $T^S := N^S$. This contradicts our supposition. Consequently, $S' \not\leq_T S$.

b) Let $R_{\mathcal{K}}^S$ be an S-TM as follows. The input to $R_{\mathcal{K}}^S$ is the code $\langle T^S \rangle$ of an arbitrary T^S. The $R_{\mathcal{K}}^S$ starts simulating T^S on $\langle T^S \rangle$, and *if* it halts (i.e., T^S would halt on $\langle T^S \rangle$), then it outputs a YES. So $R_{\mathcal{K}}^S$ is a *recognizer* of the set $\{\langle T^S \rangle \mid T^S$ halts on input $\langle T^S \rangle \} = S'$. Hence, S' is S-c.e.   □

From Theorems 12.1 and 12.2a it follows that $S <_T S'$, for any set S. Therefore, S and S' are not $\equiv_T$-equivalent, but still comparable T-degrees.

Corollary 12.1. *Let S be an arbitrary set. Then: $deg(S) < deg(S')$.*

By taking $S := \mathcal{K}$ in the above corollary, we obtain $\deg(\mathcal{K}) < \deg(\mathcal{K}')$.

NB *We have discovered that there is a* T*-degree that is higher than* $deg(\mathcal{K})$. *Hence, there exist decision problems that are more difficult than the Halting Problem.*

12.3 Hierarchies of T-Degrees

Because $\mathcal{S}'$ is a set, we can apply the function $'$ to $\mathcal{S}'$ too. This leads to an even higher T-degree $\deg((\mathcal{S}')')$. Since we can repeat this as many times as we wish, it follows that there are higher and higher $<$-comparable T-degrees—and this never ends. We conclude that there is a hierarchy of at least $\aleph_0$ T-degrees and that there is no highest T-degree. Let us now see the details.

In the same fashion as we constructed the Turing jump of the set $\mathcal{S}$, we can construct the Turing jump of the set $\mathcal{S}'$: we take the oracle for $\mathcal{S}'(=\mathcal{K}^{\mathcal{S}})$, make it available to each o-TM T^*, define for the obtained $\mathcal{S}$-TMs $T^{\mathcal{S}'}(=T^{\mathcal{K}^{\mathcal{S}}})$ the corresponding halting problem, and finally obtain the associated language (set) $\mathcal{K}^{\mathcal{S}'}$. Then, we can rewrite $\mathcal{K}^{\mathcal{S}'}$ as follows:

$$\mathcal{K}^{\mathcal{S}'} = \mathcal{K}^{\mathcal{K}^{\mathcal{S}}} = (\mathcal{K}^{\mathcal{S}})' = (\mathcal{S}')'.$$

We call this set the *second Turing jump of $\mathcal{S}$* and denote it simply by $\mathcal{S}''$ or $\mathcal{S}^{(2)}$.

In the same manner we define the sets $\mathcal{S}^{(3)}, \mathcal{S}^{(4)}, \dots$ In general, we construct $\mathcal{S}^{(i+1)}$ by "elevating" $\mathcal{S}^{(i)}$ to the oracle set, making it available to all o-TMs, and collecting in $\mathcal{S}^{(i+1)}$ the codes $\langle T^{\mathcal{S}^{(i)}} \rangle$ of those $\mathcal{S}^{(i)}$-TMs that halt on their own codes:

$$\mathcal{S}^{(i+1)} = \{\langle T^{\mathcal{S}^{(i)}} \rangle \mid T^{\mathcal{S}^{(i)}} \text{ halts on input } \langle T^{\mathcal{S}^{(i)}} \rangle\}$$
$$= \{x \mid T_x^{\mathcal{S}^{(i)}} \text{ halts on input } x\}$$
$$= \{x \mid \Psi_x^{\mathcal{S}^{(i)}}(x)\!\downarrow\}.$$

Definition 12.2. (*n*th Turing Jump) The *n*th **Turing jump of the set** $\mathcal{S}$ is the set $\mathcal{S}^{(n)}$, which is inductively defined as follows:

$$\mathcal{S}^{(n)} \stackrel{\text{def}}{=} \begin{cases} \mathcal{S} & \text{if } n = 0 \\ \left(\mathcal{S}^{(n-1)}\right)' & \text{if } n \geqslant 1 \end{cases}$$

The relation between $\mathcal{S}^{(i)}$ and $\mathcal{S}^{(i+1)}$ is described by the following theorem. The proof of the theorem would run in the same way as the proofs of Lemma 12.1, Theorems 12.1 and 12.2, and Corollary 12.1. We therefore leave it as an exercise to the reader.

Theorem 12.3. *Let S be an arbitrary set. Then:*

a) $\mathcal{S}^{(n)} <_T \mathcal{S}^{(n+1)}$
b) $\mathcal{S}^{(n+1)}$ *is* $\mathcal{S}^{(n)}$*-c.e.*
c) $\deg(\mathcal{S}^{(n)}) < \deg(\mathcal{S}^{(n+1)})$

We see that each set S is the origin of an infinite hierarchy of sets:

$$S^{(0)} <_T S^{(1)} <_T S^{(2)} <_T \ldots <_T S^{(i)} <_T S^{(i+1)} <_T \ldots$$

Associated with this hierarchy is the infinite hierarchy of T-degrees:

$$\deg\big(S^{(0)}\big) < \deg\big(S^{(1)}\big) < \deg\big(S^{(2)}\big) < \ldots < \deg\big(S^{(i)}\big) < \deg\big(S^{(i+1)}\big) < \ldots$$

There are at least $\aleph_0$ T-degrees that are comparable with the relation $<$. Why have we said *at least*? The only reason is because we are cautious: at this point, nothing excludes the possibility of the existence of T-degrees *between* $\deg\big(S^{(i)}\big)$ and $\deg\big(S^{(i+1)}\big)$, for some i. Such T-degrees would by *passed over* by T-jump and hence not constructible by it.

We have thus discovered one more surprising fact:

NB *For every degree of unsolvability there is a higher degree of unsolvability. For every decision problem, even an undecidable one, there is a more difficult decision problem. There is no most difficult decision problem.*

12.3.1 The Jump Hierarchy

Until now, S denoted an arbitrary subset of $\mathbb{N}$. In this subsection we will choose for S a particular set.

To do this, we intuitively reason as follows. The power of the oracle for S will depend on the (un)decidability of the chosen set. If we choose for S a *decidable* set, then we expect that the oracle for S will help o-TMs *as little as possible*. Actually, the help offered by such an oracle will be the same as that of an ordinary TM (see Sect 11.1). We therefore expect that the difference between two successive Turing jumps of the set S, i.e., the sets $S^{(i)}$ and $S^{(i+1)}$, will be as small as possible. This should result in a denser hierarchy $\deg\big(S^{(i)}\big)$, $i = 0, 1, 2, \ldots$, that would, hopefully, reveal *all*[2] T-degrees and hence more of their properties.

So, for S we will pick a decidable set. As decidable sets are T-equivalent, and the set $\emptyset$ is decidable, we will take $S := \emptyset$. We obtain the following *jump hierarchy of sets*

$$\emptyset^{(0)} <_T \emptyset^{(1)} <_T \emptyset^{(2)} <_T \ldots <_T \emptyset^{(i)} <_T \emptyset^{(i+1)} <_T \ldots$$

and the associated *jump hierarchy of T-degrees*

$$\deg\big(\emptyset^{(0)}\big) < \deg\big(\emptyset^{(1)}\big) < \deg\big(\emptyset^{(2)}\big) < \ldots < \deg\big(\emptyset^{(i)}\big) < \deg\big(\emptyset^{(i+1)}\big) < \ldots$$

---

[2] However, we will learn in Chap. 13 that this reasoning is too optimistic. It will prove again that intuition can be misleading.

Example 12.1. (Degrees of Some Sets) The undecidable sets $\mathcal{K}_0$, $\mathcal{K}_1$, $\mathcal{F}in$, $\mathcal{C}of$, $\mathcal{T}ot$, $\mathcal{E}xt$, which we defined in Sects. 8.2 and 8.3.5, belong to the initial T-degrees of the jump hierarchy. In particular,

$$\mathcal{K}, \mathcal{K}_0, \mathcal{K}_1 \in \deg(\emptyset^{(1)})$$
$$\mathcal{F}in, \mathcal{T}ot \in \deg(\emptyset^{(2)})$$
$$\mathcal{C}of, \mathcal{E}xt \in \deg(\emptyset^{(3)})$$

This tells us that

- the following three incomputable problems are equally difficult:

$$\mathcal{D}_H = \text{``Does } T \text{ halt on input } \langle T \rangle \text{?''}$$
$$\mathcal{D}_{Halt} = \text{``Does } T \text{ halt on input } w \text{?''}$$
$$\mathcal{D}_{\mathcal{K}_1} = \text{``Is } \mathrm{dom}(\varphi) \text{ empty?''}$$

- the next two decision problems are equally difficult, yet more difficult than the above three:

$$\mathcal{D}_{\mathcal{F}in} = \text{``Is } \mathrm{dom}(\varphi) \text{ finite?''}$$
$$\mathcal{D}_{\mathcal{T}ot} = \text{``Is } \varphi \text{ total?''}$$

- the next two decision problems are equally difficult, but more difficult than the above five:

$$\mathcal{D}_{\mathcal{C}of} = \text{``Is } \varphi \text{ undefined on finitely many elements?''}$$
$$\mathcal{D}_{\mathcal{E}xt} = \text{``Can } \varphi \text{ be extended to a total computable function?''}$$

Written succinctly:

$$\mathcal{D}_H \equiv_T \mathcal{D}_{Halt} \equiv_T \mathcal{D}_{\mathcal{K}_1} <_T \mathcal{D}_{\mathcal{F}in} \equiv_T \mathcal{D}_{\mathcal{T}ot} <_T \mathcal{D}_{\mathcal{C}of} \equiv_T \mathcal{D}_{\mathcal{E}xt}$$

This situation is depicted in Fig. 12.2.

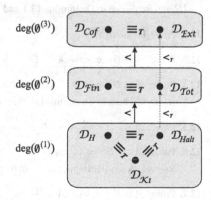

Fig. 12.2 The undecidability of some decision problems. The problems in the same T-degree are equally difficult, but more difficult than the problems in lower T-degrees.

If, for instance, $\mathcal{D}_{\mathcal{T}ot}$ were decidable, $\mathcal{D}_{\mathcal{F}in}$ and $\mathcal{D}_H$, $\mathcal{D}_{Halt}$, $\mathcal{D}_{\mathcal{K}_1}$ would also be decidable; however, $\mathcal{D}_{\mathcal{C}of}$ and $\mathcal{D}_{\mathcal{E}xt}$ would remain undecidable. □

12.4 Chapter Summary

We compare the difficulty of computational problems by relating the problems with the Turing reduction.

Using the Turing jump operator, we can construct, for an arbitrary set, a set that is at a higher T-degree than the original set.

Correspondingly, for every decision problem, even an undecidable one, there is a more difficult decision problem.

This enables us to construct the jump hierarchy of T-degrees. The hierarchy starts with the T-degree $\deg(\emptyset)$ representing decidable decision problems, and continues with infinitely many T-degrees. There is no highest T-degree in the jump hierarchy. At this point nothing suggests that there might exist T-degrees other than those that are members of the jump hierarchy, i.e., constructible by the Turing jump operator.

This means that the class of decision problems does not divide into just two subclasses, $\deg(\emptyset)$ and $\deg(\mathcal{K})$, one consisting of decidable and the other of undecidable problems, that are equally difficult as the *Halting Problem*. Instead, the subclass of undecidable decision problems is further partitioned into infinitely many subclasses consisting of more and more difficult problems.

All of this is surprising and proves once again that human experience and intuition can be deceptive.

Problems

12.1. Prove:
(a) $\emptyset' = \mathcal{K}$
(b) $\emptyset'' = \mathcal{K}^{\mathcal{K}}$
(c) $\emptyset''' = \mathcal{K}^{\mathcal{K}^{\mathcal{K}}}$

   [*Hint.* Use $\mathcal{S} = \emptyset$ in Definitions 12.1 and 12.2.]

12.2. Prove Theorem 12.3.

12.3. Prove: $\mathcal{A}$ is $\mathcal{B}$-c.e. $\Longleftrightarrow \mathcal{A} \leq_1 \mathcal{B}'$.

12.4. Prove: $\mathcal{A}$ is $\mathcal{B}$-c.e. $\wedge\ \mathcal{B} \leq_T \mathcal{C} \Longrightarrow \mathcal{A}$ is $\mathcal{C}$-c.e.

   Remark. Thus, in Theorem 11.6, the conditions for the transitivity of $\leq_T$ can be relaxed.

12.5. Prove: $\mathcal{A} \leq_T \mathcal{B} \Longleftrightarrow \mathcal{A}' \leq_1 \mathcal{B}'$.

12.6. Prove: $\mathcal{A} \equiv_T \mathcal{B} \Longrightarrow \mathcal{A}' \equiv_1 \mathcal{B}'$.

   Remark. A similar implication will be proved in the next chapter (see Theorem 13.1, p. 256).

12.7. Prove: $\mathcal{A}$ is $\mathcal{B}$-c.e. $\Longleftrightarrow \mathcal{A}$ is $\overline{\mathcal{B}}$-c.e.

Bibliographic Notes

- The *jump* operator was first informally used in Post [127] when he, as well as Kleene, struggled to formally define the abstract notion of completeness of c.e sets. Post used the Turing jump S' of a set S but did not formally define the operator $'$. So, the first published proof that for any set S there exists a set S' which is complete for S appeared in Kleene [82]. But, the jump operator was finally defined in Kleene and Post [84].
- The first results about the jump hierarchy, consisting of degrees yielded by the Turing jump operator, appeared in Kleene [79, 81, 82], Mostowski [108, 109], Post [127], and Davis [33].
- For more on degrees of unsolvability, see Lerman [93], Soare [164], Odifreddi [116], Cooper [29], Ambos-Spies and Fejer [8], and Soare [169]. See also Sacks [145], Cutland [32], and Enderton [42].

Chapter 13
The Class $\mathcal{D}$ of Degrees of Unsolvability

A structure is something that consists of parts connected together in an ordered way.

Abstract We now know that there are 1) infinitely many T-degrees; 2) the relation $<$ defined between these T-degrees; and 3) the operator Turing jump that constructs from a given set a new set at a higher T-degree. In this chapter, T-degrees will become the main object of our research. It will be useful to view T-degrees as members of a certain class, $\mathcal{D}$. We will define this class as a mathematical structure endowed with a relation and a function that we will found on $<$ and $'$, respectively. This view will simplify our expression and the investigation of the properties of the structure.

13.1 The Structure $(\mathcal{D}, \leq, ')$

Recall that a mathematical structure is a class endowed with certain relations and functions defined on the class (see the footnote on p. 35). In what follows, our intention is to define a structure whose class will contain all T-degrees, while the relation and function on this class will be based on the relation $<$ and the Turing jump operator $'$, respectively.

First, observe that instead of viewing a T-degree as a $\equiv_T$-equivalence class of subsets of $\mathbb{N}$, we can view it as a *member* of some other set. Of course, this is the *quotient set of* $\mathbb{N}$ *relative to* $\equiv_T$, i.e., the class $2^{\mathbb{N}}/\equiv_T$ of all $\equiv_T$-equivalence classes of $2^{\mathbb{N}}$. We will denote this class by $\mathcal{D}$.

Definition 13.1. (Class $\mathcal{D}$) The **class** $\mathcal{D}$ of all T-degrees is $\mathcal{D} \overset{\text{def}}{=} 2^{\mathbb{N}}/\equiv_T$.

$\mathcal{D}$ is not empty; we have seen in Chap. 12 that it has infinitely many members.

Remarks. (1) We can interpret $\mathcal{D}$ as the class of all degrees of unsolvability of decision problems. (2) It is customary to denote the members of $\mathcal{D}$ by boldface characters, e.g., a, b, c, d or 0. This notation indicates no representatives of T-degrees. When we will need a representative of a T-degree, we will indicate it explicitly, e.g., $S \in d$ or $d = \deg(S)$, saying that S is of degree d.

© Springer-Verlag Berlin Heidelberg 2015
B. Robič, *The Foundations of Computability Theory*,
DOI 10.1007/978-3-662-44808-3_13

Second, we already have a relation defined on $\mathcal{D}$; this is the relation $<$ which we introduced to compare T-degrees $\deg(\emptyset)$ and $\deg(\mathcal{K})$. It turned out that there are infinitely many T-degrees $\deg(\mathcal{S}^{(i)})$, $i \in \mathbb{N}$, and that they are linearly ordered by $<$. Now recall that linear order is just a special case of partial order (see Appendix A). We want to be able to consider other partial orders on $\mathcal{D}$, *if any*. For this reason we will "relax" the relation $<$ by replacing it with its reflexive closure $\leq$, defined as usual by $a \leq b \overset{\text{def}}{\Longleftrightarrow} a < b \vee a = b$.

Finally, we also have a function $'$, the Turing jump. But we must be careful before we apply it to the members of $\mathcal{D}$. Why? We have defined $'$ on *sets* (Definition 12.1, p. 247), and *not* on T-degrees, so it is not so obvious that $\mathcal{D}$ can inherit $'$ as it inherited the relation $\leq$. What we must clear up is the following question: If sets $\mathcal{A}$ and $\mathcal{B}$ are in the same T-degree, can $\mathcal{A}'$ and $\mathcal{B}'$ be in different T-degrees? If this could happen, then $'$ would not be well defined on $\mathcal{D}$. Luckily, the answer to the question is *no*. Thus the following theorem.

Theorem 13.1. $\mathcal{A} \equiv_T \mathcal{B} \Longrightarrow \mathcal{A}' \equiv_T \mathcal{B}'$

Proof. First, we prove that $\mathcal{A} \leq_T \mathcal{B} \Longrightarrow \mathcal{A}' \leq_T \mathcal{B}'$. So let $\mathcal{A} \leq_T \mathcal{B}$. It follows that $\mathcal{A}'$ is $\mathcal{A}$-c.e. (by Theorem 12.2 b). But then $\mathcal{A}'$ is $\mathcal{B}$-c.e. Hence, $\mathcal{A}' \leq_T \mathcal{B}'$ (by Lemma 12.1). We have thus proved $\mathcal{A} \leq_T \mathcal{B} \Longrightarrow \mathcal{A}' \leq_T \mathcal{B}'$. Second, we prove that $\mathcal{B} \leq_T \mathcal{A} \Longrightarrow \mathcal{B}' \leq_T \mathcal{A}'$. This is easy: we only have to swap $\mathcal{A}$ and $\mathcal{B}$ in the previous proof. Thus, $\mathcal{B} \leq_T \mathcal{A} \Longrightarrow \mathcal{B}' \leq_T \mathcal{A}'$. Finally, the two proved relations together result in $\mathcal{A} \equiv_T \mathcal{B} \Longrightarrow \mathcal{A}' \equiv_T \mathcal{B}'$. $\qquad\square$

Informally, this means that T-jumps of sets of the same T-degree are sets of the same T-degree. Consequently, we can extend the definition of $'$ to be a function that maps a T-degree into a (single) T-degree. Thus we can talk about the Turing jump of a *whole* T-degree. In short, $'$ is a well-defined function on the class $\mathcal{D}$. Here is the definition.

Definition 13.2. (T-jump of a T-degree) The **Turing jump of a T-degree** $d \in \mathcal{D}$ is the T-degree $d' \overset{\text{def}}{=} \deg(\mathcal{S}')$, where $\mathcal{S}$ is an arbitrary member of d.

The nth T-jump $d^{(n)}$ of a T-degree d is defined in the same fashion as for sets (see Definition 12.2, p. 249), so we omit the formal definition. Writing $\mathbf{0}^{(i)}$ instead of $\deg\big(\emptyset^{(i)}\big)$, the jump hierarchy of T-degrees is now

$$\mathbf{0}^{(0)} \leq \mathbf{0}^{(1)} \leq \mathbf{0}^{(2)} \leq \ldots \leq \mathbf{0}^{(i)} \leq \mathbf{0}^{(i+1)} \leq \ldots$$

The initial T-degrees we usually denote by $\mathbf{0} = \mathbf{0}^{(0)}$, $\mathbf{0}' = \mathbf{0}^{(1)}$, $\mathbf{0}'' = \mathbf{0}^{(2)}$, $\mathbf{0}''' = \mathbf{0}^{(3)}$.

To summarize, the class $\mathcal{D}$ has been endowed with the relation $\leq$ and the function $'$. In the following sections we will list some properties of the structure $(\mathcal{D}, \leq, ')$.

13.2 Some Basic Properties of $(\mathcal{D}, \leq, ')$

In this section will list some of the properties that concern $(\mathcal{D}, \leq, ')$ as a whole. Specifically, we will explore facts about the cardinality and structure of $(\mathcal{D}, \leq, ')$.

13.2.1 Cardinality of Degrees and of the Class $\mathcal{D}$

Concerning the cardinality, two questions are of interest:

1. *Cardinality of T-degrees.* Given any T-degree, how many sets are there in it?
2. *Cardinality of the class $\mathcal{D}$.* How many T-degrees are there in $\mathcal{D}$?

Remark. Restating the above questions in view of decision problems we obtain:
1) How many decision problems share a given degree of unsolvability?
2) How many degrees of unsolvability are there?

The first question is answered by the following theorem.

Theorem 13.2. *Every T-degree is countable.*

Recall that a set is countable if it is equinumerous to a subset of $\mathbb{N}$. Thus, the cardinality of a countable set is either a natural number or $\aleph_0$.

Proof. Let $\mathcal{B} \subseteq \mathbb{N}$ be an arbitrary set. Define $\mathrm{lcone}(\mathcal{B})$ to be the set of all sets T-reducible to $\mathcal{B}$; that is, $\mathrm{lcone}(\mathcal{B}) \stackrel{\text{def}}{=} \{\mathcal{A} \mid \mathcal{A} \leq_T \mathcal{B}\}$. The plan of the proof is this: We will prove that $\mathrm{lcone}(\mathcal{B})$ is countable; since $\deg(\mathcal{B}) \subseteq \mathrm{lcone}(\mathcal{B})$, it will then follow that $\deg(\mathcal{B})$ is countable too.

To prove that $\mathrm{lcone}(\mathcal{B})$ is countable, we must construct an injection $f : \mathrm{lcone}(\mathcal{B}) \to \mathbb{N}$. How can we do that? Since $\mathcal{A} \leq_T \mathcal{B}$, there is a $\mathcal{B}$-TM $T^{\mathcal{B}}$ capable of computing $\chi_{\mathcal{A}}(a)$ for any $a \in \mathcal{A}$. So there are countably infinitely many $\mathcal{B}$-TMs equivalent to $T^{\mathcal{B}}$, in the sense that each of them computes $\chi_{\mathcal{A}}$ (see Sect. 10.1.4). The set $\mathrm{ind}^{\mathcal{B}}(\mathcal{A})$ of all indexes of such machines is countably infinite (see Sect. 10.2.1). We now define the function $f : \mathrm{lcone}(\mathcal{B}) \to \mathbb{N}$ by letting f assign to $\mathcal{A}$ the smallest index in $\mathrm{ind}^{\mathcal{B}}(\mathcal{A})$; that is, $f(\mathcal{A}) \stackrel{\text{def}}{=} \min\{x \mid T_x^{\mathcal{B}} \text{ computes } \chi_{\mathcal{A}}\}$. (Such an index exists, as $\mathrm{ind}^{\mathcal{B}}(\mathcal{A}) \neq \emptyset$.) The function f is injective. (Otherwise, there would exist in $\mathrm{lcone}(\mathcal{B})$ two *different* sets $\mathcal{A}_1$, $\mathcal{A}_2$ that would be decided by the *same* $\mathcal{B}$-TM $T_{f(\mathcal{A}_1)}^{\mathcal{B}} = T_{f(\mathcal{A}_2)}^{\mathcal{B}}$. This would imply that $\chi_{\mathcal{A}_1} = \chi_{\mathcal{A}_2}$, which is impossible because $\mathcal{A}_1 \neq \mathcal{A}_2$.) Consequently, $\mathrm{lcone}(\mathcal{B})$ is countable. As $\deg(\mathcal{B}) \subseteq \mathrm{lcone}(\mathcal{B})$, so is $\deg(\mathcal{B})$. □

But we can do more: Kleene and Post proved that each T-degree has cardinality $\aleph_0$.

Proposition 13.1. *Every T-degree is countably infinite.*

Proof. We omit the proof. See the Bibliographic Notes to this chapter. □

We now turn to the second question. How many T-degrees exist in $\mathcal{D}$? We have seen that the jump hierarchy $\mathbf{0}^{(0)} < \mathbf{0}^{(1)} < \mathbf{0}^{(2)} < \ldots < \mathbf{0}^{(i)} < \mathbf{0}^{(i+1)} < \ldots$ contains $\aleph_0$ T-degrees. So, $\mathcal{D}$ contains at least as many T-degrees. What about other hierarchies starting in decidable sets S besides $\emptyset$? Since all decidable sets are in $\mathbf{0}$, all their Turing jumps are in $\mathbf{0}'$ (by Theorem 13.1). Thus, it seems that there can be no other hierarchy besides the jump hierarchy. On the other hand, nothing says that the Turing jump is the only way to discover new T-degrees. If other ways exist, also other degrees of unsolvability may exist. So the question is whether or not the total number of degrees of unsolvability is larger than $\aleph_0$. In other words: Is the class $\mathcal{D}$ countable or uncountable? Here is the answer.

Theorem 13.3. *The class $\mathcal{D}$ is uncountable; its cardinality is $2^{\aleph_0}$.*

Proof. There are $2^{\aleph_0}$ subsets of $\mathbb{N}$. Since each is of a certain degree of unsolvability, there can be *at most* $2^{\aleph_0}$ T-degrees in $\mathcal{D}$. Now, if there were only countably many (i.e., at most $\aleph_0$) T-degrees in $\mathcal{D}$, then—knowing (by Theorem 13.2) that each contains countably many sets— there would be countably many sets contained in *all* T-degrees. (Note that "countably many $\times$ countably many $=$ countably many"; see Appendix A.) We conclude that there must be $2^{\aleph_0}$ T-degrees in $\mathcal{D}$. □

Remarks. Let us interpret the above results in light of decision problems. There are $2^{\aleph_0}$ degrees of unsolvability (Theorem 13.3). (This is as many as there are real numbers, assuming the *Continuum Hypothesis*, p. 16.) For each degree of unsolvability there are $\aleph_0$ decision problems of that degree of unsolvability (Theorem 13.2). For instance, $\aleph_0$ decision problems are as difficult as the *Halting Problem* (see Fig. 12.2). There are $\aleph_0$ decision problems as difficult as the problem $\mathcal{D}_{\mathcal{T}ot} =$ "Is a p.c. function φ total?" A similar situation occurs for problems that are as difficult as the problem $\mathcal{D}_{\mathcal{E}xt} =$ "Can a p.c. function φ be extended to a computable one?"

Now we see that our cautiousness in Sect. 12.3.1 was well grounded: besides the $\aleph_0$ T-degrees $\mathbf{0}^{(0)} < \mathbf{0}^{(1)} < \mathbf{0}^{(2)} < \ldots < \mathbf{0}^{(i)} < \mathbf{0}^{(i+1)} < \ldots$ there are many more T-degrees. But where are they? Intuitively, we can identify two possibilities:

- there exist T-degrees that are not within the jump hierarchy;
- there exist intermediate T-degrees between a T-degree and its T-jump.

In the following, we will see that both are true.

13.2.2 The Class $\mathcal{D}$ as a Mathematical Structure

The class $\mathcal{D}$ is endowed with the relation $\leq$. Does this relation reveal any particular, distinguished order in the class $\mathcal{D}$? It is easy to prove the following theorem.

Theorem 13.4. $(\mathcal{D}, \leq)$ *is partially ordered.*

Proof. We leave it to the reader to check that $\leq$ is reflexive ($a \leq a$, for any $a \in \mathcal{D}$), transitive ($a \leq b \wedge b \leq c \Rightarrow a \leq c$, for any $a, b, c \in \mathcal{D}$), and anti-symmetric ($a \leq b \wedge b \leq a \Rightarrow a = b$, for any $a, b \in \mathcal{D}$). $\qquad\square$

Incomparable T-Degrees

Is it, perhaps, that $(\mathcal{D}, \leq)$ is even linearly ordered, as is, for example, $(\mathbb{N}, \leqslant)$? Since $(\mathcal{D}, \leq)$ is partially ordered, to answer this question we must find out whether or not every two members of $\mathcal{D}$ are $\leq$-comparable, i.e., whether $a \leq b \vee b \leq a$, for every $a, b \in \mathcal{D}$. Surprisingly, the answer is *no*. In 1954, Kleene and Post proved that there exist sets $\mathcal{A}, \mathcal{B}$, both T-reducible to the set $\emptyset'$, such that $\mathcal{A} \not\leq_T \mathcal{B} \wedge \mathcal{B} \not\leq_T \mathcal{A}$. Since $\mathcal{A}$ and $\mathcal{B}$ are $\leq_T$-*incomparable*, so are $\leq$-*incomparable* $\deg(\mathcal{A})$ and $\deg(\mathcal{B})$. (Note that $\deg(\mathcal{A}) \leq \mathbf{0}'$ and $\deg(\mathcal{B}) \leq \mathbf{0}'$.) We write $a|b$ when a, b are $\leq$-incomparable. The proof of the following Kleene-Post theorem is instructive because its idea will soon be developed further and used in *Post's Problem* (see Chap. 14).

Theorem 13.5. *There exist T-degrees a, b, such that $\mathbf{0} \leq a, b$ and $a, b \leq \mathbf{0}'$ and $a|b$.*

Proof. To prove the theorem, we must show that there exist two sets $\mathcal{A}$ and $\mathcal{B}$, such that $\mathcal{A} \leq_T \emptyset'$, $\mathcal{B} \leq_T \emptyset'$, and $\mathcal{A} \not\leq_T \mathcal{B} \wedge \mathcal{B} \not\leq_T \mathcal{A}$. Since we expect that proving the existence of $\mathcal{A}$ and $\mathcal{B}$ in one fell swoop might be too difficult (if not impossible), we will take a different approach: we will design a set of guidelines by which $\mathcal{A}$ and $\mathcal{B}$ can be constructed, at least in principle, in a systematic, algorithmic way. There are several ingredients in this method.

First, observe that the condition $\mathcal{A} \not\leq_T \mathcal{B}$ means that *no* $\mathcal{B}$-TM can decide the set $\mathcal{A}$. Hence the condition can be replaced by a conjunction $R_0 \wedge R_1 \wedge R_2 \wedge \ldots$ of countably many simpler requirements, where R_e requires that $\mathcal{A}$ cannot be decided by $T_e^{\mathcal{B}}$, the eth $\mathcal{B}$-TM. Equivalently, R_e demands that χ_A, the characteristic function of $\mathcal{A}$, not be the proper functional $\Psi_e^{\mathcal{B}}$:

$$R_e : \chi_A \neq \Psi_e^{\mathcal{B}}.$$

In the same fashion we replace the condition $\mathcal{B} \not\leq_T \mathcal{A}$ with the sequence $S_0 \wedge S_1 \wedge S_2 \wedge \ldots$, where

$$S_e : \chi_B \neq \Psi_e^{\mathcal{A}}.$$

Consequently, to prove the theorem we must show how to construct $\mathcal{A}$ and $\mathcal{B}$ such that R_e and S_e will be fulfilled for every e.

The *plan* is this: we will construct χ_A and χ_B in an infinite sequence of *stages*; at any stage s, only the current *approximations* f_s and g_s to χ_A and χ_B, respectively, will exist; we will ensure that the current f_s and g_s fulfill $R_0 \wedge \ldots \wedge R_\ell$ and $S_0 \wedge \ldots \wedge S_\ell$, respectively, for some $\ell = \ell(s)$; and we will ensure that the length $\ell(s)$ of the fulfilled conjunctions will increase monotonously from stage to stage. As a consequence, in the limit, the condition $\mathcal{A} \not\leq_T \mathcal{B} \wedge \mathcal{B} \not\leq_T \mathcal{A}$ will be fulfilled.

At any stage s, the to-be-constructed sets $\mathcal{A}$ and $\mathcal{B}$ will be approximated by sets which we denote by $\mathcal{A}_s$ and $\mathcal{B}_s$, respectively. Correspondingly, χ_A and χ_B will be approximated by f_s and g_s, respectively. So, in the limit, we expect $\mathcal{A}_\omega = \mathcal{A}$, $\mathcal{B}_\omega = \mathcal{B}$ and $f_\omega = \chi_A$, $g_\omega = \chi_B$. The construction of $\chi_{\mathcal{A}}$ and $\chi_{\mathcal{B}}$ will start at stage $s = 0$ with f_0 and g_0 representing $\mathcal{A}_0 = \mathcal{B}_0 = \emptyset$. At each next stage, we will try to fulfill first the requirement R_e, and then the requirement S_e, for some e. (Thus, at $s = 1$ we will fulfill R_0 and S_0; at $s = 2$, R_1 and S_1; and so on.) But we will try to do this in such a way that *once a requirement has been fulfilled, it will remain fulfilled forever*. (We also say that no requirement will be *injured*.) If we succeed in this plan, the requirements will become fulfilled in the order $R_0, S_0, R_1, S_1, R_2, S_2, \ldots$, which will satisfy the condition $\mathcal{A} \not\leq_T \mathcal{B} \wedge \mathcal{B} \not\leq_T \mathcal{A}$. (What about the other two conditions, $\mathcal{A} \leq_T \emptyset'$, $\mathcal{B} \leq_T \emptyset'$, that are set by the theorem? The reasons for having these will become apparent soon.)

We have come to the question about how to fulfill R_e and S_e, given that all previous Rs and Ss have been fulfilled. The answer will be given in an induction-like way by outlining what the next stage $s + 1$ should do in order to preserve the situation similar to that at stage s. So, *assume that, at stage s, we have constructed f_s and g_s in such way that*, for some n, both are defined everywhere on the initial segment $\{0, 1, \ldots, n\}$ of $\mathbb{N}$, and all the requirements $R_0, S_0, \ldots, R_{e-1}, S_{e-1}$ are fulfilled. We can represent the function f_s by a sequence of its values on the segment, that is, by the word $a_s = f_s(0) f_s(1) \ldots f_s(n) \in \{0, 1\}^*$. Of course, $f_s(i)$ tells us whether or not $i \in \mathcal{A}_s$ (when $0 \leqslant i \leqslant n$), while $f_s(i) \uparrow$ for $i > n$, as the *status* (membership) of these numbers in $\mathcal{A}$ is still open. (The same holds for g_s, which is represented by a word $b_s \in \{0, 1\}^*$.) *Then, in the next stage $s + 1$, we will try to define f_{s+1} and g_{s+1} in such way that*

- $a_s \subset a_{s+1}$ and $b_s \subset b_{s+1}$ will hold. That is, a_{s+1} is to be a *proper extension* of the word a_s, or, equivalently, a_s is to be a proper prefix of the word a_{s+1}. (Similarly for b_{s+1}.) This means that f_{s+1} is to be defined everywhere on $\{0, 1, \ldots, n, \ldots, m\}$, for some $m > n$. (Similarly for g_{s+1}.)
- R_e and S_e will be fulfilled and none of $R_0, S_0, \ldots, R_{e-1}, S_{e-1}$ will be injured.

If we attain the two objectives, the limit functions $f_\omega, g_\omega : \mathbb{N} \to \{0, 1\}$ will be *total*, and hence *characteristic functions* of the sets $\mathcal{A}_\omega = \mathcal{A}$ and $\mathcal{B}_\omega = \mathcal{B}$. These sets will fulfill all requirements R and S.

Clearly, attaining the above objectives is the crux of the proof. We will prove the following lemma.

Lemma. *Let e be an arbitrary natural number. Given a_s, b_s, there exist extensions a_{s+1}, b_{s+1} such that for any a, b which extend a_{s+1}, b_{s+1} and represent $\chi_{\mathcal{A}}$, $\chi_{\mathcal{B}}$, the $\mathcal{B}$-TM $T_e^{\mathcal{B}}$ does not decide $\mathcal{A}$.*

Proof. Let x be an arbitrary natural number for which f_s is *not* defined: $f_s(x) \uparrow$. (So x is a number for which the membership in $\mathcal{A}_s$, and hence in $\mathcal{A}$, has not been determined.) The crucial question is:

Is there a set $\mathcal{B}$ such that $T_e^{\mathcal{B}}$ would halt on input x and return either YES *or* NO? $\qquad\qquad (*)$

(*i*) If such a set $\mathcal{B}$ *does not* exist, then we can take the trivial extensions $a_{s+1} := a_s$ and $b_{s+1} := b_s$.
(*ii*) If, however, such a $\mathcal{B}$ *exists*, there is more work to do in order to construct a_{s+1} and b_{s+1}. First, we run $T_e^{\mathcal{B}}$ on x. Before halting, the machine asks the oracle finitely often whether or not a number is in $\mathcal{B}$. Let y be the largest of these numbers. Then, let b_{s+1} be the shortest extension of b_s that covers y. Now it remains to construct a_{s+1}. Note that $T_e^{\mathcal{B}}$, if run on x, would return the same answer as $T_e^{\mathcal{B}_{s+1}}$. So, to ensure that $T_e^{\mathcal{B}_{s+1}}$ (and hence $T_e^{\mathcal{B}}$) will *fail* to answer correctly the question $x \in ?\mathcal{A}_{s+1}$ (and hence the question $x \in ?\mathcal{A}$), we add x into either $\mathcal{A}_{s+1}$ or its complement $\overline{\mathcal{A}}_{s+1}$, depending on whether $T_e^{\mathcal{B}_{s+1}}$'s answer is NO or YES, respectively. As a consequence, the limit machine $T_e^{\mathcal{B}}$ will fail to decide the limit set $\mathcal{A}$.

It should be obvious that, once it has been determined which of the possibilities (*i, ii*) holds, the construction of a_{s+1} and b_{s+1} is computable in the ordinary sense. But how can we answer the question $(*)$? A systematic search for $\mathcal{B}$ is out of question. But we can obtain the answer by focusing on the program $\tilde{\delta}_e$ of the o-TM T_e^*. Each instruction of $\tilde{\delta}_e$ branches in two directions. This results in the *computation tree of $\tilde{\delta}_e$*, a tree representing all possible executions of $\tilde{\delta}_e$. Given a particular oracle set $\mathcal{B}$, the actual execution of $\tilde{\delta}_e$ on x is represented by a branch in this tree. Halting executions are represented by finite branches, and non-halting executions by infinite branches. Answers (YES, NO, $\uparrow$) are in the leaves of the tree (i.e., at the end of finite branches). We now see: there exists a set $\mathcal{B}$ for which $T_e^{\mathcal{B}}$ halts on x and returns a YES or NO *iff* the computation tree of $\tilde{\delta}_e$ has a leaf bearing the answer YES or NO. So we can apply an ordinary TM T that systematically checks the leaves of the computation tree of $\widetilde{\delta_e}$ and halts with an answer *found* as soon as a leaf with an answer YES or NO has been found. In this case we know that a set $\mathcal{B}$ with the above properties exists. *But* what if T never halts? So, we must find out whether or not T halts. Recall that this is the *Halting Problem*, which can be solved with an o-TM with the oracle set $\mathcal{K}_0$. Consequently, the question (∗) is $\mathcal{K}_0$-decidable, and the construction of a_{s+1} and b_{s+1} is $\mathcal{K}_0$-computable. (This is where the condition $\mathcal{B} \leq_T \emptyset'$ of the theorem comes from.) The lemma is proven.

Observe that we can exchange the roles of $\mathcal{A}$ and $\mathcal{B}$ in the lemma and prove that a_s, b_s can be extended so that ultimately the $\mathcal{A}$-TM $T_e^{\mathcal{A}}$ will not decide $\mathcal{B}$. Hence, applying the lemma twice, we can fulfill both requirements R_e and S_e. Now we can finally see the whole process of the construction of $\mathcal{A}$ and $\mathcal{B}$. We consider oracle Turing machines in succession, $T_0^*, T_1^*, T_2^*, ...$, and, based on the above lemma, ensure that none of them decides $\mathcal{A}$. In doing so, we alternate, for each o-TM T_e^*, the roles of the sets $\mathcal{A}$ and $\mathcal{B}$ and in this way ensure that the construction of both $\mathcal{A}$ and $\mathcal{B}$ proceeds.   □

So, there are incomparable T-degrees between $\mathbf{0}$ and $\mathbf{0}'$. Is this situation specific to the pair $\mathbf{0}$, $\mathbf{0}'$? Far from it: $\leq$-incomparable T-degrees exist between $\mathbf{d}$ and $\mathbf{d}'$, for *any* T-degree $\mathbf{d}$. Here is a generalization of the previous theorem. (See Fig. 13.1.)

Theorem 13.6. *For any T-degree $\mathbf{d}$ there exist T-degrees $\mathbf{a}, \mathbf{b}$, such that $\mathbf{d} \leq \mathbf{a}, \mathbf{b}$ and $\mathbf{a}, \mathbf{b} \leq \mathbf{d}'$ and $\mathbf{a}|\mathbf{b}$.*

Proof. The proof is a relativization of the above proof. See the Bibliographic Notes to this chapter.   □

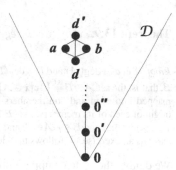

Fig. 13.1 For any T-degree $\mathbf{d}$ there exist $\leq$-incomparable T-degrees $\mathbf{a}, \mathbf{b}$

Theorems 13.3 and 13.6 suggest that there might be uncountably many $\leq$-incomparable pairs of T-degrees. That this is indeed so is stated in the next theorem.

Theorem 13.7. *There are $2^{\aleph_0}$ mutually $\leq$-incomparable T-degrees.*

Proof. See the Bibliographic Notes to this chapter.   □

Distinguished T-Degrees

Since there are $\leq$-incomparable elements in $\mathcal{D}$, the relation $\leq$ does not linearly order $\mathcal{D}$. This gives rise to a series of new questions about the existence of certain distinguished elements:

1. Are there $\leq$-minimal, $\leq$-least, $\leq$-greatest, or $\leq$-maximal elements in $\mathcal{D}$?
2. Do every two members of $\mathcal{D}$ have an $\leq$-upper bound or even the $\leq$-lub?[1]
3. Do every two members of $\mathcal{D}$ have a $\leq$-lower bound or even the $\leq$-glb?[2]
4. Is $(\mathcal{D}, \leq)$ a lattice, and if so, of what kind?

Let us answer these questions.

There can be no $\leq$-*greatest* and no $\leq$-*maximal* element, because the Turing jump constructs, for arbitrary T-degree d, a higher T-degree d'. But, there is a $\leq$-*least* element in $(\mathcal{D}, \leq)$.

Theorem 13.8. *There is a $\leq$-least T-degree in $(\mathcal{D}, \leq)$; this is $\mathbf{0}$.*

Proof. Since $\emptyset$ is decidable, $\emptyset \leq_T \mathcal{A}$ for every set $\mathcal{A}$. Hence, $\mathbf{0} \leq a$ for every $a \in \mathcal{D}$. □

This is not surprising. The decidability of a decidable set is insensitive to the decidability of other sets. Thus, the degrees of true unsolvability are in $\mathcal{D} - \{\mathbf{0}\}$.

Is $(\mathcal{D}, \leq)$ a lattice? Well, to be a lattice, any two T-degrees a, b must have a $\leq$-lub and a $\leq$-glb. The first requirement is satisfied, as the following theorem states.

Theorem 13.9. *Any two T-degrees have a $\leq$-least upper bound.*

Proof. Let $a = \deg(\mathcal{A})$ and $b = \deg(\mathcal{B})$ be arbitrary T-degrees. Consider the *join* $\mathcal{A} \oplus \mathcal{B}$ of $\mathcal{A}$ and $\mathcal{B}$, that is, the set $\mathcal{A} \oplus \mathcal{B} \overset{\text{def}}{=} \{2x \mid x \in \mathcal{A}\} \cup \{2y+1 \mid y \in \mathcal{B}\}$. The members of $\mathcal{A}$ and $\mathcal{B}$ are injectively mapped into even and odd members of the join, respectively. Informally, $\mathcal{A} \oplus \mathcal{B}$ remembers the origin of each of its members. $\mathcal{A} \oplus \mathcal{B}$ is the $\leq_T$-lub of $\mathcal{A}$ and $\mathcal{B}$. To prove this, we must check that 1) $\mathcal{A} \leq_T \mathcal{A} \oplus \mathcal{B}$ and $\mathcal{B} \leq_T \mathcal{A} \oplus \mathcal{B}$, and 2) $\mathcal{A} \oplus \mathcal{B} \leq_T \mathcal{C}$, for any $\leq_T$-upper bound $\mathcal{C}$ of $\mathcal{A}, \mathcal{B}$. (This we leave as an exercise.) It follows that $\deg(\mathcal{A} \oplus \mathcal{B})$ is the $\leq$-lub of $a = \deg(\mathcal{A})$ and $b = \deg(\mathcal{B})$. □

We denote the $\leq$-least upper bound of $a, b \in \mathcal{D}$ by $a \vee b$. In particular, if $a = \deg(\mathcal{A})$ and $b = \deg(\mathcal{B})$, then $a \vee b = \deg(\mathcal{A} \oplus \mathcal{B})$. Informally, the set $\mathcal{A} \oplus \mathcal{B}$ remembers every member of $\mathcal{A}$ and every member of $\mathcal{B}$ by keeping track of the origin of each of its members. Thus, each representative of $a \vee b$ bears more information about its contents than any representative of a or b. Finally, of course, any finite set of T-degrees has a $\leq$-least upper bound.

In contrast to the above theorem, the $\leq$-glb of two degrees need not always exist. This was first proved by Kleene and Post in 1954.

---
[1] least upper bound
[2] greatest lower bound

Theorem 13.10. *There is a pair of T-degrees that have no $\leq$-greatest lower bound.*

Proof. We omit the proof. See the Bibliographic Notes to this chapter. □

We must conclude that $(\mathcal{D}, \leq)$ is not a lattice. However, because of the existence of least upper bounds, we say that $(\mathcal{D}, \leq)$ is an *upper semi-lattice*.

Remarks. Let us look at the above results in light of decision problems. There are pairs of decision problems such that none is more difficult than the other (Theorem 13.6). For any finite set of undecidable decision problems there is a "superproblem" (i.e., a decision problem whose solution would make all problems in the set decidable) which is the easiest among all "superproblems" of the set (Theorem 13.9).

13.2.3 Intermediate T-Degrees

Can there be *intermediate degrees* between $\mathbf{0}^{(i)}$ and $\mathbf{0}^{(i+1)}$, that is, degrees that are passed over when T-jump takes $\mathbf{0}^{(i)}$ to $\mathbf{0}^{(i+1)}$? The answer to this question has already been obtained by Theorems 13.5 and 13.6. But we can ask further: How many T-degress can be passed over by a T-jump? The following theorem, which was also proved by Kleene and Post in 1954, tells us that for each degree d there are *infinitely many* pairwise $\leq$-incomparable degrees between d and d'. (See Fig. 13.2.)

Theorem 13.11. *For any T-degree d and $n \geq 1$, there are pairwise $\leq$-incomparable T-degrees $c_1, \ldots, c_n$ such that $d < c_k < d'$, for $k = 1, \ldots, n$.*

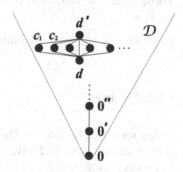

Fig. 13.2 For any d, there are infinitely many pairwise $\leq$-incomparable T-degrees c_k that are passed over by T-jump from d to d'

Proof. See the Bibliographic Notes to this chapter. □

The above theorem guarantees the existence of $\leq$-*incomparable* T-degrees between d and d'. Can it happen that, for some d, there are T-degrees between d and d' that are comparable or even *linearly* ordered by $\leq$? If so, can such a sequence of intermediate T-degrees have infinitely many members? The answer to both questions is yes. (See Fig. 13.3.)

Theorem 13.12. *For any T-degree d and $n \geqslant 1$, there are T-degrees $c_1, \ldots, c_n$ such that $d < c_1 < \ldots < c_n < d'$.*

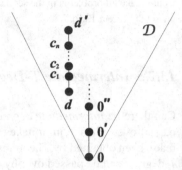

Fig. 13.3 For any d, there are infinitely many linearly ordered T-degrees c_i that are passed over by T-jump from d to d'

Proof. We omit the proof. See the Bibliographic Notes to this chapter.   □

Moreover, Kleene and Post proved that linearly ordered intermediate T-degrees are *dense*, in the sense that if a and b are any such T-degrees, then there is a T-degree c between them. This is stated in the next theorem.

Theorem 13.13. *If $d < a < b < d'$, then there is a T-degree c such that $a < c < b$.*

Proof. We omit the proof. See the Bibliographic Notes to this chapter.   □

13.2.4 Cones

It may seem that because of Theorems 13.3 and 13.7 there is not much left for $\leq$-comparable elements in $\mathcal{D}$. But infinite sets contain equipollent proper subsets (see Appendix), so there still can be many elements of $\mathcal{D}$ that are $\leq$-comparable. This gives rise to the question: How many members of $\mathcal{D}$ are $\leq$-comparable to a given $d \in \mathcal{D}$?

Let us pick an arbitrary $d \in \mathcal{D}$ and consider all the elements of $\mathcal{D}$ that are $\le$-comparable to d. We divide these elements into two sets, the set of elements that are higher than (or equal to) d, and the set of elements that are lower than (or equal to) d. These sets we call the *upper cone* and *lower cone* of d, respectively. See Fig. 13.4. Here is the definition.

Definition 13.3. (Upper and Lower Cone) The **upper cone** of a T-degree d is the set $\mathrm{ucone}(d) \overset{\text{def}}{=} \{x \in \mathcal{D} \mid d \le x\}$, and the **lower cone** of d is the set $\mathrm{lcone}(d) \overset{\text{def}}{=} \{x \in \mathcal{D} \mid x \le d\}$.

Fig. 13.4 The upper and lower cone of a T-degree d. Any T-degree that is $\le$-comparable to d is in d's upper or lower cone

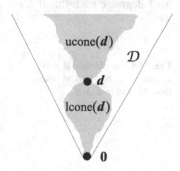

The next two theorems tell us what the population is of each of the two cones.

Theorem 13.14. *The upper cone* $\mathrm{ucone}(d)$ *is uncountable, for any T-degree d.*

Proof. We omit the proof. See the Bibliographic Notes to this chapter. □

Since we can construct with the Turing jump only $\aleph_0$ T-degrees, Theorem 13.14 tells us that the Turing jump cannot uncover all T-degrees above a given T-degree. We have seen one such method in the proof of Theorem 13.6, and another will be described in the next chapter.

Theorem 13.15. *The lower cone* $\mathrm{lcone}(d)$ *is countable, for any T-degree d.*

Proof. We omit the proof. See the Bibliographic Notes to this chapter. □

13.2.5 Minimal T-Degrees

By Theorem 13.8, the T-degree $\mathbf{0}$ is lower than any other T-degree $\mathbf{d}$. Pick an arbitrary $\mathbf{d} (\neq \mathbf{0})$. There may exist a T-degree $\mathbf{c}$ which is strictly between $\mathbf{0}$ and $\mathbf{d}$, that is, $\mathbf{0} < \mathbf{c} < \mathbf{d}$. (For instance, take $\mathbf{d} = \mathbf{0}''$ and $\mathbf{c} = \mathbf{0}'$; or, take $\mathbf{d} = \mathbf{0}'$ and $\mathbf{c} = \mathbf{a}$ in Theorem 13.5.)

The following question immediately arises: Does there exist a T-degree $\mathbf{d} (\neq \mathbf{0})$ such that no T-degree is strictly between $\mathbf{0}$ and $\mathbf{d}$? If such a $\mathbf{d}$ exists, it is called *minimal*. (See Fig. 13.5.) Here is the official definition.

Definition 13.4. (Minimal Degree) A T-degree $\mathbf{d}$ is **minimal** if $\mathbf{d} \neq \mathbf{0}$ and there is no T-degree $\mathbf{c}$ such that $\mathbf{0} < \mathbf{c} < \mathbf{d}$.

Fig. 13.5 An element $\mathbf{d} \neq \mathbf{0}$ is minimal if no element other than $\mathbf{0}$ is lower than $\mathbf{d}$

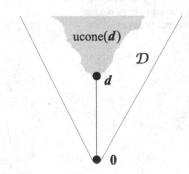

In 1956, Spector[3] proved that minimal degrees actually exist. He proved the existence of a minimal degree below $\mathbf{0}''$.

Theorem 13.16. *There exists a minimal T-degree $\mathbf{d}$; in addition, $\mathbf{d} \leq \mathbf{0}''$.*

Proof. We omit the proof. See the Bibliographic Notes to this chapter.   □

In 1961, Sacks[4] proved the existence of minimal degrees that are even below $\mathbf{0}'$.

Remarks. Let us interpret the above results in light of decision problems. For every degree of unsolvability there are $\aleph_0$ decision problems of that difficulty (Proposition 13.1). There are uncountably many degrees of unsolvability that can have a decision problem (Theorem 13.2). There exist pairs of decision problems such that neither problem is easier (or, more difficult) than the

---

[3] Clifford Spector, 1931–1961, American mathematician.

[4] Gerald Enoch Sacks, 1933, American mathematician and logician.

other (Theorems 13.5 and 13.6). Actually, there are uncountably many pairs of such incomparable decision problems (Theorem 13.7). For any finite set of decision problems there exists a "superproblem" whose decidability would imply the decidability of every decision problem in the set. In addition, there is no easier superproblem for this set of decision problems (Theorem 13.9). For any decision problem, there are uncountably many more difficult decision problems (Theorem 13.14), and only countably many easier decision problems (Theorem 13.15). There are undecidable decision problems for which there exist no easier decision problems besides the decidable ones (Theorem 13.16).

13.3 Chapter Summary

Every T-degree is countably infinite. The class of all T-degrees is uncountable and is partially ordered by the relation $\leq$. The Turing jump operator maps a T-degree into a higher T-degree. There exist T-degrees that are incomparable with the relation $\leq$; in fact, there are uncountably many pairs of such incomparable degrees. The T-degree of decidable decision problems, $\mathbf{0}$, is the $\leq$-least T-degree in the class. Any two T-degrees have a $\leq$-least upper bound, but not necessarily a $\leq$-greatest lower bound. Between any T-degree and its T-jump there are infinitely many pairwise incomparable T-degrees and infinitely many linearly ordered T-degrees. The upper cone of any T-degree contains uncountably many T-degrees, and its lower cone contains only countably many T-degrees. There are minimal degrees, some of which are below the T-degree $\mathbf{0}''$, or even below $\mathbf{0}'$.

Problems

13.1. Prove: $(\mathcal{D}, \leq)$ is partially ordered.

13.2. Prove: $\mathcal{A} \leq_T \mathcal{B} \Longrightarrow \mathcal{A}' \leq_T \mathcal{B}'$.

13.3. Prove: There are sets $\mathcal{A}$ and $\mathcal{B}$ such that $\mathcal{A} <_T \mathcal{B}$ and $\mathcal{A}' \equiv_T \mathcal{B}'$.

Remark. Thus, the converse of the implication $\mathcal{A} \equiv_T \mathcal{B} \Longrightarrow \mathcal{A}' \equiv_T \mathcal{B}'$ (Theorem 13.1) does *not* hold; we may have $\mathcal{A} \not\equiv_T \mathcal{B}$ and still $\mathcal{A}' \equiv_T \mathcal{B}'$.

13.4. Prove: The Turing jump operator $'$ defined on the class $\mathcal{D}$ is *not* injective.

[*Hint.* See Problem 13.3.]

13.5. Complete the proof of Theorem 13.9.

Bibliographic Notes

- The abstraction from degrees of undecidability of *sets* to *degrees* as such was introduced in Kleene and Post [84]. This paper revealed many important facts about degrees of unsolvability:

the T-jump operator was defined on T-degrees [84, §1]; the existence of infinitely many $\leq$-incomparable T-degrees between any two $d \neq 0$ and d' was proved [84, §2]; $(\mathcal{D}, \leq)$ was proved to be an upper semi-lattice [84, §1] but not a lattice [84, §4]; the class $\mathcal{D}$ was proved to be uncountable; the cardinality of each T-degree was proved to be $\aleph_0$; the fact that each T-jump passes over infinitely many degrees was proved; the existence of such a set of T-degrees, which is dense, was proved.

- In describing the idea of the proof of Theorem 13.5 we have leaned on Soare [164, Chap. VI] and Shen and Vereshchagin [149].
- The existence of minimal T-degrees was proved in Spector [170], and the existence of minimal T-degrees below $0'$ in Sacks [144].
- The reader can also find proofs of Theorems 13.6, 13.7, 13.12, and 13.15 in Soare [164].
- The degrees of unsolvability also cover (in order of appearance) Rogers [139], Cutland [32], Odifreddi [116], Cooper [29], Ambos-Spies and Fejer [8], Enderton [42], and Weber [186].
- A survey of the results on $(\mathcal{D}, \leq, ')$ to the end of the 1970s is Simpson [156], to the mid-1980s Lerman [93], and from the early 1980s on Shore [151]. A survey of current knowledge of this area is Slaman [158] and Soare [169].

Chapter 14
C.E. Degrees and the Priority Method

*If something has priority over other things, it is regarded as
being more important than them and is dealt with first.
An injury is damage done to somebody's body.*

Abstract Among the Turing degrees, the so-called computably enumerable (c.e.)
degrees are all important. This is because they stem from c.e. sets, the sets that often
spring up in practice. In this chapter we will present the basic facts of c.e. degrees.
We will then describe *Post's Problem*, a problem about c.e. degrees that was posed
by Emil Post in 1944. After a series of attempts by Post and others, the problem
was finally solved in 1956 by Muchnik and Friedberg. They simultaneously and
independently devised a method, called the *Priority Method*, and applied it to solve
the problem. We will describe *Post's Problem* and the *Priority Method*.

14.1 C.E. Turing Degrees

Undecidable c.e. sets are interesting for several reasons. The first reason is that most
of the undecidable decision problems that have arisen in practice, i.e., outside pure
Computability Theory, are represented by undecidable c.e. sets. (See Sect. 8.3 for a
number of such problems that emerged in various areas of science.) The second reason is that, roughly speaking, any such set is in a way "close" to the decidable sets,
in the sense that it is more amenable to recognition than any other, non-c.e. set. Put
differently, for any such set the membership problem is semi-decidable since there
exists an algorithm that is capable of recognizing any member of the set, though
incapable of recognizing every non-member of the set. (See Sect. 8.2.)

Remarks. An undecidable c.e. set is mirrored in the corresponding decision problem. There exists
an algorithm capable of solving any positive instance of the problem, but no algorithm is capable
of also solving any negative instance. Informally, such a decision problem is just semi-solvable.

As a consequence, much of the research has been and still is devoted to c.e. sets and
their degrees of unsolvability. We will describe some of this research in this chapter.

© Springer-Verlag Berlin Heidelberg 2015   269
B. Robič, *The Foundations of Computability Theory*,
DOI 10.1007/978-3-662-44808-3_14

Completeness and c.e. Degrees

In 1944, Post proved that the set $\mathcal{K}$ has the following interesting property: every c.e. set is 1-reducible to $\mathcal{K}$. (See Definition 9.2 (p. 213) and Problem 9.3 (p. 214).) Since 1-reducibility is also m-reducibility (Sect. 9.2.4), and hence T-reducibility, it followed that every c.e. set is T-reducible to $\mathcal{K}$. Post called such a set *complete* relative to the T-reducibility. We now define the concept of completeness more generally, i.e., relative to an arbitrary reduction, $\leq_C$. (See discussion on $\leq_C$ in Sect. 9.2.1.)

Definition 14.1. (Complete Set) Let $\leq_C$ be an arbitrary reducibility. A c.e. set $\mathcal{S}$ is said to be C-**complete** if $\mathcal{A} \leq_C \mathcal{S}$ for every c.e. set $\mathcal{A}$.

So, Post was aware that $\mathcal{K}$ is T-complete. As this played an important role in his further research, we state it as a theorem.

Theorem 14.1. *The set $\mathcal{K}$ is T-complete.*

Just as c.e. sets are important, so are their degrees of unsolvability. These we introduce in the next definition.

Definition 14.2. (c.e. Degree) A T-degree is **computably enumerable** (**c.e.**) if it contains a c.e. set.

Remark. It is customary to call a T-degree that is computably enumerable just a *c.e. degree*.

Since $\emptyset$ and $\mathcal{K}$ are c.e. sets (see Sect. 8.2.1), both **0** and **0**$'$ are c.e. degrees.

14.2 Post's Problem

In 1944, Post asked whether there are any c.e. degrees strictly between **0** and **0**$'$? Here is the citation from his paper of 1944:

> A primary problem in the theory of recursively enumerable sets is the problem of determining the degrees of unsolvability of the unsolvable decision problems thereof. We shall early see that for such problems there is certainly a highest degree of unsolvability. Our whole development largely centers on the single question of whether there is, among these problems, a lower degree of unsolvability than that, or whether they are all of the same degree of unsolvability.

Remarks. "Recursively enumerable sets" is the old name for c.e. sets. The problems of Post's interest are decision problems associated with undecidable c.e. sets, that is, undecidable semi-decidable decision problems. Clearly, the highest degree of unsolvability of such problems is $0'$. Post asks whether or not any such problem can be of lower degree of unsolvability than $0'$.

This question became known as *Post's Problem.*

Definition 14.3. (Post's Problem) Is there a c.e. degree c such that $0 < c < 0'$?

We already know that there are T-degrees strictly between 0 and $0'$ (Theorem 13.5). However, the T-degrees constructed in the proof of this theorem need not be c.e. This is why Post embarked on solving this problem. In the next section we will describe his approach. Although Post's approach proved to be only partially successful, it nevertheless brought many new ideas and, more importantly, motivated researchers to search for different methods that would lead to a solution to the problem.

14.2.1 Post's Attempt at a Solution to Post's Problem

To solve this problem positively, Post wanted to prove that there is a c.e. set $\mathcal{A}$ such that $\emptyset <_T \mathcal{A} <_T \mathcal{K}$. How would he do that? He had the following idea: If he managed to prove that there exists an undecidable c.e. set $\mathcal{A}$ such that $\mathcal{K} \nleq_T \mathcal{A}$, then $\emptyset <_T \mathcal{A} <_T \mathcal{K}$ would follow. In other words, Post wanted to find an *undecidable c.e.* set that is *T-incomplete.* So he set a list of goals—called *Post's Program*—that should be attained in order to solve the problem.

Post's Program had three goals:

A. define a *property* of a c.e. set;
B. prove that any set with this property is *undecidable* and *T-incomplete*;
C. prove that there *exist* c.e. sets with this property.

He started modestly, with the m-reducibility $\leq_m$, which is more special than $\leq_T$, the Turing reducibility. Then, if he succeeded in fulfilling his program with $\leq_m$, he would try to generalize the proof to the reducibility $\leq_T$.

Suppose that $\mathcal{K} \leq_m \mathcal{A}$ for some c.e. set $\mathcal{A}$. Then, there is a computable function r such that $r(\mathcal{K}) \subseteq \mathcal{A}$ and $r(\overline{\mathcal{K}}) \subseteq \overline{\mathcal{A}}$ (see Problem 9.1b on p. 213). Post's idea was to achieve $\mathcal{K} \nleq_m \mathcal{A}$ by defining $\mathcal{A}$ so that $\overline{\mathcal{A}}$ would be unable to contain $r(\overline{\mathcal{K}})$, for any computable function r. Thus, the complement $\overline{\mathcal{A}}$ would be too "sparse" to accommodate $r(\overline{\mathcal{K}})$, in the sense that it would lack certain objects which, in contrast, abound in $\overline{\mathcal{K}}$. Post hoped that this property of $\mathcal{A}$ would prevent the m-reduction of $\mathcal{K}$ to $\mathcal{A}$.

How should he define this property of "having a sparse complement"? Post analyzed the properties of the set $\mathcal{K}$ and discovered that the complement $\overline{\mathcal{K}}$ contains infinitely many c.e. subsets. (See the details in Box 14.1). So, if $\overline{\mathcal{A}}$ had *no* c.e. subsets at all, it might have been unable to host $r(\overline{\mathcal{K}})$, for any computable function r. This led him to the definition of *simple* sets. Informally, a c.e. set $\mathcal{A}$ is simple if $\overline{\mathcal{A}}$ is "sparse," in the sense that, although infinite, $\overline{\mathcal{A}}$ does not contain any infinite c.e. set. Here is the official definition.

Definition 14.4. (Simple Set) A set $\mathcal{A}$ is **simple** if $\mathcal{A}$ is c.e. and $\overline{\mathcal{A}}$ is infinite but $\overline{\mathcal{A}}$ contains no infinite c.e. set.

Next, in accordance with goal B of his program, it was necessary to prove that a simple set $\mathcal{A}$ would actually do the job, i.e., guarantee that $\mathcal{K} \not\leq_m \mathcal{A}$. In other words, a simple set would be m-incomplete. That this is indeed so, Post proved in the following theorem.

Theorem 14.2. *If a set $\mathcal{A}$ is simple, then $\mathcal{A}$ is not m-complete.*

It remained to be proved that a simple set actually exists (goal C). Using diagonalization, Post proved the following theorem.

Theorem 14.3. *There exists a simple set $\mathcal{A}$.*

Proof. We omit the proof. See the Bibliographic Notes to this chapter.   □

In summary, Post discovered that there exist c.e. sets that are strictly between $\emptyset$ and $\mathcal{K}$ in the $\leq_m$ ordering. So he could write the following corollary.

Corollary 14.1. *There is a c.e. set $\mathcal{A}$, such that $\emptyset <_m \mathcal{A} <_m \mathcal{K}$.*

Yet, this was not the final solution to his problem, because the question was whether the statement of Corollary 14.1 holds for the Turing reducibility $\leq_T$.

In order to generalize the above result to the ordering $\leq_T$, Post defined a more general notion of reducibility, the *bounded truth-table reducibility* $\leq_{btt}$. (See the Bibliographic Notes to this chapter for definitions of the notions that we mention in this paragraph.) Post was able to prove that if $\mathcal{A}$ is simple, then $\mathcal{A}$ is btt-incomplete (i.e., $\mathcal{K} \not\leq_{btt} \mathcal{A}$). Thus, there is a c.e. set $\mathcal{A}$ such that $\emptyset <_{btt} \mathcal{A} <_{btt} \mathcal{K}$. This success led him to try with the still more general notion of *truth-table reducibility* $\leq_{tt}$ (that is, $\leq_{btt}$ with no bounds). However, it turned out that if $\mathcal{A}$ is simple, the relation

$K \nleq_{tt} A$ *cannot* be proved. Since this indicated that the notion of the simple set was too weak, he defined the more powerful notion of the *hyper-simple set*. He was then able to prove that (goal B) if A is hyper-simple, then A is tt-incomplete (i.e., $K \nleq_{tt} A$), and (goal C) a hyper-simple set actually exists. It followed that there is a c.e. set A, such that $\emptyset <_{tt} A <_{tt} K$.

The reducibilities $\leq_1, \leq_m, \leq_{btt}$, and $\leq_{tt}$ are called *strong*. This is because they are obtained by imposing additional conditions on the Turing reducibility $\leq_T$. We now see that Post aimed to gradually relax the strength of the additional conditions and eventually obtain the Turing reducibility; during this he would define, for each new weaker condition C, a new kind of c.e. sets (goal A), prove that such sets meet the condition $\emptyset <_C A <_C K$ (goal B), and prove the existence of such sets (goal C).

Unfortunately, his progress stalled after success with the truth-table reducibility ($C = tt$), and this was still far from the Turing reducibility. To continue, he defined *hyperhyper-simple* sets, but he couldn't prove their existence.

Box 14.1 (Creative Sets).

When Post analyzed the set K and its complement $\overline{K}$, he discovered that K possesses an interesting property, which he then called *creativity*. Here is the definition of a creative set.

Definition 14.5. (Creative Set) A set C is **creative** if C is c.e. and there is a p.c. function φ such that $(\forall x)[W_x \subseteq \overline{C} \implies \varphi(x)\downarrow \wedge \varphi(x) \in \overline{C} - W_x]$.

What does that mean? First, observe that W_x denotes the xth c.e. set. (To see that, combine Definition 6.4 (p. 125) and Problem 6.7 (p. 140).) The function φ produces, for any W_x, an element $\varphi(x)$ *witnessing* that $\overline{C} \neq W_x$ (as $\varphi(x) \in \overline{C} - W_x$). We call φ the *production function* of the creative set C. Since we can effectively produce, for any W_x, a counterexample for the assertion $\overline{C} = W_x$, we say that $\overline{C}$ is *effectively* non-c.e. Hence, C is an effectively undecidable set.

The set $\overline{C}$ contains an infinite c.e. set; moreover, it contains infinitely many infinite c.e. sets (see Problem 14.1).

Since the set K is creative (see Problem 14.3), $\overline{K}$ is effectively non-c.e. and contains infinitely many infinite c.e. sets. Now Post's next step was obvious: for the searched-for set A, take a *non-creative* set. A non-creative set he called a *simple* set (see Definition 14.4).

In summary, the ultimate goal of *Post's Program* for solving *Post's Problem* was to define an appropriate *structural* property of c.e. sets that would guarantee the existence, undecidability, and incompleteness of c.e. sets having this property. As we have seen, the structural property that attracted Post was "sparseness of the complement." Could Post have succeeded in attaining his program had he continued his investigation in this direction? The answer is no. In 1965, Yates[1] proved that the structural property of having a "sparse complement," and the new hyperhyper-simple sets, could not lead to a positive solution to *Post's Problem*.

---

[1] C. E. M. Yates, British mathematician, logician, and computer scientist.

14.3 The Priority Method and Priority Arguments

Nevertheless, today we know that the answer to *Post's Problem* is positive: there is a c.e. set $\mathcal{A}$ such that $\emptyset <_T \mathcal{A} <_T \mathcal{K}$, and there is a c.e. degree a such that $0 < a < 0'$. How was the problem resolved?

Post posed his problem and described his attempt at a solution in 1944. Ten years later, in 1954, when Kleene and Post published their seminal paper containing many other discoveries and ideas (see Chap. 13), the problem was still open. In the paper, Kleene and Post also introduced the *method of finite extensions*, which they used to prove the existence of $\leq$-incomparable *T*-degrees (see Theorem 13.5). This attracted the attention of several researchers. In 1956, two of them, Friedberg[2] and Muchnik[3], simultaneously and independently upgraded the finite extensions method into a subtler one, the *Finite-Injury Priority Method*, as it is called today. By applying it, Friedberg and Muchnik obtained a positive answer to *Post's Problem*.

14.3.1 The Priority Method in General

We will now describe the *Priority Method* in general. Let P be a property sensible of sets. Is there a c.e. set with the property P? To answer the question affirmatively, we can embark on the construction of such a set. We can try to construct a c.e. set S with the property P by adhering to the guidelines described in the following.

1. The set S will be constructed step by step, in an infinite sequence of *stages*. Each stage, say i, will construct a finite set S_i that will be an approximation of the set S. Informally, S_i will be obtained by adding new elements into S_{i-1} and/or banning certain elements from entering S_i. Intuitively, we want the sets S_i to monotonously grow as i increases, and eventually (in the limit) develop into the set S. The plan is to define the stages in such a way that two objectives will be achieved:

Objective 1: $S_{i-1} \subseteq S_i$, for every $i \geqslant 1$;

Objective 2: $\bigcup_{i=1}^{\infty} S_i = S$.

That is, each S_i should be a better approximation of S than S_{i-1}, so that $\lim_{i \to \omega} S_i = S$. But, isn't our plan unrealistic? How can we approximate an unknown object such as S? If S exists, it will become fully known only after infinitely many stages of the construction; and yet, we intend to obtain, at each stage i, a *better* approximation S_i of S. How will we evaluate the quality of S_i? Will S_i correctly extend S_{i-1}? This information will be needed to avoid missing the objectives.

---

[2] Richard Michael Friedberg, American mathematician.

[3] Albert Abramovič Mučnik, 1934, Russian mathematician.

2. The above situation is resolved by making a radical turn in our view of what is approximated during the construction. Instead of approximating the unknown set S, we approximate the known property P. The idea is that the properties of each finite set S_i should approximate the property P; that is, they should satisfy, *at least partially*, the property P. To implement this idea, we must somehow *atomize* P, i.e., break it into a set of primitive properties. Only then will a set S_i be capable of fulfilling a part of P. The primitive properties are called the *requirements*. Clearly, it is sensible to atomize P in such a way that any set fulfilling *all* the requirements will fulfill the whole of P. (Such would be the set S.) To ensure this, we must break P into a *conjunction* of requirements. So we have developed the first two guidelines:

G1: Write P as a conjunction $R_0 \wedge R_1 \wedge R_2 \ldots$ of countably many requirements R_i.
G2: Do G1 in such a way that a set has the property $P \Longleftrightarrow$ the set fulfills every R_i.

3. What is the requirement and how is it fulfilled? In order to keep it as simple as possible, a requirement is only allowed to specify that certain numbers must be added (i.e., *enumerated*) into the set S_i and/or that certain other numbers must be kept out of this set (for some i). So, a requirement will be fulfilled by carrying out finitely many instructions of the form

$$x \in !\, S_i \quad (\text{add } x \text{ into } S_i)$$

or

$$x \notin !\, S_i \quad (\text{ban } x \text{ from } S_i).$$

Note that $x \notin !\, S_i$ does not mean that x is deleted from S_i; it means that x is *kept out of S_i*. A requirement R is *initiated* (tried to be fulfilled) at the stage when it *receives attention*. At that stage, say i, R can be fulfilled simply by ensuring the presence or absence of a particular number in the set S_i, i.e., by imposing either $x \in !\, S_i$ or $x \notin !\, S_i$, where x is a *candidate*—a number whose *status* (potential membership in S_j) has not been considered, for any $j < i$. In general, there are infinitely many candidates and we must choose one of them. This is summarized in the following guideline:

G3: At stage i, a requirement R is fulfilled by deciding, for a candidate x, whether $x \in !\, S_i$ or $x \notin !\, S_i$.

Remark. Fulfilling R uncovers just a small part of the information about S, namely, the status of the candidate number. The current information about the contents of S, which has been gathered by fulfilling the requirements by the end of stage i, is represented by S_i. Since S_i fulfills more requirements than S_{i-1}, it meets P better than S_{i-1}, and hence better approximates S.

4. It is now obvious that we must arrange the construction in such way that each and every requirement will eventually be considered. If we succeed in finding such an arrangement, then each and every requirement will get a chance to be fulfilled, so that *all* the requirements *may* eventually become fulfilled. In that case, they will be fulfilled by the limit set $\lim_{i \to \omega} S_i$. According to *G2*, this set will have the property P, so it will be the searched-for set S.

An intuitively appealing guideline for considering the requirements would be: consider R_0, then consider R_1, then consider R_2, and so on. Actually, this arrangement worked in the proof of Theorem 13.5 (p. 259), and the reason for this is easy to find: *a requirement, once fulfilled, always remained fulfilled.* The proving method where this holds is called the *Finite Extension Method*, in short the *FEM*. But the FEM is not considered to be a true *Priority Method*, because it cannot deal with the more general situation, which we describe in the following.

5. Unfortunately, considering the requirements in a simple succession may not suffice. Why? By any stage only finitely many requirements can be fulfilled, which leaves infinitely many Rs to be considered and fulfilled in the rest of the construction. But, the future requirements may not be independent from those already fulfilled; they may interact with each other. So, we are faced with a permanent lack of information about future Rs—and this may have significant consequences.

In particular, it may turn out, at any stage i, that it is impossible to fulfill the current requirement R. How can that happen? There can be two reasons. First, at some previous stage $j < i$, a requirement R' was fulfilled in a *wrong* way; that is, R' was fulfilled by banning a candidate y from S_j, i.e., setting $y \notin ! S_j$, and this decision now, at stage i, *prevents* us from fulfilling R. In short, we banned from S_j a candidate that we shouldn't have—and we could not anticipate that at stage j. The second reason can be that R' and R are contradictory requirements and cannot both be fulfilled. (Isn't there also a third possibility? Cannot $y \in ! S_j$ be a wrong decision? Since the set S is to be c.e., we assume that the decision $y \in ! S_j \subset S$ is well grounded and can be made effectively, so there will be no need to revoke it.)

We see that there should be a possibility of returning to a previously fulfilled requirement and fulfilling it in some other way that would allow us to proceed with the construction. But, things are more complicated than that: changing a decision about S_j (e.g., from $y \notin ! S_j$ to $y \in ! S_j$) may affect the sets $S_{j+1}, \ldots, S_{i-1}$ (which were constructed from S_j), so also the requirements fulfilled at stages $j+1, \ldots, i-1$ may have to be reconsidered and fulfilled in some other way.

6. How can the bewildering situation described in **5** be controlled so that all the changes will be systematically enforced? Here, Friedberg and Muchnik introduced a new ingredient in the method: they assumed that different requirements have different *priorities*. Informally, R's priority is used to represent the *importance* of R. Here is the new guideline.

G4: With different requirements associate different priorities.

It proves to be useful to index the Rs in such a way that their priorities decrease as the index increases; that is, R_k is of higher priority than R_{k+1}, for any $k \geqslant 0$. So, a requirement with a lower index has higher priority and is therefore more important. From now on we will assume that such an indexing has been done.

7. The construction is arranged so that, at each stage, the next initiated require-ment is the one that has the highest priority among the requirements which require attention at that stage. We say that such a requirement *receives attention*.

G5: At any stage initialize the highest-priority R that requires attention at that stage.

8. How did Friedberg and Muchnik apply the concept of priorities to resolve the situation described in paragraph **5**? Recall the situation: to fulfill the current R, a pre-viously fulfilled R' should be reconsidered and fulfilled in some other way. Whether or not this is allowed will depend on the priorities of R and R' in the following way:

- R *has higher priority than* R'. If R' was fulfilled by wrongly choosing $y \notin !S_j$, we are allowed to return to R', revoke that choice, and change it to $y \in !S_j$. This will enable the fulfillment of R. However, it will also turn R' unfulfilled again. We say that R' has been *injured* by R. (We will have to cure R', that is, fulfill R' in some other way that does not affect R. We will do that by considering some other candidate number y' and choosing either $y' \in !S_j$ or $y' \notin !S_j$.)
- R *has a lower priority than* R'. In this case R is not allowed to injure R'.

This is summarized in the next guideline.

G6: A requirement can be injured only by a higher-priority requirement.

9. Friedberg and Muchnik's next key assumption was that an injured requirement R will be reconsidered *after* all the injured requirements having higher priority than R are cured. Here is the guideline.

G7: An injured R will be initiated after all higher-priority injured Rs are cured.

If R_k is injured, then there can be at most k more important injured requirements: $R_0, \ldots, R_{k-1}$. It will take finitely many steps to cure all of them, so an attempt at R_k's recovery is guaranteed to start after a finite number of stages.

10. Although only finitely many requirements can injure an R, there might still exist a requirement that would injure R infinitely many times. In such a case, R would never stop being injured and could not recover once and for all. Could the entire conjunction $R_0 \wedge R_1 \wedge R_2 \ldots$ then be fulfilled? To prevent such a situation, Friedberg and Muchnik assumed that each requirement can be injured only finitely many times. So, they introduced the following guideline.

G8: A requirement can be injured only finitely many times.

This concludes the general description of the *Priority Method* for the construction of a set S with property P. However, there are several variations of the method. For example, when the method adheres to all of the guidelines $G1, \ldots, G8$, it is called the *Finite-Injury Priority Method* (*FIPM*). When $G8$ cannot be assumed, we obtain the *Infinite-Injury Priority Method* (*IIPM*).

A proof using any kind of the *Priority Method* is called a *priority argument*. Priority arguments have been classified into a hierarchy based on their complexity.

14.3.2 The Friedberg-Muchnik Solution to Post's Problem

Using their *Priority Method*, Friedberg and Muchnik proved that there exist *c.e.* sets $\mathcal{A}$ and $\mathcal{B}$ such that $\mathcal{A} \not\leq_T \mathcal{B} \wedge \mathcal{B} \not\leq_T \mathcal{A}$. (As $\mathcal{A}, \mathcal{B}$ are c.e., also $\mathcal{A} \leq_T \emptyset'$ and $\mathcal{B} \leq_T \emptyset'$.) It immediately followed that $\deg(\mathcal{A})$ and $\deg(\mathcal{B})$ are $\leq$-incomparable c.e.-degrees— and *Post's Problem* was solved positively. Here is the Friedberg-Muchnik theorem.

Theorem 14.4. *There exist incomparable c.e.-degrees.*

Proof. We describe the idea of the proof; for the details see the Bibliographic Notes to this chapter. Friedberg-Muchnik's general goal was to extend the proof of Theorem 13.5 (p. 259) to the case of c.e. sets. So the requirements stated in that proof remain the same (now they are all denoted by Rs):

$$R_{2e} : \chi_{\mathcal{A}} \neq \Psi_e^{\mathcal{B}}$$
$$R_{2e+1} : \chi_{\mathcal{B}} \neq \Psi_e^{\mathcal{A}}$$

Fulfilling a Single Requirement. To fulfill R_{2e} we first associate with it a candidate number, an x whose status (membership in $\mathcal{A}$) is still open. Then we look for a stage $s+1$ such that $\Psi_e^{\mathcal{B}_s}(x){\downarrow} = 0$. If there is no such stage, then this means that $x \notin \mathcal{A}$ and $\Psi_e^{\mathcal{B}}(x){\uparrow} \vee \Psi_e^{\mathcal{B}}(x){\downarrow} \neq 0$, implying that x fulfills R_{2e}. If, however, such an $s+1$ exists, then R_{2e} will *require attention* at that stage.

When, at stage $s+1$, R_{2e} actually *receives attention*, we do the following:

- add x into $\mathcal{A}_{s+1}$ (and, hence, into $\mathcal{A}$);
- protect the construction. We do this by trying to restrain too-small numbers y from later entering $\mathcal{B}$ by any requirement of lower priority than R_{2e}. More specifically, we choose new candidates for all (lower-priority) requirements $R_k, k > 2e$, and initialize them. This ensures that only (higher-priority) requirements $R_k, k < 2e$, can later injure R_{2e} by adding some small y into $\mathcal{B}$.

This fulfills the requirement R_{2e}. If later a requirement of higher priority than R_{2e} enumerates into $\mathcal{B}$ and injures R_{2e}, then R_{2e} is initialized and must be fulfilled again by another candidate. (To fulfill a requirement R_{2e+1} we use the same strategy as for R_{2e} but with the roles of $\mathcal{A}$ and $\mathcal{B}$ reversed.)

Fulfilling All Requirements. We saw that, occasionally, we must choose new candidate numbers for some requirements. There are two restrictions on this: first, for any R_k, we can choose another candidate for R_k only finitely often; and second, candidates for different requirements must be distinct. The latter restriction is met by choosing all candidates for a requirement R_k from the set $\mathbb{N}^{[k]} \stackrel{\text{def}}{=} \{\langle n, k \rangle \mid n \in \mathbb{N}\}$.

Construction of $\mathcal{A}$ and $\mathcal{B}$. Initially, we have $\mathcal{A}_0 = \mathcal{B}_0 = \emptyset$. At stage $s > 0$ we do the following. Let R_k be the highest-priority unfulfilled requirement and let r be the stage at which R_k was the last time initialized; of course, $r < s$. Then we determine x in the following way:

- if $k = 2e$: Let x be the least $x \in \mathbb{N}^{[k]} - \mathcal{A}_{s-1}$ such that $x > r$ and $\Psi_e^{\mathcal{B}_{s-1}}(x) = 0$;
- if $k = 2e+1$: Let x be the least $x \in \mathbb{N}^{[k]} - \mathcal{B}_{s-1}$ such that $x > r$ and $\Psi_e^{\mathcal{A}_{s-1}}(x) = 0$.

In the first case we add x into $\mathcal{A}$, and in the second case we add x into $\mathcal{B}$. (Observe that $x < s$.) We then declare R_k fulfilled and initialize all requirements of lower priority (as described above).

It can be proved that, for every k, requirement R_k receives attention at most finitely often, is injured at most $2^k - 1$ times, and is eventually fulfilled forever.   □

14.3.3 Priority Arguments

We have seen that Friedberg and Muchnik's solution to *Post's Problem* is somewhat difficult to follow. This is not a coincidence. Today, when the *Priority Method* is the main technique for establishing results about c.e. sets, it is known that priority arguments are usually very complex and sophisticated. First, the requirements and the strategy by which they are fulfilled must be carefully constructed to produce the required result—and this must be done for each problem separately. In addition, requirements can be first fulfilled and later injured, so the membership of a number in the constructed set can be first determined and later undone.

However, as the results obtained by priority arguments have been multiplying, attempts to make priority arguments more systematic and intelligible have also started appearing. The aim of these attempts is to isolate the general principles that are common to the existing priority arguments (and, hopefully, to those yet to be constructed). Ultimately, the attempts should lead to a *framework* what would offer a uniform approach to the construction of priority arguments.

While waiting for such a systematic simplification of priority arguments, it has become desirable to prove results without priority arguments, or to see if results proved with priority arguments can also be proved without them. For example, in 1986, Kučera[4] devised a proof of *Post's Problem* without using the priority method. Kučera's proof is involved too, but the resulting set is less artificial.

14.4 Some Properties of C.E. Degrees

When it became known that there are more than just two c.e. degrees, research into c.e. degrees started to flourish. Many of the results were (and are being) obtained by priority arguments. We now briefly list some of the first results. For further results see the Bibliographic Notes to this chapter.

1. *Every c.e. degree is $\leq \mathbf{0}'$.*
2. *Not every T-degree which is $< \mathbf{0}'$ is a c.e. degree.*
3. *Density Theorem: Between any two c.e. degrees there is a third c.e. degree.*
4. *There are two c.e. degrees with no glb in the c.e. degrees.*
5. *There is a pair of nonzero c.e. degrees whose glb is $\mathbf{0}$.*
6. *Nondiamond Theorem: There is no pair of c.e. degrees whose glb is $\mathbf{0}$ and lub is $\mathbf{0}'$.*

The above theorems were proved in the mid-1960s. In particular, 3 was proved in 1964 by Sacks; 4 and 5 were proved in 1966 by Lachlan[5] and Yates; and 6 was proved in 1966 by Lachlan. See Bibliographic Notes to this chapter for the details.

---

[4] Antonín Kučera, Czech mathematician, logician, and computer scientist.

[5] Alistair H. Lachlan, Canadian mathematician, logician, and computer scientist.

14.5 Chapter Summary

A c.e. set is said to be Turing complete (T-complete) if every c.e. set is T-reducible to it. The set $\mathcal{K}$ is T-complete. A T-degree is said to be c.e. if it contains a c.e. set. Both $\mathbf{0}$ and $\mathbf{0}'$ are c.e. degrees. *Post's Problem* asks whether or not there exists a c.e. degree c that is strictly between $\mathbf{0}$ and $\mathbf{0}'$, that is, $\mathbf{0} < c < \mathbf{0}'$.

Post attempted a solution to his problem by devising a program which is now called *Post's Program*. His aim was to define a structural property of c.e. sets that would guarantee the existence, undecidability, and Turing incompleteness of c.e. sets having this property. In trying to achieve that, Post defined various reducibilities and special kinds of c.e. sets, but did not succeed in attaining his program.

A positive solution to *Post's Problem* was obtained in 1965 by Friedberg and Muchnik, who devised and used a new method called the *Priority Method*. In this method, a conjunction of simple requirements must be fulfilled. As more and more requirements are fulfilled, a larger and larger part of the set to be constructed is uncovered. However, the requirements are interrelated, so the requirements that have already been fulfilled may become unfulfilled again, thus temporarily concealing a part of the uncovered set. Such injured requirements must be fulfilled again in some other way, and the process repeats. Although involved, the *Priority Method* is today the main technique for establishing results about c.e. sets.

Proofs that use this method are called priority arguments. There are attempts to isolate the general principles common to the priority arguments and to integrate them into a framework that would offer a uniform approach to the construction of priority arguments.

At the same time, priority-free proofs are searched for because the sets constructed by them are less artificial. One such is Kučera's priority-free proof of *Post's Problem*.

Problems

14.1. Prove: If $\mathcal{C}$ is a creative set, then $\overline{\mathcal{C}}$ contains an infinite c.e. subset.

[*Hint*. Let φ be the production function of $\mathcal{C}$. Let n be an index of the empty set, i.e., $\emptyset = \mathcal{W}_n$. Consider the set $\mathcal{W} = \{x_1, x_2, \ldots\}$, where x_i is defined inductively as follows: $\mathcal{W}_{x_1} = \{\varphi(n)\}$ and $\mathcal{W}_{x_{i+1}} = \mathcal{W}_{x_i} \cup \{\varphi(x_i)\}$. So, the construction of the set $\mathcal{W}$ starts with $\mathcal{W}_n = \emptyset$ as the first current set and then, in each step, adds to the current set its witness to obtain the next current set.]

14.2. Prove: If $\mathcal{C}$ is a creative set, then $\overline{\mathcal{C}}$ contains infinitely many infinite c.e. subsets.

[*Hint*. See Problem 14.1 and use any $n \in \text{ind}(\emptyset)$.]

14.3. Prove: $\mathcal{K}$ is a creative set.

Bibliographic notes

- Post formulated the (now called) *Post's Problem* in [124]. See also Post [127].
- That $\mathcal{K}$ (actually $\mathcal{K}_0$) is 1-complete and hence also m-complete was proved in Post [124].
- The existence of *simple* sets was proved in Post [124]. See also Cooper [29, Chap. 6], Weber [186, Chap. 5], and Soare [164, Chap. V].
- *Strong reducibilities* $\leq_m, \leq_{htt}, \leq_{tt}$ were introduced in Post [124]. For more on these reducibilities see Odifreddi [115].
- That *Post's Program* could not succeed with the structural property of "sparseness of the complement" was proved in Yates [189].
- The solution to *Post's Problem* appeared in Muchnik [110, 111] and Friedberg [48]. In describing the solution we followed Nies [113] and Soare [164]. See also Cooper [29].
- A wealth of information about the ideas, techniques, and applications of the *priority argument* is Cooper [29, Chap. 12]. For the *Finite Injury* and *Infinite Injury Priority Method*, see Soare [164, Chaps. VII and VIII]. A framework for developing priority arguments is described in Lerman [94].
- A solution to *Post's Problem*, which does not use the priority method and therefore has no injury to the requirements, was obtained in 1986 by Kučera [89]. See Cooper [29, Chap. 15], Nies [113, Chap. 4], and Soare [164, Chap. VII] for the expositions of this proof.
- Theorems 3,4,5, and 6 from Sect. 14.4 were proved by Sacks [146], Lachlan [90, 91], and Yates [190], respectively.
- For an overview of c.e. sets, see Soare [166], and for an overview of c.e. degrees, see Shore [152].

Chapter 15
The Arithmetical Hierarchy

For every sensible question there is an answer; for every answer there is a sensible question.

Abstract In this chapter we will introduce a different view of sets of natural numbers. Sometimes such a set can be defined by a property of its members, where the property is expressed by a formula of *Formal Arithmetic*. Sets defined by formulas of the same complexity constitute an *arithmetical class*. Different complexities of formulas give rise to different arithmetical classes. There is also an ordering between these classes, so they form the so-called *Arithmetical hierarchy*. We will show that the *Arithmetical hierarchy* is closely connected with the *Jump hierarchy*.

15.1 Decidability of Relations

Before we delve deeper into the main subject of this chapter, we must prepare the ground by defining a few new notions and proving some basic facts about them.

A k-ary *relation* on a set S is a subset $\mathcal{R}$ of S^k. If $\mathcal{R}$ is a k-ary relation on S, it is customary to write $R(a_1, \ldots, a_k)$ to indicate that $(a_1, \ldots, a_k) \in \mathcal{R}$. When $k = 1$ we say that R is a *property* defined on S; when $k = 2$ we call R a *binary* relation on S. Just like any other set, the set $\mathcal{R}$ can also be decidable, semi-decidable, undecidable, or can have any other sensible property of sets. In this case we say that R has (or does not have) such a property. Here are the definitions that we will need.

Definition 15.1. (Decidable Relation) A k-ary relation R on a set S is **decidable** (or **semi-decidable**, or **undecidable**) if the corresponding set $\mathcal{R} \subseteq S^k$ is decidable (or semi-decidable, or undecidable).

Remarks. 1) Instead of decidable (semi-decidable, undecidable) relation, we may say computable (c.e., incomputable) relation. 2) If R is decidable, we can decide, for any $(a_1, \ldots, a_k) \in S^k$, whether or not $R(a_1, \ldots, a_k)$. If R is undecidable but still semi-decidable, such a deciding is guaranteed to return an answer (a YES) only for k-tuples that are in $\mathcal{R}$. Otherwise, there is no such guarantee.

© Springer-Verlag Berlin Heidelberg 2015
B. Robič, *The Foundations of Computability Theory*,
DOI 10.1007/978-3-662-44808-3_15

We will now fix the set $\mathcal{S}$ to $\mathcal{S} = \mathbb{N}$. Relations on $\mathbb{N}$ are said to be *arithmetical*.

Example 15.1. (Relation R_{Halt}) Let us define an arithmetical relation $R_{Halt}(e,x,s)$ as follows:

$R_{Halt}(e,x,s) \equiv$ "Turing machine T_e halts on input x in at most s steps and returns a result."

This relation is decidable, because we can decide, for any $(e,x,s) \in \mathbb{N}^3$, whether or not $R_{Halt}(e,x,s)$. To appreciate this, recall from Sect. 7.2 that we can construct T_e for any natural e, start T_e on any natural x, and wait until a natural number is output or the first s steps have been completed. $\quad\square$

Let us now take an arbitrary decidable relation $R(x,y)$ on $\mathbb{N}$ and define the set $\mathcal{A}$ to be $\mathcal{A} = \{x \in \mathbb{N} \mid \exists y R(x,y)\}$. So $\mathcal{A}$ consists of those numbers that are R-related to some number. What can be said about the decidability of $\mathcal{A}$? Here is the answer.

Theorem 15.1. *A set $\mathcal{A} \subseteq \mathbb{N}$ is c.e. iff $\mathcal{A} = \{x \in \mathbb{N} \mid \exists y R(x,y)\}$ for some decidable relation R on $\mathbb{N}$.*

Proof. ($\Leftarrow$) Let R be an arbitrary decidable relation on $\mathbb{N}$. Let $D_{\mathcal{R}}$ be a decider of the corresponding set $\mathcal{R}$, and $G_{\mathbb{N}^2}$ a pair generator (see Sect. 6.3.4 and Box 6.3). Then we can construct a generator $G_{\mathcal{A}}$ of the set $\mathcal{A}$ as follows. $G_{\mathcal{A}}$ repeats the following sequence of steps: (1) it calls $G_{\mathbb{N}^2}$ to generate the next (x,y); (2) it calls $D_{\mathcal{R}}$ to see if (x,y) is in $\mathcal{R}$; (3) if $(x,y) \in \mathcal{R}$ then it generates (outputs) x. Since $\mathcal{A}$ can be generated by a TM, it is a c.e. set. ($\Rightarrow$) If $\mathcal{A}$ is c.e., then it is the domain of a p.c. function, φ_e (Problem 6.7). So $\mathcal{A} = \{x \in \mathbb{N} \mid \exists s R_{Halt}(e,x,s)\}$, with R_{Halt} from Example 15.1. $\quad\square$

15.2 The Arithmetical Hierarchy

In the 1940s, Kleene and Mostowski[1] were exploring the sets of natural numbers that are defined as $\{x \in \mathbb{N} \mid F(x)\}$, where $F(x)$ is a formula of *Formal Arithmetic*, and x is a free individual variable in F (see Sect. 3.2).

Kleene was investigating how the syntactical complexity of $F(x)$ affects the decidability of the set $\{x \in \mathbb{N} \mid F(x)\}$. The obvious question was how to measure the syntactical complexity of $F(x)$. Here, Kleene leaned on Kuratowski[2] and Tarski,[3] who discovered that every formula can be transformed into *prenex normal form* (*pnf*), i.e., a logically equivalent formula consisting of a string of quantifiers followed by a quantifier-free formula. He could therefore assume that $F(x)$ is of the form $Q_1 y_1 \ldots Q_k y_k R(x, y_1, \ldots, y_k)$, where $y_i \neq y_j$ for $i \neq j$, Q_i is $\forall$ or $\exists$, and R is a decidable arithmetical relation (predicate). In addition, adjacent quantifiers of the same kind can be contracted and replaced by a single quantifier of that kind (see Box 15.1). After all possible contractions have been performed, the resulting pnf has a sequence of *alternating* quantification symbols, i.e., $F(x)$ is either of the form

[1] Andrzej Mostowski, 1913-1975, Polish mathematician.

[2] Kazimierz Kuratowski, 1896–1980, Polish mathematician and logician.

[3] Alfred Tarski, 1901–1983, Polish mathematician and logician.

$$\exists y_1 \forall y_2 \exists y_3 \ldots Q y_n R(x, y_1, y_2, \ldots, y_n),$$

where Q is $\exists$ if n is odd, and $\forall$ if n is even; or it is of the form

$$\forall y_1 \exists y_2 \forall y_3 \ldots Q y_n R(x, y_1, y_2, \ldots, y_n),$$

where Q is $\forall$ if n is odd, and $\exists$ if n is even.

Box 15.1 (Contraction of Quantifiers in Prenex Normal Forms).

We describe the contraction of quantifiers on an example. Let $\exists y_1 \exists y_2 \forall y_3 \forall y_4 \forall y_5 P(y_1, y_2, y_3, y_4, y_5)$ be a formula. We introduce two individual variables $u = (y_1, y_2)$ and $v = (y_3, y_4, y_5)$. Recall that the projection function π_i^k returns the ith component of a k-tuple (Box 5.1, p. 74). Using u, v, π^2 and π^3 we transform the formula into an equivalent formula $\exists u \forall v P(\pi_1^2(u), \pi_2^2(u), \pi_1^3(v), \pi_2^3(v), \pi_3^3(v))$. This is the contracted pnf. Generally, given a formula $Q_1 y_1 \ldots Q_k y_k P(y_1, \ldots, y_k)$, we first partition $Q_1 y_1 \ldots Q_k y_k$ into maximal subsequences of adjacent quantifiers of the same kind; then we introduce, for each subsequence, a new individual variable (tuple); next, we replace each subsequence by the corresponding quantifier of that kind; and use projection functions in the predicate P.

So, Kleene could assume that $F(x)$ is already in the contracted prenex normal form. Any set $\{x \in \mathbb{N} \mid F(x)\}$ defined by such a distinguished $F(x)$ he called *arithmetical*.

Definition 15.2. (Arithmetical Set) A set $\mathcal{A}$ is an **arithmetical set** if $\mathcal{A} = \{x \in \mathbb{N} \mid F(x)\}$, such that, for some $n \geqslant 0$, the predicate $F(x) = \exists y_1 \forall y_2 \exists y_3 \ldots Q y_n R(x, y_1, y_2, \ldots, y_n)$ or $F(x) = \forall y_1 \exists y_2 \forall y_3 \ldots Q y_n R(x, y_1, y_2, \ldots, y_n)$, and R is a decidable relation.

Then he defined the syntactical complexity of $F(x)$ as the number n of quantification symbols in $F(x)$. Depending on n and the first quantification symbol of $F(x)$, he classified the arithmetical sets $\{x \in \mathbb{N} \mid F(x)\}$ into various classes. These classes he called *arithmetical* and denoted them by Σ_n, Π_n, and Δ_n. Here is the definition.

Definition 15.3. (Arithmetical Classes) The **arithmetical classes** Σ_n, Π_n, and Δ_n are defined as follows:

$\Sigma_n = $ class of all sets $\{x \in \mathbb{N} \mid F(x)\}$, where $F(x) = \exists y_1 \forall y_2 \exists y_3 \ldots Q y_n R(x, y_1, \ldots, y_n)$ for some decidable arithmetical relation R;

$\Pi_n = $ class of all sets $\{x \in \mathbb{N} \mid F(x)\}$, where $F(x) = \forall y_1 \exists y_2 \forall y_3 \ldots Q x_n R(x, y_1, \ldots, y_n)$ for some decidable arithmetical relation R;

$\Delta_n = $ class of all sets $\{x \in \mathbb{N} \mid F(x)\}$ that are in $\Sigma_n \cap \Pi_n$.

Example 15.2. ($\mathcal{K}$ is in Σ_1) The set $\mathcal{K}$ is in Σ_1, because $\mathcal{K} \stackrel{\text{def}}{=} \{x \mid \varphi_x(x)\downarrow\} = \{x \mid \exists s R_{Halt}(x,x,s)\}$, where R_{Halt} is the (decidable) relation from Example 15.1.   □

Example 15.3. ($\mathcal{K}_0$ is in Σ_1) This is because $\mathcal{K}_0 \stackrel{\text{def}}{=} \{\langle e,x \rangle \mid \varphi_e(x)\downarrow\} = \{\langle e,x \rangle \mid \exists s R_{Halt}(e,x,s)\} = \{\langle e,x \rangle \mid \exists s R_{Halt}(\langle e,x \rangle_1, \langle e,x \rangle_2, s)\}$, where $\langle e,x \rangle_1 = e$, $\langle e,x \rangle_2 = x$, and R_{Halt} is from Example 15.1. □

We now justify the title of this chapter, the *Arithmetical Hierarchy*. So, is there a hierarchy of arithmetical classes? Yes; in 1943, Kleene proved the following.

Theorem 15.2. *For any $n \geqslant 0$, the following hold:*

     a) $\Sigma_n \subset \Sigma_{n+1}$ b) $\Pi_n \subset \Pi_{n+1}$

     c) $\Sigma_n \subset \Pi_{n+1}$ d) $\Pi_n \subset \Sigma_{n+1}$

            e) $\Delta_n \subset \Sigma_n$

            f) $\Delta_n \subset \Pi_n$

Proof idea. In the first part we prove, for each of the above relations, that the left-hand side is related to the right-hand side with the relation $\subseteq$. To do this, we introduce a *dummy* variable y_{n+1}. In the second part we prove, for each $\mathcal{X} \subseteq \mathcal{Y}$ of the relations $\Sigma_n \subseteq \Sigma_{n+1}$, $\Pi_n \subseteq \Pi_{n+1}$, $\Sigma_n \subseteq \Pi_{n+1}$, $\Pi_n \subseteq \Sigma_{n+1}$, $\Delta_n \subseteq \Sigma_n$, $\Delta_n \subseteq \Pi_n$, that $\mathcal{X} \neq \mathcal{Y}$. We use *diagonalization* to prove that there is a set that is in $\mathcal{Y}$ but not in $\mathcal{X}$. The diagonal argument is a generalization of the argument that we used in proving the existence of undecidable c.e. sets. (See Box 15.2 for further details.)

Box 15.2 (Proof of Theorem 15.2).

     a) Let S be an arbitrary element of the arithmetical class Σ_n. Then $S = \{x \in \mathbb{N} \mid F(x)\}$, where $F(x) = \exists y_1 \forall y_2 \exists y_3 \ldots Q y_n R(x, y_1, \ldots, y_n)$ for some decidable arithmetical relation R. Now define a new relation $R'(x, y_1, \ldots, y_n, y_{n+1}) \stackrel{\text{def}}{=} R(x, y_1, \ldots, y_n) \wedge (y_{n+1} = y_{n+1})$ and a new formula $F'(x) \stackrel{\text{def}}{=} \exists y_1 \forall y_2 \exists y_3 \ldots Q y_n Q' y_{n+1} R'(x, y_1, \ldots, y_n, y_{n+1})$, where Q' denotes the alternated Q. Finally, observe that $S = \{x \in \mathbb{N} \mid F'(x)\} \in \Sigma_{n+1}$.

     b) The proof is similar to the proof of a), except that $F(x) = \forall y_1 \exists y_2 \forall y_3 \ldots Q y_n R(x, y_1, \ldots, y_n)$ and $F'(x) \stackrel{\text{def}}{=} \forall y_1 \exists y_2 \forall y_3 \ldots Q y_n Q' y_{n+1} R'(x, y_1, \ldots, y_n, y_{n+1})$.

     c) Let $S \in \Sigma_n$ and $S = \{x \in \mathbb{N} \mid F(x)\}$, where $F(x) = \exists y_1 \forall y_2 \exists y_3 \ldots Q y_n R(x, y_1, \ldots, y_n)$ for some decidable arithmetical relation R. Observe that $F(x) = \exists y_1 \forall y_2 \exists y_3 \ldots Q y_n R(x, y_1, \ldots, y_n) = \forall y_{n+1} \exists y_1 \forall y_2 \exists y_3 \ldots Q y_n [R(x, y_1, \ldots, y_n) \wedge (y_{n+1} = y_{n+1})]$, where we have introduced a new variable y_{n+1}. Now define a new relation $R'(x, y_1, \ldots, y_n, y_{n+1}) \stackrel{\text{def}}{=} R(x, y_1, \ldots, y_n) \wedge (y_{n+1} = y_{n+1})$ and a new formula $F'(x) \stackrel{\text{def}}{=} \forall y_{n+1} \exists y_1 \forall y_2 \exists y_3 \ldots Q y_n R'(x, y_1, \ldots, y_n, y_{n+1})$. Then, after appropriate renaming of the variables y_i, we see that $S = \{x \in \mathbb{N} \mid F'(x)\} \in \Pi_{n+1}$.

     d) The proof is similar to the proof of c), except that $F(x) = \forall y_1 \exists y_2 \forall y_3 \ldots Q x_n R(x, y_1, \ldots, y_n)$ and $F'(x) \stackrel{\text{def}}{=} \exists y_1 \forall y_2 \exists y_3 \ldots Q' y_n Q y_{n+1} R'(x, y_1, \ldots, y_n, y_{n+1})$.

     e,f) $\Delta_n \subseteq \Sigma_n$ follows directly from the definition of Δ_n. The same holds for $\Delta_n \subseteq \Pi_n$.

We demonstrate the second part of the proof on cases e) and f). We will construct a set $\mathcal{P}$ such that $\mathcal{P} \in \Sigma_n - \Pi_n$ and $\overline{\mathcal{P}} \in \Pi_n - \Sigma_n$. Since $\Delta_n = \Sigma_n \cap \Pi_n$, parts e) and f) of the theorem will follow.

Each element of Σ_1 is a c.e. set and hence the domain of a p.c. function (see Problem 6.7). But p.c. functions can be effectively enumerated (see Proposition 6.1 and Definition 6.4). It follows that we can effectively enumerate the elements of Σ_1. We can also enumerate the elements of Π_1. This is because if $\mathcal{B} \in \Pi_1$, then $\mathcal{B} = \overline{\mathcal{A}_e}$ for some $\mathcal{A}_e \in \Sigma_1$, and we can rename $\mathcal{B}$ as $\mathcal{B}_e$. From the two enumerations we can construct enumerations of elements of Σ_2 and Π_2 and then of Σ_n and Π_n for higher n. Thus we can speak, for any $n \geqslant 1$, of the eth set of the class Σ_n or the class Π_n.

Now define a set $\mathcal{S}$ as follows: $\mathcal{S} \stackrel{\text{def}}{=} \{\langle e,x\rangle \in \mathbb{N} \,|\, e\text{th set in } \Sigma_n \text{ contains } x\}$. The set $\mathcal{S}$ is in Σ_1, because we can write $\mathcal{S} = \{\langle e,x\rangle \in \mathbb{N} \,|\, \exists s R_{Halt}(e,x,s)\}$, where R_{Halt} is the relation from Example 15.1. Consequently, $\mathcal{S}$ is in Σ_n, for any $n \geqslant 1$. (Notice that $\mathcal{S}$ is related to the universal set (language) $\mathcal{K}_0$; see Definition 8.5 on p. 167.)

Based on $\mathcal{S}$ we finally define the set $\mathcal{P}$ as follows: $\mathcal{P} \stackrel{\text{def}}{=} \{x \in \mathbb{N} \,|\, \langle x,x\rangle \in \mathcal{S}\}$. The set $\mathcal{P}$ is in Σ_1, because $\mathcal{P} = \{x \in \mathbb{N} \,|\, \exists s R_{Halt}(x,x,s)\}$. ($\mathcal{P}$ is related to the diagonal set $\mathcal{K}$; see Definition 8.6.) Hence, $\mathcal{P} \in \Sigma_n$. However, $\mathcal{P} \notin \Pi_n$. (*Suppose the contrary:* $\mathcal{P} \in \Pi_n$. Then it would follow that $\overline{\mathcal{P}} \in \Sigma_n$, so $\overline{\mathcal{P}}$ would be the e'th set in Σ_n for some e'. Then, by definition of $\mathcal{P}$, we would have that $e' \in \overline{\mathcal{P}} \Leftrightarrow \langle e',e'\rangle \in \mathcal{S}$. But, by definition of $\mathcal{S}$, we would also have $e' \in \overline{\mathcal{P}} \Leftrightarrow e' \notin \mathcal{P} \Leftrightarrow \langle e',e'\rangle \notin \mathcal{S}$, which would be a contradiction.) So, $\mathcal{P} \in \Sigma_n - \Pi_n$. Similarly we find that $\overline{\mathcal{P}} \in \Pi_n - \Sigma_n$.

Informally, Theorem 15.2e,f tell us that not every arithmetical set can be defined in both ways, that is, both as a member of Σ_n and as a member of Δ_n. Next, for any $n \geqslant 0$, there are arithmetical sets $\{x \in \mathbb{N} \,|\, F(x)\}$ that cannot be defined by properties $F(x)$ having just n alternating quantifiers (Theorem 15.2a, b, c, d).

The inclusions between the arithmetical classes are depicted in Fig. 15.1.

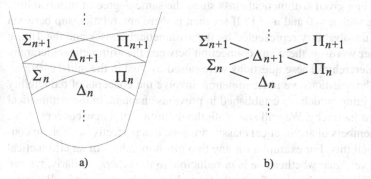

Fig. 15.1 a) The inclusions between the arithmetical classes at two successive levels; b) Hasse diagram representing the classes of two successive levels ordered by inclusion

To get some feeling for the arithmetical classes, let us see what kinds of sets are gathered in the classes at the lowest levels of the hierarchy. For $n = 0$, any formula $F(x)$ is just a decidable unary relation $R(x)$, so the corresponding set $\{x \in \mathbb{N} \,|\, F(x)\}$ is decidable. Obviously, this is true of the members of Π_0 and Δ_0 too. Thus,

$$\Sigma_0 = \Pi_0 = \Delta_0 = \text{ class of all decidable sets.}$$

Now take $n = 1$. The class Σ_1 contains all the sets defined by $\{x \in \mathbb{N} \mid \exists y_1 R(x, y_1)\}$. By Theorem 15.1 any such set is c.e., so

$$\Sigma_1 = \textit{class of all c.e. sets.}$$

What about the class Π_1? By definition it is the class of all the sets that are defined by $\{x \in \mathbb{N} \mid \forall y_1 R(x, y_1)\}$. But any such set is the complement of some c.e. set—namely, the set $\{x \in \mathbb{N} \mid \exists y_1 \neg R(x, y_1)\}$—and is therefore called *co-c.e.* Consequently,

$$\Pi_1 = \textit{class of all co-c.e. sets.}$$

The class Δ_1 is, by definition, equal to $\Sigma_1 \cap \Pi_1$, so it contains all the sets that are both c.e. and co-c.e. But, by Theorem 7.3 (p. 144), such sets are decidable. It follows that

$$\Delta_1 = \textit{class of all decidable sets.}$$

Note that, for any of the classes $\Sigma_0, \Sigma_1, \Pi_0, \Pi_1, \Delta_0, \Delta_1$, the sets in the class have the same degree of unsolvability. We will see shortly that this is not a coincidence.

15.3 The Link with the Jump Hierarchy

What about the arithmetical classes $\Sigma_n, \Pi_n, \Delta_n, n = 2, 3, \ldots$, which reside on higher levels of the arithmetical hierarchy? What kinds of sets are gathered in these classes? Do all the sets in a given arithmetical class share the same degree of unsolvability, as was the case with $n = 0$ and $n = 1$? If so, then is there any relationship between the degrees of unsolvability represented by the arithmetical classes and the Turing degrees. In other words, is there any connection between the arithmetical hierarchy and the jump hierarchy? These questions were raised by Post in the mid-1940s.

To answer the questions we must somehow involve the concepts of reducibility and the Turing jump, which we established in previous chapters, in the arithmetical classes and their hierarchy. We will start with the definition of a new concept.

Since the members of arithmetical classes are sets, it is perfectly sensible to consider their reducibility. For example, for any two given members of an arithmetical class we can investigate whether one is *m*-reducible to the other. Similarly, we can distinguish a member of the class from other members of the class, and call it *complete* in that class if every member of the class is *m*-reducible to it. Here is the definition.

Definition 15.4. (Σ_n-Complete Set) A set $\mathcal{A}$ is Σ_n-**complete** if $\mathcal{A} \in \Sigma_n$ and $\mathcal{X} \leq_m \mathcal{A}$ for every $\mathcal{X} \in \Sigma_n$. Similarly are defined Π_n-**complete** and Δ_n-**complete** sets.

Some clues of what might be the relationship between the jump and arithmetical hierarchy will give us the class Σ_1:

a) We know that $\mathcal{K} \in \Sigma_1$ (Example 15.2). We also know that $\mathcal{K}$ is m-complete (Problem 9.2, p. 214). So $\mathcal{K}$ is Σ_1-complete. Recall that $\mathcal{K} = \emptyset'$ (Definition 12.1, p. 247). It follows that $\emptyset'$ *is* Σ_1-*complete*. Observing this we now speculate and ask:

Is it for *any* $n \geqslant 1$ true that $\emptyset^{(n)}$ is Σ_n-complete?

b) Let $\mathcal{A} \in \Sigma_1$. Then $\mathcal{A}$ is c.e. and there is an ordinary TM recognizing $\mathcal{A}$. The recognizer is equivalent to an o-TM with the oracle set $\emptyset$, say $T^\emptyset$. So, $\mathcal{A}$ is $\emptyset$-c.e. Observing this may lead us to the next bold speculation:

Is it for *any* $n \geqslant 0$ true that Σ_{n+1} consists of $\emptyset^{(n)}$-c.e. sets?

Indeed, in 1948, Post announced that the answer to both questions is *yes*. This is stated in the following theorem, which is usually called *Post's Theorem*.

Theorem 15.3. *Let $\mathcal{A} \subseteq \mathbb{N}$ and $n \geqslant 0$. Then:*

a) $\emptyset^{(n)}$ *is* Σ_n-*complete for* $n \geqslant 1$;

b) $\mathcal{A} \in \Sigma_{n+1} \Longleftrightarrow \mathcal{A}$ *is* $\emptyset^{(n)}$-*c.e.*

c) $\mathcal{A} \in \Delta_{n+1} \Longleftrightarrow \mathcal{A} \leqslant_T \emptyset^{(n)}$.

Proof idea. First, part b is proven by induction on n; this is then used to prove parts a and c. For details see the Bibliographic Notes to this chapter.   □

Since the theorem is important, we invest some time in commenting on it:

1. Part a reveals the link between the concepts of the Turing jump and the class Σ_n in terms of m-reducibility. For $n = 1$ it tells us that the set $\emptyset' (= \mathcal{K}^\emptyset = \mathcal{K})$ is in Σ_1 and that any set $\mathcal{X} \in \Sigma_1$ is m-reducible to $\mathcal{K}$. Well, we already knew that. But, for $n = 2$, it tells us that the second jump $\emptyset'' (= (\emptyset')' = \mathcal{K}' = \mathcal{K}^\mathcal{K})$ is in Σ_2 and every set $\mathcal{X} \in \Sigma_2$ is m-reducible to $\mathcal{K}^\mathcal{K}$. And for $n = 3$ it says that $\emptyset''' (= \mathcal{K}^{\mathcal{K}^\mathcal{K}}) \in \Sigma_3$ and any set in Σ_3 is m-reducible to $\emptyset'''$; that is, $\emptyset'''$ is Σ_3-complete.

2. Part b reveals the connection between any two consecutive classes Σ_n and Σ_{n+1} in terms of their relative computability. Specifically, Σ_{n+1} contains exactly the sets that would become c.e. if there existed an oracle for the set $\emptyset^{(n)}$. For example, the sets in Σ_1 are $\emptyset$-c.e. (that is, c.e.), which is nothing new. But the sets in Σ_2 are $\emptyset'$-c.e. (that is, $\mathcal{K}$-c.e.), so they can be recognized by o-TMs with the oracle set $\mathcal{K}$. Similarly, any set in Σ_3 is $\emptyset''$-c.e., so it can be recognized with an o-TM with the oracle set $\mathcal{K}^\mathcal{K}$.

3. Part c reveals the link between the concepts of the arithmetical class Δ_{n+1} and the Σ_n-complete set in terms of Turing reducibility. Indeed, it tells us that any set of the class Δ_{n+1} can be T-reduced to $\emptyset^{(n)}$, the Σ_n-complete set.

15.4 Practical Consequences: Proving the Incomputability

The link between the jump hierarchy and the arithmetical hierarchy gives rise to yet another method for proving the undecidability of sets and, consequently, of decision problems. Recall that a decision problem $\mathcal{D}$ is represented by the corresponding formal language $L(\mathcal{D})$, a subset of the standard universe Σ^* (see Sect. 8.1.2). But $L(\mathcal{D})$ is associated, via a bijection $f : \Sigma^* \to \mathbb{N}$, with the set $f(L(\mathcal{D})) \subseteq \mathbb{N}$ (see Sect. 6.3.6). Now, *if* this set is arithmetical, i.e., $f(L(\mathcal{D})) = \{x \in \mathbb{N} \,|\, F(x)\}$ with $F(x)$ complying with the Definition 15.2, then the membership of $f(L(\mathcal{D}))$ in any of the arithmetical classes $\Sigma_n, \Pi_n, \Delta_n$ uncovers the degree of undecidability of the set $L(\mathcal{D})$—and, therefore, of the decision problem $\mathcal{D}$.

So, the method is as follows. Given a decision problem $\mathcal{D}$, try to construct an arithmetical description of the set $f(L(\mathcal{D}))$, that is, try to construct a predicate $F(x)$ that complies with Definition 15.2, such that $f(L(\mathcal{D})) = \{x \in \mathbb{N} \,|\, F(x)\}$.

The steps of the method are then as follows.

Method. (Proof by Arithmetical Hierarchy) The degree of undecidability of a decision problem $\mathcal{D}$ can be found as follows:

1. Consider the set $L(\mathcal{D})$.
2. Try to prove that $f(L(\mathcal{D}))$ is arithmetical, i.e., write $f(L(\mathcal{D})) = \{x \in \mathbb{N} \,|\, F(x)\}$, where the predicate $F(x)$ is either $F(x) = \exists y_1 \forall y_2 \exists y_3 \ldots Q y_n R(x, y_1, y_2, \ldots, y_n)$ or $F(x) = \forall y_1 \exists y_2 \forall y_3 \ldots Q y_n R(x, y_1, y_2, \ldots, y_n)$, and R is a decidable relation, for some $n \geqslant 0$.
3. If step 2 succeeded and n is minimal, then $L(\mathcal{D})$ (and hence $\mathcal{D}$) belongs to $\mathbf{0}^{(n)}$.

We now give some examples.

Example 15.4. (*Halting Problem $\mathcal{D}_{Halt}$*) The language of $\mathcal{D}_{Halt}$ is $L(\mathcal{D}_{Halt}) = \{\langle T, w \rangle \,|\, T \text{ halts on } w\}$ $= \mathcal{K}_0$. We can rewrite it as $\mathcal{K}_0 = \{\langle T, w \rangle \,|\, \exists s : T \text{ halts on } w \text{ in at most } s \text{ steps and returns a result}\}$. The relation $R(T, w, s) \overset{\text{def}}{=}$ "T halts on w in at most s steps and returns a result" is decidable. So, $\mathcal{K}_0 = \{\langle T, w \rangle \,|\, \exists s R(T, w, s)\}$. The corresponding subset of $\mathbb{N}$ is $\mathcal{K}_0 = \{x \,|\, \exists s R_{Halt}(\pi_1^2(x), \pi_2^2(x), s)\}$, where π_1^2, π_2^2 are projection functions and R_{Halt} is from Example 15.1. Thus, $\mathcal{K}_0 \in \Sigma_1$. $\qquad \square$

Example 15.5. (Empty Proper Set, $\mathcal{D}_{Emp}$) The proper set of a Turing machine T is the set $L(T) \overset{\text{def}}{=} \{w \in \Sigma^* \,|\, T \text{ accepts } w\}$ (see Definition 6.8, p. 131), and the decision problem $\mathcal{D}_{Emp}$ asks whether or not $L(T) = \emptyset$ (see Sect. 8.3.1). The language of $\mathcal{D}_{Emp}$ is $L(\mathcal{D}_{Emp}) = Emp = \{\langle T \rangle \,|\, L(T) = \emptyset\}$. We can restate it: $Emp = \{\langle T \rangle \,|\, \forall w \forall s : T \text{ does not accept } w \text{ in } s \text{ steps}\}$. The relation $R(T, w, s) \overset{\text{def}}{=}$ "T does not accept w in s steps" is decidable (just run T on w for at most s steps). So, $Emp = \{\langle T \rangle \,|\, \forall w \forall s R(T, w, s)\}$. The corresponding subset of $\mathbb{N}$ is $Emp = \{x \,|\, \forall t \forall s R(x, t, s)\}$, where $x = f(\langle T \rangle)$ and $t = f(w)$. The prefix $\forall t \forall s$ can be contracted into $\forall y$, where $t = \pi_1^2(y)$ and $s = \pi_2^2(y)$, so $Emp = \{x \,|\, \forall y R(x, \pi_1^2(y), \pi_2^2(y))\}$. Hence, $Emp \in \Pi_1$.

Example 15.6. (Finite Proper Set) This is the decision problem $\mathcal{D} \equiv$ "Is $L(T)$ finite?" Its language is $L(\mathcal{D}) = \{\langle T \rangle \,|\, L(T) \text{ is finite}\}$. Clearly, $L(T)$ is finite if and only if there is an upper bound on the length of its words. That is, $L(T)$ is finite *iff* there exists an ℓ such that, for any $w \in L(T)$, $|w| \leqslant \ell$. Therefore, $L(\mathcal{D}) = \{\langle T \rangle \mid \exists \ell \forall w (w \in L(T) \Rightarrow |w| \leqslant \ell)\}$. Using the equivalence $(A \Rightarrow B) \Leftrightarrow (\neg A \vee B)$, we obtain $L(\mathcal{D}) = \{\langle T \rangle \mid \exists \ell \forall w (w \notin L(T) \vee |w| \leqslant \ell)\}$. A word is not in $L(T)$ *iff* T does not accept it, regardless of the number s of steps performed. Hence, $L(\mathcal{D}) = \{\langle T \rangle \mid \exists \ell \forall w \forall s ((T \text{ does not accept } w \text{ in } s \text{ steps}) \vee |w| \leqslant \ell)\}$. Now define the relation $R(T, w, s, \ell) \stackrel{\text{def}}{=}$ "$(T$ does not accept w in s steps$)$ $\vee$ $|w| \leqslant \ell$." The relation R is decidable (run T on w for s steps, and if w has been accepted, compare $|w|$ with ℓ). Hence, $L(\mathcal{D})$ can be expressed as $L(\mathcal{D}) = \{\langle T \rangle \mid \exists \ell \forall w \forall s R(T, w, s, \ell)\}$. After contracting $\forall w \forall s$ into $\forall t$ we obtain $L(\mathcal{D}) = \{\langle T \rangle \mid \exists \ell \forall t R(T, \pi_1^2(t), \pi_2^2(t), \ell)\}$. The corresponding subset of $\mathbb{N}$ is in Σ_2. □

Example 15.7. (Finite Function Domain, $\mathcal{D}_{\mathcal{F}in}$) This is the problem "Is dom(φ) finite?" Its language is $\mathcal{F}in = \{e \mid \text{dom}(\varphi_e(x)) \text{ is finite}\}$. Leaning on Example 15.6, we can easily prove that $\mathcal{F}in \in \Sigma_2$. □

Example 15.8. (Function Totality, $\mathcal{D}_{\mathcal{T}ot}$) "Is a computable function φ total?" This is the decision problem $\mathcal{D}_{\mathcal{T}ot}$. Its language is $L(\mathcal{D}_{\mathcal{T}ot}) = \mathcal{T}ot = \{e \mid \forall x \varphi_e(x) \!\downarrow\}$ and can be rewritten as $\mathcal{T}ot = \{e \mid \forall x \exists s R_{Halt}(e, x, s)\}$; see Example 15.1. Thus, $\mathcal{T}ot \in \Pi_2$. □

Figure 15.2 shows the initial part of the arithmetical hierarchy for $n = 1, 2, 3, 4$. The corresponding Σ_n-complete sets are $\emptyset'$, $\emptyset''$, $\emptyset'''$, $\emptyset^{(4)}$, and the Π_n-complete sets are $\overline{\emptyset'}$, $\overline{\emptyset''}$, $\overline{\emptyset'''}$, $\emptyset^{(4)}$. We mention without proof that some of these are the sets $\mathcal{K}, \mathcal{F}in, \mathcal{I}nf, \mathcal{C}of, \mathcal{T}ot$ (and their complements) that we introduced in Sect. 8.3.5 and used in Example 12.1 (p. 251). Actually, the following holds:

$$\emptyset''' \equiv_m \mathcal{C}of \qquad \overline{\emptyset'''} \equiv_m \overline{\mathcal{C}of}$$
$$\emptyset'' \equiv_m \mathcal{F}in \qquad \overline{\emptyset''} \equiv_m \mathcal{T}ot \equiv_m \mathcal{I}nf$$
$$\emptyset' \equiv_m \mathcal{K} \qquad \overline{\emptyset'} \equiv_m \overline{\mathcal{K}}$$

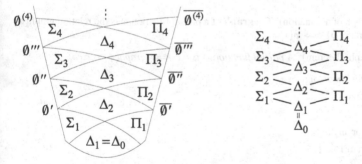

Fig. 15.2 The lowest levels of the arithmetical hierarchy: $(n = 1)$ Σ_1 is the class of c.e. sets; Π_1 is the class of co-c.e. sets; Δ_1 is the class of decidable sets; $\emptyset'$ is Σ_1-complete; $\overline{\emptyset'}$ is Π_1-complete; $(n = 2)$ Σ_2 is the class of $\emptyset'$-c.e. sets; Π_2 is the class of $\emptyset'$-co c.e. sets; Δ_2 is the class of $\emptyset'$-decidable sets; $\emptyset''$ is Σ_2-complete; $\overline{\emptyset''}$ is Π_2-complete; $(n = 3)$ Σ_3 is the class of $\emptyset''$-c.e. sets; Π_3 is the class of $\emptyset''$-co-c.e. sets; Δ_3 is the class of $\emptyset''$-decidable sets; $\emptyset'''$ is Σ_3-complete; $\overline{\emptyset'''}$ is Π_3-complete; $(n = 4)$ Σ_4 is the class of $\emptyset'''$-c.e. sets; Π_4 is the class of $\emptyset'''$-co-c.e. sets; Δ_4 is the class of $\emptyset'''$-decidable sets; $\emptyset^{(4)}$ is Σ_4-complete; and $\overline{\emptyset^{(4)}}$ is Π_4-complete

15.5 Chapter Summary

A k-ary relation R is decidable if there is a Turing machine capable of deciding, for any given k-tuple of elements $(a_1, a_2, \ldots, a_k)$, whether or not $R(a_1, a_2, \ldots, a_k)$. A set $\mathcal{A}$ is arithmetical if it can be represented as $\mathcal{A} = \{x \in \mathbb{N} \mid F(x)\}$ such that, for some $n \geqslant 0$, the predicate $F(x) = \exists y_1 \forall y_2 \exists y_3 \ldots Q y_n R(x, y_1, y_2, \ldots, y_n)$ or $F(x) = \forall y_1 \exists y_2 \forall y_3 \ldots Q y_n R(x, y_1, y_2, \ldots, y_n)$, and R is a decidable relation on $\mathbb{N}$. The sets defined by the properties $F(x) = \exists y_1 \forall y_2 \exists y_3 \ldots Q y_n R(x, y_1, y_2, \ldots, y_n)$ are in the arithmetical class Σ_n, and the sets defined by $F(x) = \forall y_1 \exists y_2 \forall y_3 \ldots Q y_n R(x, y_1, y_2, \ldots, y_n)$ are in the class Π_n. The intersection of Σ_n and Π_n is the class Δ_n. The classes Σ_n and Π_n are proper subclasses of both Σ_{n+1} and Π_{n+1}, and Δ_n is a proper subclass of Σ_n and Π_n. A member $\mathcal{A}$ of Σ_n is Σ_n-complete if any member of Σ_n is m-reducible to $\mathcal{A}$; and a member $\mathcal{B}$ of Π_n is Π_n-complete if any member of Π_n is m-reducible to $\mathcal{B}$.

There is a connection between the arithmetical hierarchy and the jump hierarchy. Specifically, a) the set $\emptyset^{(n)}$ is Σ_n-complete (for $n \geqslant 1$); b) the class Σ_{n+1} consists of exactly $\emptyset^{(n)}$-c.e. sets; and c) any member of Δ_{n+1} is T-reducible to $\emptyset^{(n)}$. In particular, Δ_1 contains decidable sets, Σ_1 contains c.e. sets, and Π_1 contains co-c.e. sets. Thus, the jump hierarchy is interleaved with the arithmetical hierarchy.

We can use the arithmetical hierarchy in proving the undecidability of a set (or a decision problem). To do this, we must express the set (or the language of the problem) as an arithmetical set. Then, its degree of undecidability is obtained from the index of the lowest arithmetical class that contains it.

Problems

Definition 15.5. (Graph of a Function) The **graph** of a partial function φ is the relation $\mathrm{graph}(\varphi)$ defined by $(x, y) \in \mathrm{graph}(\varphi) \Longleftrightarrow \varphi(x) = y$.

Theorem 15.4. (Graph Theorem) *A partial function φ is p.c.* $\Longleftrightarrow \mathrm{graph}(\varphi)$ *is c.e.*

15.1. Prove the *Graph Theorem* above.

15.2. Prove:

(a) $\Sigma_n \subseteq \Sigma_m \cap \Pi_m$, for any $m > n$.
(b) $\Pi_n \subseteq \Sigma_m \cap \Pi_m$, for any $m > n$.

   [*Hint.* Use Theorem 15.2.]

15.3. Prove:

(a) $\mathcal{A} \in \Sigma_n \Longrightarrow \mathcal{A} \times \mathcal{A} \in \Sigma_n$
(b) $\mathcal{A}, \mathcal{B} \in \Sigma_n \Longrightarrow \mathcal{A} - \mathcal{B} \in \Delta_{n+1}$

15.4. Prove:

(a) $\mathcal{A}, \mathcal{B} \in \Sigma_n \Longrightarrow \mathcal{A} \cup \mathcal{B} \in \Sigma_n$
(b) $\mathcal{A}, \mathcal{B} \in \Sigma_n \Longrightarrow \mathcal{A} \cap \mathcal{B} \in \Sigma_n$
(c) $\mathcal{A}, \mathcal{B} \in \Pi_n \Longrightarrow \mathcal{A} \cup \mathcal{B} \in \Pi_n$
(d) $\mathcal{A}, \mathcal{B} \in \Pi_n \Longrightarrow \mathcal{A} \cap \mathcal{B} \in \Pi_n$

   [*Hint.* $\forall y_1 \exists y_2 \ldots R \wedge \forall z_1 \exists z_2 \ldots S = \forall y_1 \forall z_1 \exists y_2 \exists z_2 \ldots (R \wedge S).$]

15.5. Prove:

(a) $\mathcal{A} \leq_m \mathcal{B} \wedge \mathcal{B} \in \Sigma_n \implies \mathcal{A} \in \Sigma_n$

(b) $\mathcal{A} \leq_m \mathcal{B} \wedge \mathcal{B} \in \Pi_n \implies \mathcal{A} \in \Pi_n$

[*Hint.* Let $x \in \mathcal{A} \Leftrightarrow f(x) \in \mathcal{B}$ where f is a computable function. If $R(x, y_1, \ldots, y_n)$ is a decidable relation, so is $R(f(x), y_1, \ldots, y_n)$.]

Bibliographic Notes

- The *prenex normal form* of a formula and the transformation to obtain it were described by Tarski and Kuratowski. For the description of the Tarski-Kuratowski transformation algorithm see Mendelson [105, Chap. 2] or Rogers [139, Chap. 14].
- The *arithmetical hierarchy* was first studied by Kleene [81] and Mostowski [108].
- The link between the jump hierarchy and the arithmetical hierarchy appeared in Post [127]. For the proof of Theorem 15.3, see Cooper [29, Chap. 10]. There, the reader can also find the proofs that sets $\mathcal{K}, \mathcal{F}in, \mathcal{I}nf, \mathcal{C}of, \mathcal{T}ot$ and their complements are Σ_n- or Π_n-complete for small n.

Chapter 16
Further Reading

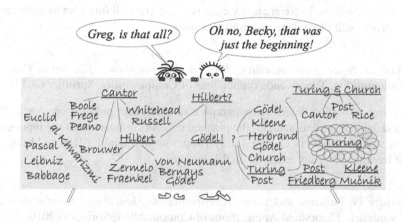

This text has been designed so that it can serve as a stepping stone to a more advanced study of *Computability Theory*, or as an introduction to *Computational Complexity Theory*. Here are some suggestions for further reading.

S. Barry Cooper, *Computability Theory*, Chapman & Hall/CRC Mathematics (2004).

Cooper's monograph consists of three parts. Being now acquainted with the fundamental concepts of *Computability Theory*, you should have no problems with the first part. But carefully reading it, you will complement your current knowledge and view many issues from different perspectives. The second part starts with oracle computations, which should be easy for you, and proceeds to topics fundamental to *Computational Complexity Theory*. The third part brings in advanced topics about degree structures, forcing, determinacy, and applications to mathematics and science. There are many examples and exercises.

© Springer-Verlag Berlin Heidelberg 2015
B. Robič, *The Foundations of Computability Theory*,
DOI 10.1007/978-3-662-44808-3_16

Rebecca Weber, *Computability Theory*, Student Mathematical Library, vol. 62, American Mathematical Society (2012).

Weber's monograph, similarly to Cooper's, will complement your current knowledge and present it from other perspectives. The second half of the monograph will give you additional explanation of methods, tools, and arithmetical hierarchy, and the last chapter will give you a taste of various areas of *Computability Theory* where research is currently active.

Robert I. Soare, *Recursively Enumerable Sets and Degrees*, Springer (1987).

Soare's monograph is a concise survey of *Computability Theory* during the periods 1931–1943, 1944–1960, and 1961–1987. It will deepen and broaden the fundamental concepts that you are now acquainted with, and bring you deep into *Post's Problem*, oracle constructions, and finitary and infinitary methods for constructing c.e. sets and degrees. There are many exercises and you will find a lot of information there. Proofs will often demand additional work.

Robert I. Soare, *Computability Theory and Applications: The Art of Classical Computability*, Theory and Applications of Computability, Springer (2015).

At the time of writing, Soare's monograph is scheduled for publication shortly. Publicly available information about the planned contents indicates that the monograph will survey the results in *Computability Theory* up to the mid-2010s and bring you to the frontier of current research on this theory.

Rodney G. Downey and Denis R. Hirschfeldt, *Algorithmic Randomness and Complexity*, Theory and Applications of Computability, Springer (2010).

Downey-Hirschfeldt's monograph will introduce you to a research area that has been flourishing since the late 1990s. It will explain to you how relative computability, information content, and randomness interact.

André Nies, *Computability and Randomness*, Oxford Logic Guides, Oxford University Press (2009).

Nies's monograph will explain to you how *Computability Theory* is used in the study of randomness of sets of natural numbers; conversely, it will show you how ideas originating from randomness are used to enrich *Computability Theory*. You will find many advanced topics that will extend your current knowledge.

Appendix A
Mathematical Background

In this appendix, we review the basic notions, concepts and facts of logic, set theory, algebra, analysis, and formal language theory that are used throughout this book. For further details see, for example, [105, 134, 138] for logic, [61, 56, 88] for set theory, [40, 74, 119] for algebra, [51] for formal languages, and [77, 142] for mathematical analysis.

Propositional Calculus P

Syntax

- An *expression* of **P** is a finite sequence of symbols. Each symbol denotes either an individual constant or an individual variable, or it is a logic connective or a parenthesis. *Individual constants* are denoted by $a, b, c, \ldots$ (possibly indexed). *Individual variables* are denoted by $x, y, z, \ldots$ (possibly indexed). *Logic connectives* are $\vee, \wedge, \Rightarrow, \Leftrightarrow$ and $\neg$; they are called the disjunction, conjunction, implication, equivalence, and negation, respectively. *Punctuation marks* are parentheses.
- Not every expression of **P** is well-formed. An expression of **P** is *well formed* if it is either 1) an individual-constant or individual-variable symbol, or 2) one of the expressions $F \vee G$, $F \wedge G$, $F \Rightarrow G$, $F \Leftrightarrow G$, and $\neg F$, where F and G are well-formed expressions of **P**. A well-formed expression of **P** is called a *sentence*.

Semantic

- The intended meanings of the logic connectives are: "or" ($\vee$), "and" ($\wedge$), "implies" ($\Rightarrow$), "if and only if" ($\Leftrightarrow$), "not" ($\neg$).
- Let $\{\top, \bot\}$ be a set. The elements $\top$ and $\bot$ are called *logic values* and stand for "true" and "false," respectively. Often, 1 and 0 are used instead of $\top$ and $\bot$, respectively.
- Any sentence has either the truth value $\top$ or $\bot$. A sentence is said to be *true* if its truth value is $\top$, and *false* if its truth value is $\bot$. Individual constants and individual variables obtain their truth values by assignment. When logic connectives combine sentences into new sentences, the truth value of the new sentence is determined by truth values of its component sentences. Specifically, let E and F be sentences. Then:

  - $\neg E$ is true if E is false, and $\neg E$ is false if E is true.
  - $E \vee F$ is false if both E and F are false; otherwise $E \vee F$ is true.
  - $E \wedge F$ is true if both E and F are true; otherwise $E \wedge F$ is false.
  - $E \Rightarrow F$ is false if E is true and F is false; otherwise $E \Rightarrow F$ is true.

© Springer-Verlag Berlin Heidelberg 2015
B. Robič, *The Foundations of Computability Theory*,
DOI 10.1007/978-3-662-44808-3

– E ⇔ F is true if both E and F are either true or false; else, E ⇔ F is false.

• The following hold: ¬(E ∨ F) ⇔ (¬E) ∧ (¬F) and ¬(E ∧ F) ⇔ (¬E) ∨ (¬F).

First-Order Logic L

Syntax

• An *expression* of **L** is a finite sequence of symbols, where each symbol is an individual-constant symbol (e.g., a, b, c), an individual-variable symbol (e.g., x, y, z), a logic connective (∨, ∧, ⇒, ⇔, ¬), a function symbol (e.g., f, g, h), a predicate symbol (e.g., P, Q, R), a quantification symbol (∀, ∃), or a punctuation mark (e.g., colon, parenthesis). (Predicates are also called *relations*.)

• We are only interested in the well-formed expressions of **L**. To define these, we need two definitions. First, a *term* is either 1) an individual-constant or individual-variable symbol, or 2) a function symbol applied to terms (e.g., f(a, x)). Second, an *atomic formula* is a predicate symbol applied to terms (e.g., P(y, f(a, x))). Finally, we say that an expression of **L** is *well formed* if it is either 1) an atomic formula, or 2) one of the expressions F ∨ G, F ∧ G, F ⇒ G, F ⇔ G, ¬F, ∀τF, and ∃τF, where F and G are well-formed expressions of **L** and τ is an individual-variable symbol. A logic expression of **L** that is well formed is called a *formula*.

Semantic

• The intended meanings of the quantification symbols are: "for all" (∀), "exists" (∃). For the meanings of logic connectives, see *Propositional Calculus* **P** above.

• The truth value of a formula is determined as follows. Let E and F be formulas. Then:

– ∀τF is true if F is true for every possible assignment of a value to τ.

– ∃τF is true if F is true for at least one possible assignment of a value to τ.

– For the truth values of F ∨ G, F ∧ G, F ⇒ G, F ⇔ G, ¬F, see *Propositional Calculus* **P** above.

Sets

Basics

• Given any objects $a_1, \ldots, a_n$, the *set* containing $a_1, \ldots, a_n$ as its only elements is denoted by $\{a_1, \ldots, a_n\}$. More generally, given a property P, the set of those elements having the property P is written as $\{x \mid P(x)\}$. If an element x is in a set $\mathcal{A}$, we say that x is a *member* of $\mathcal{A}$ and write $x \in \mathcal{A}$; otherwise, we write $x \notin \mathcal{A}$ and say that x is not a member of $\mathcal{A}$. The set with no members is called the *empty set* and denoted by $\emptyset$.

• For sets $\mathcal{A}$ and $\mathcal{B}$, we say that $\mathcal{A}$ is a *subset* of $\mathcal{B}$, written $\mathcal{A} \subseteq \mathcal{B}$, if each member of $\mathcal{A}$ is also a member of $\mathcal{B}$. A set $\mathcal{A}$ is a *proper subset* of $\mathcal{B}$, written $\mathcal{A} \subsetneq \mathcal{B}$, if $\mathcal{A} \subseteq \mathcal{B}$ but there is a member of $\mathcal{B}$ not in $\mathcal{A}$. Instead of $\subsetneq$ we also write $\subset$.

• Sets $\mathcal{A}$ and $\mathcal{B}$ are *equal*, written $\mathcal{A} = \mathcal{B}$, if $\mathcal{A} \subseteq \mathcal{B}$ and $\mathcal{B} \subseteq \mathcal{A}$.

• Given a set $\mathcal{A} = \{x_\iota \mid \iota \in \mathcal{I}\}$, the set $\mathcal{I}$ is called the *index set* of $\mathcal{A}$.

• By $(a_1, \ldots, a_n)$, or also by $\langle a_1, \ldots, a_n \rangle$, we denote the *ordered n-tuple* of objects $a_1, \ldots, a_n$. When $n = 2$, the n-tuple is called the ordered *pair*. Two ordered n-tuples $(a_1, \ldots, a_n)$ and $(b_1, \ldots, b_n)$ are *equal*, denoted by $(a_1, \ldots, a_n) = (b_1, \ldots, b_n)$, if $a_i = b_i$ for $i = 1, \ldots, n$.

Operations on Sets

- The *union* of sets $\mathcal{A}$ and $\mathcal{B}$, written as $\mathcal{A} \cup \mathcal{B}$, is the set of elements that are members of at least one of $\mathcal{A}$ and $\mathcal{B}$.
- The *intersection* of sets $\mathcal{A}$ and $\mathcal{B}$, written as $\mathcal{A} \cap \mathcal{B}$, is the set of elements that are members of both $\mathcal{A}$ and $\mathcal{B}$. We say that $\mathcal{A}$ and $\mathcal{B}$ are *disjoint* if $\mathcal{A} \cap \mathcal{B} = \emptyset$.
- The *difference* of sets $\mathcal{A}$ and $\mathcal{B}$, written as $\mathcal{A} - \mathcal{B}$, is the set of those members of $\mathcal{A}$ which are not in $\mathcal{B}$.
- If $\mathcal{A} \subseteq \mathcal{B}$, then the *complement of $\mathcal{A}$ with respect to $\mathcal{B}$* is the set $\mathcal{B} - \mathcal{A}$.
- The *power set* of a set $\mathcal{A}$ is the set of all subsets of $\mathcal{A}$ and is denoted by $2^{\mathcal{A}}$.
- The *Cartesian product* of a finite sequence of sets $\mathcal{A}_1, \ldots, \mathcal{A}_n$ is the set of all ordered n-tuples $(a_1, \ldots, a_n)$, where $a_i \in \mathcal{A}_i$ for each i. In this case it is denoted by $\mathcal{A}_1 \times \ldots \times \mathcal{A}_n$. If $\mathcal{A}_1 = \ldots = \mathcal{A}_n = \mathcal{A}$, the Cartesian product is denoted by $\mathcal{A}^n$. By convention, $\mathcal{A}^1$ stands for $\mathcal{A}$.

Relations

Basics

- An n-ary *relation* on a set $\mathcal{A}$ is a subset of $\mathcal{A}^n$. When $n = 2$ we say that the relation is *binary*, or in short, a *relation*. If R is a relation, we write xRy to indicate that $(x,y) \in R$. A 1-ary relation on $\mathcal{A}$ is a subset of $\mathcal{A}$, and is called a *property* on $\mathcal{A}$.
- A relation R on $\mathcal{A}$ is:

  - *reflexive* if xRx for each $x \in \mathcal{A}$.
  - *irreflexive* if xRx for no $x \in \mathcal{A}$.
  - *symmetric* if xRy implies yRx, for arbitrary $x,y \in \mathcal{A}$.
  - *asymmetric* if xRy implies that not yRx, for arbitrary $x,y \in \mathcal{A}$.
  - *anti-symmetric* if xRy and yRx imply $x = y$, for arbitrary $x,y \in \mathcal{A}$.
  - *transitive* if xRy and yRz imply xRz, for arbitrary $x,y,z \in \mathcal{A}$.

Ordered Sets

- A *preordered set* is a pair $(\mathcal{A},R)$, where $\mathcal{A}$ is a set and R a binary relation on $\mathcal{A}$, such that (i) R is reflexive, and (ii) R is transitive. In this case, we say that R is a *preorder* on $\mathcal{A}$. Two elements $x,y \in A$ are *incomparable* by R (in short, *R-incomparable*) if neither xRy nor yRx.
- A *partially ordered* set is a pair $(\mathcal{A},R)$, where $\mathcal{A}$ is a set and R a binary relation on $\mathcal{A}$, such that (i) R is reflexive, (ii) R is transitive, and (iii) R is anti-symmetric. In this case, we say that R is a *partial order* on $\mathcal{A}$. A partial order is often denoted by $\leq, \leqslant, \preceq, \preccurlyeq$ or any other symbol indicating the properties of this order.
- Let $(\mathcal{A}, \preccurlyeq)$ be a partially ordered set. The relation $\prec$ on $\mathcal{A}$ is the *strict partial order* corresponding to $\preccurlyeq$ if $a \prec b \Leftrightarrow a \preccurlyeq b \land a \neq b$, for arbitrary $a,b \in \mathcal{A}$. We say that $\prec$ is the *irreflexive reduction* of $\preccurlyeq$. Conversely, $\preccurlyeq$ is the *reflexive closure* of $\prec$, since $a \preccurlyeq b \Leftrightarrow a \prec b \lor a = b$.
- Let $(\mathcal{A}, \preccurlyeq)$ be a partially ordered set and $a,b \in \mathcal{A}$. When $a \preccurlyeq b$, we say that a is *smaller than or equal to* (or *lower than or equal to*) b. Correspondingly, we say that b is *larger than or equal to* (or *higher than or equal to*) a. When $a \prec b$, we say that a is *smaller than* (or *lower than*, or *below*) b. Correspondingly, we say that b is *larger than* (or *higher than*, or *above*) a.
- Let $(\mathcal{A}, \preccurlyeq)$ be a partially ordered set and $a,b,c,d \in \mathcal{A}$. Then we say:

  - a is $\preccurlyeq$-*minimal* if $x \preccurlyeq a$ implies $x = a$ for $\forall x \in \mathcal{A}$ (nothing in $\mathcal{A}$ is smaller than a).
  - b is $\preccurlyeq$-*least* if $b \preccurlyeq x$ for $\forall x \in \mathcal{A}$ (b is smaller than any other in $\mathcal{A}$).
  - c is $\preccurlyeq$-*maximal* if $c \preccurlyeq x$ implies $x = c$ for $\forall x \in \mathcal{A}$ (nothing in $\mathcal{A}$ is greater than c).
  - d is $\preccurlyeq$-*greatest* if $x \preccurlyeq d$ for $\forall x \in \mathcal{A}$ (d is greater than any other in $\mathcal{A}$).

When the relation $\preccurlyeq$ is understood, we can drop the prefix "$\preccurlyeq$-". The least and greatest elements are called the *zero* (0) and *unit* (1) element, respectively.

- Let $(\mathcal{A}, \preccurlyeq)$ be a partially ordered set, $\mathcal{B} \subseteq \mathcal{A}$, and $u, v, w, z \in \mathcal{A}$. Then we say:

  - u is a $\preccurlyeq$-*upper bound* of $\mathcal{B}$ if $x \preccurlyeq u$ for all $x \in \mathcal{B}$.
  - v is a $\preccurlyeq$-*least upper bound* (or $\preccurlyeq$-*lub*) of $\mathcal{B}$ if v is a $\preccurlyeq$-upper bound of $\mathcal{B}$ and $v \preccurlyeq u$ for every $\preccurlyeq$-upper bound u of $\mathcal{B}$.
  - w is a $\preccurlyeq$-*lower bound* of $\mathcal{B}$ if $w \preccurlyeq x$ for all $x \in \mathcal{B}$.
  - z is a $\preccurlyeq$-*greatest lower bound* (or $\preccurlyeq$-*glb*) of $\mathcal{B}$ if z is a $\preccurlyeq$-lower bound of $\mathcal{B}$ and $w \preccurlyeq z$ for every $\preccurlyeq$-lower bound w of $\mathcal{B}$.

  When the relation $\preccurlyeq$ is understood, we can drop the prefix "$\preccurlyeq$-".

- A *lattice* is a partially ordered set $(\mathcal{A}, \preccurlyeq)$ in which any two elements have an lub and a glb. The lub of $a, b \in \mathcal{A}$ is denoted by $a \vee b$, and the glb by $a \wedge b$. An *upper semi-lattice* is a partially ordered set $(\mathcal{A}, \preccurlyeq)$ in which any two elements have an lub (but not necessarily a glb).
- A *linearly* (or *totaly*) *ordered* set is a partially ordered set $(\mathcal{A}, \preccurlyeq)$ such that for all $x, y \in \mathcal{A}$ either $x \preccurlyeq y$ or $y \preccurlyeq x$. In this case we say that $\preccurlyeq$ is a *linear order* on $\mathcal{A}$.
- A *well-ordered* set is a linearly ordered set $(\mathcal{A}, \preccurlyeq)$ such that every non-empty subset of $\mathcal{A}$ has a $\preccurlyeq$-least element. We say that such a $\preccurlyeq$ is a *well-order* on $\mathcal{A}$.
- Associated with every well-ordered set $(\mathcal{A}, \preccurlyeq)$ is the corresponding *Principle of Complete Mathematical Induction*: If P is a property such that, for any $b \in \mathcal{A}$, $P(b)$, whenever $P(a)$ for all $a \in \mathcal{A}$ such that $a \preccurlyeq b$, then $P(x)$ for all $x \in \mathcal{A}$. When $\mathcal{A}$ is infinite, a proof using this principle is called a proof by *transfinite induction*.

Equivalence Relations

- A relation R on $\mathcal{A}$ is an *equivalence relation* if (i) R is reflexive, (ii) R is symmetric, and (iii) R is transitive. In this case, the *R-equivalence class* of $a \in \mathcal{A}$ is the set $\{x \in \mathcal{A} \mid xRa\}$. Elements of the R-equivalence class of a are said to be *R-equivalent* to a. If C is an equivalence class, any element of C is called a *representative* of the class C.
- A *partition* of $\mathcal{A}$ is any collection $\{\mathcal{A}_i \mid i \in \mathcal{I}\}$ of nonempty subsets of $\mathcal{A}$ such that (i) $\mathcal{A} = \bigcup_{i \in \mathcal{I}} \mathcal{A}_i$, and (ii) $\mathcal{A}_i \cap \mathcal{A}_j = \emptyset$, for all $i, j \in \mathcal{I}$ with $i \neq j$. So, $\mathcal{A}$ is the disjoint union of the sets in the partition.
- Any equivalence relation on $\mathcal{A}$ is associated with a partition of $\mathcal{A}$, and vice versa. If R is an equivalence relation on $\mathcal{A}$, then the associated partition of $\mathcal{A}$ is called the *quotient set of $\mathcal{A}$ relative to R* and is denoted by $\mathcal{A}/R$. The members of $\mathcal{A}/R$ are the R-equivalence classes of $\mathcal{A}$. The function $f : \mathcal{A} \to \mathcal{A}/R$ that associates with each element $a \in \mathcal{A}$ the R-equivalence class of a is called the *natural map* of $\mathcal{A}$ relative to R.
- An equivalence relation is often denoted by $\sim, \simeq, \equiv$, or any other symbol indicating the properties of this relation.

Functions

Basics

- A total *function f from $\mathcal{A}$ into $\mathcal{B}$* is a triple $(\mathcal{A}, \mathcal{B}, f)$ where $\mathcal{A}$ and $\mathcal{B}$ are nonempty sets and for every $x \in \mathcal{A}$ there is a unique member, denoted by $f(x)$, of $\mathcal{B}$. We call $\mathcal{A}$ the *domain* of f and denote it by dom(f). The set $\mathcal{B}$ we call the *co-domain* of f and denote it by codom(f). We usually write $f : \mathcal{A} \to \mathcal{B}$ instead of $(\mathcal{A}, \mathcal{B}, f)$. A function is also called a *mapping*.
- In specifying a definition of $f : \mathcal{A} \to \mathcal{B}$ we say that f is *well-defined* if we are assured that f is single-valued, i.e., with each member of $\mathcal{A}$, f associates a *unique* member of $\mathcal{B}$.
- When the domain of a function consists of ordered n-tuples, the function is said to be *of n arguments*. A (total) *function of n arguments on a set $\mathcal{S}$* is a function f whose domain is $\mathcal{S}^n$. We write $f(a_1, \dots, a_n)$ instead of $f(\langle a_1, \dots, a_n \rangle)$.

- Let $f : \mathcal{A} \to \mathcal{B}$ and $\mathcal{C} \subseteq \mathcal{A}$. The *image* of $\mathcal{C}$ under f is a set denoted by $f(\mathcal{C})$ and defined by $f(\mathcal{C}) = \{f(x) \mid x \in \mathcal{C}\}$. In particular, $f(\mathcal{A})$ is called the *range* of f and denoted by $\mathrm{rng}(f)$.
- A function $f : \mathcal{A} \to \mathcal{B}$ is:

   - *injective* if $f(x) \neq f(y)$ whenever $x \neq y$; we also say that such an f is an *injection*.
   - *surjective* if $f(\mathcal{A}) = \mathcal{B}$; we also say that such an f is a *surjection*.
   - *bijective* if it is injective and surjective; we also say that such an f is a *bijection*.

- An element $a \in \mathcal{A}$ is called the *fixed point* of a function $f : \mathcal{A} \to \mathcal{A}$ if $f(a) = a$.
- Let $f : \mathcal{A} \to \mathcal{B}$ and $\mathcal{C} \subseteq \mathcal{A}$. Then a function $g : \mathcal{C} \to \mathcal{B}$ is the *restriction* of f to $\mathcal{C}$ if $g(x) = f(x)$ for each $x \in \mathcal{C}$. The restriction of f to $\mathcal{C}$ is denoted by $f|_{\mathcal{C}}$. In that case f is the *extension* of g to $\mathcal{A}$.
- Let $f : \mathcal{A} \to \mathcal{B}$ and $g : \mathcal{C} \to \mathcal{D}$ be functions and $f(\mathcal{A}) \subseteq \mathcal{C}$. Then the *composite function* (or *composition*) of f and g, denoted by $g \circ f$, is the function $g \circ f : \mathcal{A} \to \mathcal{D}$ defined by $(g \circ f)(x) = g(f(x))$, for each $x \in \mathcal{A}$.
- Let $\mathcal{A}$ and $\mathcal{U}$ be sets, and let $\mathcal{A} \subseteq \mathcal{U}$. The *characteristic function* of $\mathcal{A}$ is the function $\chi_{\mathcal{A}} : \mathcal{U} \to \{0,1\}$ such that $\chi_{\mathcal{A}}(u) = 1$ if $u \in \mathcal{A}$ and $\chi_{\mathcal{A}}(u) = 0$ if $u \notin \mathcal{A}$.
- Let $\mathcal{A}$ and $\mathcal{B}$ be nonempty sets. Then the set of all functions having the domain $\mathcal{A}$ and co-domain $\mathcal{B}$ is denoted by $\mathcal{B}^{\mathcal{A}}$.

Cardinality

- Two sets $\mathcal{A}$ and $\mathcal{B}$ are said to be *equinumerous* (or *equipollent* or of the same *power*), which is denoted by $\mathcal{A} \simeq \mathcal{B}$, if there exists a bijection $f : \mathcal{A} \to \mathcal{B}$. In that case we say that $\mathcal{A}$ and $\mathcal{B}$ have the *same cardinal number*. The cardinal number of a set $\mathcal{A}$ is denoted by $|\mathcal{A}|$. The relation $\simeq$ is an equivalence relation.
- A cardinal number $|\mathcal{A}|$ is *smaller* than a cardinal number $|\mathcal{B}|$, written $|\mathcal{A}| < |\mathcal{B}|$, if there is an injection $f : \mathcal{A} \to \mathcal{B}$, but $\mathcal{A}$ and $\mathcal{B}$ are not equinumerous.
- *Cantor's Theorem* states that $|\mathcal{A}| < |2^{\mathcal{A}}|$, for any set $\mathcal{A}$.
- A set $\mathcal{A}$ is

   - *finite* if either $\mathcal{A} = \emptyset$ or $\mathcal{A} \simeq \{1, 2, \ldots, n\}$ for some natural n;
   - *infinite* if it is not finite;
   - *countable* (or *enumerable*, or *denumerable*) if $\mathcal{A} \simeq \mathcal{B}$ for some $\mathcal{B} \subseteq \mathbb{N}$; when $\mathcal{B} = \mathbb{N}$, the set $\mathcal{A}$ is said to be *countably infinite*;
   - *uncountable* if it is not countable.

- If a set $\mathcal{A}$ is infinite, then there is $\mathcal{B} \subsetneq \mathcal{A}$ such that $\mathcal{B} \simeq \mathcal{A}$.
- Any subset of a countable set is countable. The union of countably many countable sets is countable. The Cartesian product of two countable sets is countable.
- Let n be a natural number, $\aleph_0 = |\mathbb{N}|$, and $c = |\mathbb{R}|$ the cardinality of continuum. Then: $\aleph_0 + n = \aleph_0$, $\aleph_0 + \aleph_0 = \aleph_0$, $n \cdot \aleph_0 = \aleph_0$, $\aleph_0^n = \aleph_0$, $c + \aleph_0 = c$, and $\aleph_0 \cdot c = c$.
- A *sequence* is a function f defined on $\mathbb{N}$, the set of natural numbers. If we write $f(n) = x_n$, for $n \in \mathbb{N}$, we also denote the sequence f by $\{x_n\}$, or by $x_0, x_1, x_2, \ldots$ When $x_n \in \mathcal{A}$ for all $n \in \mathbb{N}$, we say that $\{x_n\}$ is a *sequence of elements of* $\mathcal{A}$. The elements of any at most countable set can be arranged in a sequence.
- The cardinality of $\mathcal{B}^{\mathcal{A}}$, the set of all functions mapping $\mathcal{A}$ into $\mathcal{B}$, is $|\mathcal{B}|^{|\mathcal{A}|}$.

Operations and Algebraic Structures

- An *n-ary operation* on a set $\mathcal{A}$ is a function $\star : \mathcal{A}^n \to \mathcal{A}$. When $n = 2$, we say that the operation is *binary*. In this case we write $a \star b$ instead of $\star(a, b)$ When $n = 1$, the operation is said to be *unary* and we write $a^{\star}$ instead of $\star(a)$.
- A binary operation on a set $\mathcal{A}$ is:

   - *associative* if $a \star (b \star c) = (a \star b) \star c$, for all $a, b, c \in \mathcal{A}$.
   - *commutative* if $a \star b = b \star a$, for all $a, b \in \mathcal{A}$.

- A *semigroup* is a pair $(\mathcal{A}, \star)$, where $\star$ is an associative binary operation on $\mathcal{A}$.
- A *group* is a semigroup $(\mathcal{A}, \star)$ satisfying the following requirements:
  - there exists an element $e \in \mathcal{A}$ such that $a \star e = e \star a = a$, for all $a \in \mathcal{A}$ (e is called an *identity* of $\mathcal{A}$);
  - for each $a \in \mathcal{A}$ there exists an element $a^{-1} \in \mathcal{A}$ such that $a \star a^{-1} = a^{-1} \star a = e$ (a^{-1} is called an *inverse* of a).

Natural Numbers

- Natural numbers are $0, 1, 2, \ldots$. The set of all natural numbers is denoted by $\mathbb{N}$. The cardinal number of $\mathbb{N}$ is denoted by $\aleph_0$ (aleph zero).
- A *prime* is a natural number greater than 1 that has no positive divisors other than 1 and itself. There are infinitely many primes. A natural number greater than 1 that is not a prime is called a *composite*.
- The *Fundamental Theorem of Arithmetic* states: Any positive integer $(\neq 1)$ can be expressed as a product of primes; this expression is unique except for the order in which the primes occur. Thus, any positive integer $n (\neq 1)$ can be written as $p_1^{\alpha_1} p_2^{\alpha_2} \ldots p_r^{\alpha_r}$, where $p_1, p_2, \ldots, p_r$ are primes satisfying $p_1 < p_2 < \ldots < p_r$, and $\alpha_1, \alpha_2, \ldots, \alpha_r$ are positive integers.
- The *Principle of Mathematical Induction* is: Any subset of $\mathbb{N}$ that contains 0 and, for every natural k, contains $k + 1$ whenever it contains k, is equal to $\mathbb{N}$.
- The set $(\mathbb{N}, <)$ is well ordered. It is also denoted by ω.
- The *Principle of Complete Mathematical Induction*: Any subset of ω that, for every natural k, contains k whenever it contains every natural $i < k$ is equal to ω.
- The set of all subsets of $\mathbb{N}$, i.e., the set $2^{\mathbb{N}}$, is uncountable. Its cardinality is $2^{\aleph_0}$. This is equal to $c = |\mathbb{R}|$, the cardinality of continuum.
- Functions $f : \mathbb{N} \to \mathbb{N}$, $k \geq 1$, are called *numerical*.
- The set $\mathbb{N}^{\mathbb{N}}$ of all functions $f : \mathbb{N} \to \mathbb{N}$ is uncountable: $|\mathbb{N}^{\mathbb{N}}| = 2^{\aleph_0} = c$. In particular, the set $\{0, 1\}^{\mathbb{N}}$ of all characteristic functions $\chi : \mathbb{N} \to \{0, 1\}$ is equinumerous to the set $2^{\mathbb{N}}$. Since each χ is identified with an infinite sequence of 0s and 1s, the set of all infinite binary sequences is also uncountable.
- The *join* of two sets $\mathcal{A}, \mathcal{B} \subseteq \mathbb{N}$ is the set denoted by $\mathcal{A} \oplus \mathcal{B}$ and defined by $\mathcal{A} \oplus \mathcal{B} \overset{\text{def}}{=} \{2x \,|\, x \in \mathcal{A}\} \cup \{2y + 1 \,|\, y \in \mathcal{B}\}$. Informally, $\mathcal{A} \oplus \mathcal{B}$ "remembers" every member of $\mathcal{A}$ and every member of $\mathcal{B}$.

Formal Languages

Basics

- An *alphabet* Σ is a finite non-empty set of abstract *symbols*.
- A *word* of length $k \geq 0$ *over the alphabet* Σ is a finite sequence $x_1, \ldots, x_k$ of symbols in Σ. A word $x_1, \ldots, x_k$ is usually written without commas, i.e., as $x_1 \ldots x_k$.
- The length of a word w is denoted by $|w|$. The word of length zero is called the *empty word* and denoted by ε.
- If $w = x_1 \ldots x_k$ is a word, then the word $w^R = x_k \ldots x_1$ is called the *reversal of* w.
- Two words $x_1 \ldots x_r$ and $y_1 \ldots y_s$ over the alphabet Σ are *equal*, written $x_1 \ldots x_r = y_1 \ldots y_s$, if $r = s$ and $x_i = y_i$ for each i.
- Let x and y be words over the alphabet Σ. The word x is a *subword* of y if $y = uxv$ for some words u and v. The word x is a *proper* subword of y if x is a subword of x, but $x \neq y$.

- Let x and y be words over the alphabet Σ. The word x is a *prefix* of y, written $x \sqsubseteq y$, if $y = xv$ for some word v. The word x is a *proper* prefix of y, written $x \sqsubset y$, if x is a prefix of x, but $x \neq y$.
- The *set of all words, including ε, over the alphabet Σ* is denoted by Σ^*.
- The set Σ^* is countably infinite.
- Each subset $\mathcal{L} \subseteq \Sigma^*$ is called a *formal language* (or *language* in short).

Operations on Languages

- If $x = x_1 \ldots x_r$ and $y_1 \ldots y_s$ are words, then xy, called the *concatenation* of x and y, is the word $x_1 \ldots x_r y_1 \ldots y_s$.
- For languages $\mathcal{L}_1$ and $\mathcal{L}_2$, the *concatenation* (or *product*) of $\mathcal{L}_1$ and $\mathcal{L}_2$ is a language denoted by $\mathcal{L}_1\mathcal{L}_2$ and defined by $\mathcal{L}_1\mathcal{L}_2 = \{xy \mid x \in \mathcal{L}_1 \wedge y \in \mathcal{L}_2\}$.
- For a language $\mathcal{L}$ let $\mathcal{L}^0 = \{\varepsilon\}$ and, for each $n \geqslant 1$, let $\mathcal{L}^n = \mathcal{L}^{n-1}\mathcal{L}$. The *Kleene star of $\mathcal{L}$* is the language denoted by $\mathcal{L}^*$ and defined by $\mathcal{L}^* = \bigcup_{i=0}^{\infty} \mathcal{L}^i$. Similarly, *Kleene plus of $\mathcal{L}$* is the language denoted by $\mathcal{L}^+$ and defined by $\mathcal{L}^+ = \bigcup_{i=1}^{\infty} \mathcal{L}^i$. In particular, for the alphabet Σ, the language Σ^n contains all words of length n over Σ, and Σ^* contains all words over Σ.

Orders on Languages

- Let $\leq$ be a linear order on the alphabet Σ. A *lexicographic order* $\leq_{\text{lex}}$ on Σ^n, induced by $\leq$, is the order in which $x_1 \ldots x_n <_{\text{lex}} y_1 \ldots y_n$ if there is a $j, 1 \leq j \leq n$, such that $x_i = y_i$ for each $i = 1, \ldots, j-1$, but $x_j < y_j$.
- A *shortlex order* on a language $\mathcal{L} \subseteq \Sigma^*$ is the order in which words of $\mathcal{L}$ are primarily ordered by their increasing length, and words of the same length are then lexicographically ordered. The shortlex order is a well-order on Σ^* and, consequently, on $\mathcal{L}$.

Appendix B
Notation Index

© Springer-Verlag Berlin Heidelberg 2015
B. Robič, *The Foundations of Computability Theory*,
DOI 10.1007/978-3-662-44808-3

References

1. Aaronson, S.: Quantum Computing Since Democritus. Cambridge University Press (2013)
2. Ackermann, W.: On Hilbert's Construction of the Real Numbers. Mathematische Annalen **99**, 118–133 (1928). (Reprint in: van Heijenoort, 1999)
3. Adams, R.: An Early History of Recursive Functions and Computability: From Gödel to Turing. Docent Press (2011)
4. Adler, A.: Some Recursively Unsolvable Problems in Analysis. Proceedings of the American Mathematical Society **22**, 523–526 (1969)
5. Aho, A.V., Hopcroft, J.E., Ullman, J.D.: The Design and Analysis of Computer Algorithms, 3rd edn. Addison-Wesley (1974)
6. Aho, A.V., Lam, M.S., Sethi, R., Ullman, J.: Compilers: Principles, Techniques, and Tools, 2nd edn. Addison-Wesley (2006)
7. Ambos-Spies, K.: Polynomial Time Reducibilities and Degrees. In: E.R. Griffor (ed.) Handbook of Computability Theory, pp. 683–706. Elsevier/North-Holland (1999)
8. Ambos-Spies, K., Fejer, P.F.: Degrees of Unsolvability (2006). Unpublished paper. http://www.cs.umb.edu/~fejer/articles/History_of_Degrees.pdf
9. Baaz, M., Papadimitriou, C.H., Putnam, H.W., Scott, D.S., Harper, C.L. (eds.): Kurt Gödel and the Foundations of Mathematics: Horizons of Truth. Cambridge University Press (2011)
10. Barendregt, H.: The Lambda Calculus. Its Syntax and Semantics. College Publications (2012)
11. Barwise, J.: Handbook of Mathematical Logic. Studies in Logic and the Foundations of Mathematics. Elsevier/North-Holland (1977)
12. Benacerraf, P., Putnam, H. (eds.): Philosophy of Mathematics, 2nd edn. Cambridge University Press (1984)
13. Berger, R.: The Undecidability of the Domino Problem. Memoirs of the American Mathematical Society **66** (1966)
14. Boole, G.: An Investigation of the Laws of Thought: On Which are Founded the Mathematical Theories of Logic and Probabilities. Macmillan (1854). (Cambridge University Press, reissue edn. 2009)
15. Boolos, G.S., Burgess, J.P., Jeffrey, R.C.: Computability and Logic, 4th edn. Cambridge University Press (2002)
16. Boolos, G.S., Jeffrey, R.C.: Computability and Logic, 3rd edn. Cambridge University Press (1989)
17. Boone, W.W.: The Word Problem. Annals of Mathematics **70**(2), 207–265 (1959)
18. Brouwer, L.E.J.: On the Significance of the Principle of Excluded Middle in Mathematics, Especially in Function Theory (1923). (Reprint in: van Heijenoort, 1999)
19. Burali-Forti, C.: Una questione sui numeri transfiniti *and* Sulle classi ben ordinate. Rendiconti Circ. Mat. Palermo **11**, 154–164, 260 (1897). (A Question on Transfinite Numbers *and* On Well-Ordered Classes. Reprint in: van Heijenoort, 1999)

© Springer-Verlag Berlin Heidelberg 2015
B. Robič, *The Foundations of Computability Theory*,
DOI 10.1007/978-3-662-44808-3

20. Cantor, G.: Über eine Eigenschaft des Inbegriffes aller reellen algebraischen Zahlen. J. Math. **77**, 258–262 (1874)
21. Chaitin, G.: Meta Maths: The Quest for Omega. Atlantic Books (2006)
22. Chang, C.C., Keisler, H.J.: Model Theory, 3rd edn. Dover Books on Mathematics. Dover Publications (2012)
23. Church, A.: A Set of Postulates for the Foundation of Logic (second paper). Annals of Mathematics **34**, 839–864 (1933)
24. Church, A.: A Note on the Entscheidungsproblem. Journal of Symbolic Logic **1**, 40–41 (1936). (Reprint in: Davis, 1965)
25. Church, A.: An Unsolvable Problem of Elementary Number Theory. American Journal of Mathematics **58**, 345–363 (1936). (Reprint in: Davis, 1965)
26. Cohen, P.: Set Theory and the Continuum Hypothesis. W.A. Benjamin (1966)
27. Cook, S.A., Reckhow, R.A.: Time Bounded Random Access Machines. Journal of Computer and System Sciences **7**(4), 354–375 (1973)
28. Cooper, K., Torczon, L.: Engineering a Compiler, 2nd edn. Morgan Kaufmann (2011)
29. Cooper, S.B.: Computability Theory. Chapman & Hall/CRC Mathematics (2004)
30. Cooper, S.B., van Leeuwen, J. (eds.): Alan Turing: His Work and Impact. Elsevier (2013)
31. Copeland, B.J., Posy, C.J., Shagrir, O. (eds.): Computability: Turing, Gödel, Church, and Beyond. The MIT Press (2013)
32. Cutland, N.J.: Computability: An Introduction to Recursive Function Theory. Cambridge University Press (1980)
33. Davis, M.: On the Theory of Recursive Undecidability. Ph.D. thesis, Princeton University (1950)
34. Davis, M.: Computability and Unsolvability. McGraw-Hill (1958)
35. Davis, M.: The Undecidable. Raven Press Books (1965)
36. Davis, M.: Hilbert's Tenth Problem Is Unsolvable. The American Mathematical Monthly **80**(3), 233–269 (1973)
37. Davis, M.: Turing Reducibility. Notices of the AMS **53**(10), 1218–1219 (2006)
38. Davis, M., Sigal, R., Weyuker, E.J.: Computability, Complexity, and Languages, Fundamentals of Theoretical Computer Science, 2nd edn. Morgan Kaufmann (1994)
39. Downey, R.G., Hirschfeldt, D.R.: Algorithmic Randomness and Complexity. Theory and Applications of Computability. Springer (2010)
40. Dummit, D.S., Foote, R.M.: Abstract Algebra. Prentice-Hall (1991)
41. Ebbinghaus, H.D., Flum, J., Thomas, W.: Mathematical Logic, 2nd edn. Undergraduate Texts in Mathematics. Springer (1994)
42. Enderton, H.B.: Computability Theory: An Introduction to Recursion Theory. Academic Press (2011)
43. Epstein, R.L., Cornielli, W.A.: Computability: Computable Functions, Logic, and the Foundations of Mathematics, 3rd edn. Advanced Reasonong Forum (2008)
44. Etzion-Petruschka, Y., Harel, D., Myers, D.: On the Solvability of Domino Snake Problems. Theoret. Comput. Sci. **131**, 243–269 (1994)
45. Fernández, M.: Models of Computation: An Introduction to Computability Theory. Undergraduate Topics in Computer Science. Springer (2009)
46. Feynman, R.P.: Feynman Lectures on Computation. Penguin Books (1999)
47. Frege, G.: Begriffsschrift, eine der arithmetischen nachgebildete Formelsprache des reinen Denkens. Halle A/S, Louis Nebert, Halle (1879). (Begriffsschrift, a Formula Language, Modeled upon that of Arithmetic, for Pure Thought. Reprint in: van Heijenoort, 1999)
48. Friedberg, R.M.: Two Recursively Enumerable Sets of Incomparable Degrees of Unsolvability. In: Proc. Nat'l Acad. Sci. USA, vol. 43, pp. 236–238 (1957)
49. Gandy, R.: The Confluence of Ideas in 1936. In: R. Herken (ed.) The Universal Turing Machine: A Half-Century Survey, pp. 55–111. Oxford University Press (1988)
50. George, A., Velleman, D.J.: Philosophies of Mathematics. Willey-Blackwell (2001)
51. Ginsburg, S.: Algebraic and Automata-Theoretic Properties of Formal Languages, *Fundamental Studies in Computer Science*, vol. 2. North-Holland (1975)

52. Gödel, K.: Die Vollständigkeit der Axiome des logischen Funktionenkalküls. Monatsh. Math. Physik **37**, 349–360 (1930). (The Completeness of the Axioms of the Functional Calculus of Logic. Reprint in: van Heijenoort, 1999)
53. Gödel, K.: Über formal unentscheidbare Sätze der Principia Mathematica und verwandter Systeme I. Monatsh. Math. Physik **38**, 173–198 (1931). (On Formally Undecidable Propositions of Principia Mathematica and Related Systems I. Reprint in: van Heijenoort, 1999)
54. Gödel, K.: On Undecidable Propositions of Formal Mathematical Systems (1934). (Lectures at Institute for Advanced Study, Princeton. Reprint in: Davis, 1965)
55. Gödel, K.: On Formally Undecidable Propositions of Principia Mathematica and Related Systems. Dover Publications (1992). (Republication with the introduction by R. B. Braithwaite)
56. Goldrei, D.: Classic Set Theory. Chapman & Hall/CRC Mathematics (1996)
57. Goldreich, O.: Computational Complexity: A Conceptual Perspective. Cambridge University Press (2008)
58. Goldreich, O.: P, NP, and NP-Completeness: The Basics of Computational Complexity. Cambridge University Press (2010)
59. Griffor, E.R. (ed.): Handbook of Computability Theory, *Studies in Logic and the Foundations of Mathematics*, vol. 140. Elsevier/North-Holland (1999)
60. Halava, V., Harju, T., Hirvensalo, M.: Undecidability Bounds for Integer Matrices Using Claus Instances. International Journal of Foundations of Computer Science **18**(5), 931–948 (2007)
61. Halmos, P.R.: Naive Set Theory. Van Nostrand (1960)
62. Harel, D., Feldman, Y.: Algorithmics: The Spirit of Computing, 3rd edn. Pearson (2004)
63. van Heijenoort, J. (ed.): From Frege to Gödel: A Source Book in Mathematical Logic, 1879-1931. Harvard University Press (1999)
64. Heyting, A.: Intuitionism: An Introduction, 3rd edn. Studies in Logic and the Foundations of Mathematics. Elsevier/North-Holland (1971)
65. Hilbert, D.: Über die Grundlagen der Logik und der Aritmetik. Verhandlungen des Dritten Internationalen Mathematiker-Kongress in Heidelberg, 8.-13. August 1904 pp. 174–185 (1905). (On the Foundations of Logic and Arithmetic. Reprint in: van Heijenoort, 1999)
66. Hilbert, D.: Grundlagen der Geometrie, 7th edn. Teubner-Verlag, Leipzig, Berlin (1930)
67. Hilbert, D., Ackermann, W.: Grundzüge der theoretischen Logik. Springer (1928). (*Principles of Mathematical Logic*, Chelsea, 1950)
68. Roger Hindley, J., Jonathan P. Seldin: Lambda-Calculus and Combinatory Logic: An Introduction, 2nd edn. Cambridge University Press (2008)
69. Hodges, A.: Alan Turing: The Enigma. Vintage Books (1992)
70. Holmes, M.R., Forster, T., Libert, T.: Alternative Set Theories. In: D.M. Gabbay, Akihito Kanamori, John Woods (eds.) Handbook of the History of Logic, vol. 6. Sets and Extensions in the Twentieth Century, pp. 559–632. Elsevier (2012)
71. Homer, S., Selman, A.L.: Computability and Complexity Theory, 2nd edn. Springer (2011)
72. Hopcroft, J.E., Ullman, J.D.: Introduction to Automata Theory, Languages, and Computation. Addison-Wesley (1979)
73. Irvine, A.D.: Bertrand Russell's Logic. In: D.M. Gabbay, John Woods (eds.) Handbook of the History of Logic, vol. 5. Logic from Russell to Church, pp. 1–28. Elsevier (2009)
74. Jacobson, N.: Basic Algebra I, 2nd edn. Dover Publications (2009)
75. Jech, T.J.: The Axiom of Choice. Studies in Logic and the Foundations of Mathematics. Elsevier/North-Holland (1973)
76. Jech, T.J.: Set Theory, The Third Millennium, revised and expanded edn. Springer Monographs in Mathematics. Springer (2011)
77. Johnsonbaugh, R., Pfaffenberger, W.E.: Foundations of Mathematical Analysis. Dover Books on Mathematics. Dover Publications (2010)
78. Kanamori, A.: Set Theory from Cantor to Cohen. In: D.M. Gabbay, Akihito Kanamori, John Woods (eds.) Handbook of the History of Logic, vol. 6. Sets and Extensions in the Twentieth Century, pp. 1–72. Elsevier (2012)

79. Kleene, S.C.: General Recursive Functions of Natural Numbers. Mathematische Annalen **112**(5), 727–742 (1936). (Reprint in: Davis, 1965)

80. Kleene, S.C.: On Notations for Ordinal Numbers. Journal of Symbolic Logic **3**, 150–155 (1938)

81. Kleene, S.C.: Recursive Predicates and Quantifiers. Transactions of the American Mathematical Society **53**, 41–73 (1943)

82. Kleene, S.C.: Introduction to Metamathematics. Elsevier/North-Holland (1952)

83. Kleene, S.C.: Mathematical Logic. Wiley (1967)

84. Kleene, S.C., Post, E.: The Upper Semi-lattice of Degrees of Recursive Unsolvability. Annals of Mathematics **59**(3), 379–407 (1954)

85. Kozen, D.C.: Automata and Computability. Springer (1997)

86. Kozen, D.C.: Theory of Computation. Springer (2006)

87. Kunen, K.: Set Theory, An Introduction to Independence Proofs. Studies in Logic and the Foundations of Mathematics. Elsevier (1980)

88. Kunen, K.: The Foundations of Mathematics. Studies in Logic. College Publications (2009)

89. Kučera, A.: An Alternative, Priority-Free, Solution to Post's Problem. In: Proc. Mathematical Foundations of Computer Science (Bratislava, 1986), *Lecture Notes in Computer Science*, vol. 233, pp. 493–500. Springer (1986)

90. Lachlan, A.H.: The Impossibility of Finding Relative Complements for Recursively Enumerable Degrees. The Journal of Symbolic Logic **31**(3), 434–454 (1966)

91. Lachlan, A.H.: Lower Bounds for Pairs of Recursively Enumerable Degrees. Proceedings of the London Mathematical Society **16**(3), 537–569 (1966)

92. Lee, J.M.: Introduction to Topological Manifolds, 2nd edn. Graduate Texts in Mathematics. Springer (2011)

93. Lerman, M.: Degrees of Unsolvability: Local and Global Theory. Perspectives in Mathematical Logic. Springer (1983)

94. Lerman, M.: A Framework for Priority Arguments, *Lecture Notes in Logic*, vol. 34. Cambridge University Press (2010)

95. Levy, A.: Basic Set Theory. Dover Publications (2002)

96. Lewis, H.R., Papadimitriou, C.H.: Elements of the Theory of Computation. Prentice-Hall (1981)

97. Machover, M.: Set Theory, Logic and Their Limitations. Cambridge University Press (1996)

98. Machtey, M., Young, P.: An Introduction to the General Theory of Algorithms. Elsevier/North-Holland (1978)

99. Manin, Y.I.: A Course in Mathematical Logic. Graduate Texts in Mathematics. Springer (1977)

100. Markov, A.A.: The Theory of Algorithms. Trudy Math. Inst. Steklova **38**, 176–189 (1951). (A.M.S. transl. 15:1–14,1960)

101. Markov, A.A.: The Theory of Algorithms. Trudy Math. Inst. Steklova **42**, 3–375 (1954)

102. Markov, A.A.: Unsolvability of the Problem of Homeomorphy. Proc. Int. Cong. Math. pp. 300–306 (1958). (Russian)

103. Matiyasevič, Y.V.: Enumerable Sets are Diophantine. Dokl. Akad. Nauk SSSR **191**, 279–282 (1970). (Russian. English translation in: Soviet Math. Doklady, 11:354–357, 1970)

104. Matiyasevič, Y.V.: Diophantine Representation of Enumerable Predicates. Izv. Akad. Nauk SSSR **Ser. Mat. 35**, 3–30 (1971). (Russian)

105. Mendelson, E.: Introduction to Mathematical Logic. The University Series in Undergraduate Mathematics. Van Nostrand (1964)

106. Moore, G.H.: Zermelo's Axiom of Choice: Its Origins, Development & Influence. Dover Publications (2013)

107. Moschovakis, J.R.: The Logic of Brouwer and Heyting. In: D.M. Gabbay, John Woods (eds.) Handbook of the History of Logic, vol. 5. Logic from Russell to Church, pp. 77–126. Elsevier (2009)

108. Mostowski, A.: On Definable Sets of Positive Integers. Fundamenta Mathematicae **34**, 81–112 (1947)

109. Mostowski, A.: A Classification of Logical Systems. Studia Philosophica 4, 237–274 (1951)
110. Muchnik, A.A.: On the Unsolvability of the Problem of Reducibility in the Theory of Algorithms. Dokl. Akad. Nauk SSSR 108, 194–197 (1956). (Russian)
111. Muchnik, A.A.: Solution of Post's Reduction Problem and of Certain Other Problems in the Theory of Algorithms. Trudy Moskov. Mat. Obšč. 7, 391–405 (1958)
112. Nagel, E., Newman, J.R.: Gödel's Proof. Routledge (1958)
113. Nies, A.: Computability and Randomness. Oxford Logic Guides. Oxford University Press (2009)
114. Novikov, P.S.: On the Algorithmic Unsolvability of the Word Problem in Group Theory. Izvestiya Akademii Nauk 18, 485–524 (1954). (Translated in: A.M.S. transl 9:1–124,1958)
115. Odifreddi, P.: Reducibilities. In: E.R. Griffor (ed.) Handbook of Computability Theory, pp. 89–120. Elsevier/North-Holland (1999)
116. Odifreddi, P.G.: Classical Recursion Theory, Studies in Logic and the Foundations of Mathematics, vol. 125. Elsevier/North-Holland (1989)
117. Peano, G.: Arithmetices Principia, Novo Methodo Exposita (1889). (The Principles of Arithmetic, Presented by a New Method. Reprint in: van Heijenoort, 1999)
118. Péter, R.: Recursive Functions, 3rd edn. Academic Press (1967)
119. Pinter, C.C.: A Book of Abstract Algebra, 2nd edn. Dover Publications (2010)
120. Post, E.: Introduction to a General Theory of Elementary Propositions. American Journal of Mathematics 43, 163–185 (1921). (Reprint in: van Heijenoort, 1999)
121. Post, E.: Finite Combinatory Processes – Formulation I. Journal of Symbolic Logic 1, 103–105 (1936). (Reprint in: Davis, 1965)
122. Post, E.: Absolutely Unsolvable Problems and Relatively Undecidable Propositions: Account of an Anticipation (1941). (Submitted for publ. and rejected. Reprint in: Davis, 1965)
123. Post, E.: Formal Reductions of the General Combinatorial Decision Problem. American Journal of Mathematics 65, 197–215 (1943)
124. Post, E.: Recursively Enumerable Sets of Positive Integers and Their Decision Problems. Bulletin of the American Mathematical Society 50, 284–316 (1944). (Reprint in: Davis, 1965)
125. Post, E.: A Variant of a Recursively Unsolvable Problem. Bulletin of the American Mathematical Society 52, 264–268 (1946)
126. Post, E.: Recursive Unsolvability of a Problem of Thue. Journal of Symbolic Logic 12, 1–11 (1947). (Reprint in: Davis, 1965)
127. Post, E.: Degrees of Recursive Unsolvability. (Preliminary report). Bulletin of the American Mathematical Society 54, 641–642 (1948). Abstract
128. Potter, M.: Set Theory and its Philosophy. Oxford University Press (2004)
129. Presburger, M.: Über die Vollständigkeit eines gewissen Systems der Arithmetik ganzer Zahlen, in welchem die Addition als einzige Operation hervortritt. Comptes Rendus du Premier Congrès des Mathématiciens des Pays Slaves pp. 92–101 (1929). Warsaw, Poland
130. Priest, G.: Paraconsistent Logic. In: D.M. Gabbay, Franz Guenthner (eds.) Handbook of Philosophical Logic, vol. 6., 2nd edn., pp. 287–393. Kluwer Academic Publishers (2002)
131. Prijatelj, N.: Matematične strukture I., II., III. del. DMFA, Ljubljana (1985). (Mathematical Structures, vol. I, II, III. In Slovenian)
132. Prijatelj, N.: Osnove matematične logike, 1., 2., 3. del. DMFA, Ljubljana (1994). (Fundamentals of Mathematical Logic, vol. 1, 2, 3. In Slovenian)
133. Radó, T.: On Non-Computable Functions. The Bell System Technical Journal 41(3), 877–884 (1962)
134. Rautenberg, W.: A Concise Introduction to Mathematical Logic, 3rd edn. Universitext. Springer (2009)
135. Rice, H.G.: Classes of Recursively Enumerable Sets and Their Decision Problems. Transactions of the American Mathematical Society 74(2), 358–366 (1953)
136. Rich, E.: Automata, Computability, and Complexity, Theory and Applications. Pearson Prentice-Hall (2008)
137. Richardson, D.: Some Undecidable Problems Involving Elementary Functions of a Real Variable. The Journal of Symbolic Logic 33(4), 514–520 (1968)

138. Robbin, J.W.: Mathematical Logic: A First Course. W.A. Benjamin (1969)
139. Rogers, H.: Theory of Recursive Functions and Effective Computability. McGraw-Hill (1967)
140. Rogozhin, Y.: Small Universal Turing Machines. Theoretical Computer Science **168**(2), 215–240 (1996)
141. Rosenberg, A.L.: The Pillars of Computation Theory: State, Encoding, Nondeterminism. Universitext. Springer (2009)
142. Rudin, W.: Principles of Mathematical Analysis, 3rd edn. McGraw-Hill (1976)
143. Russell, B.: Letter to Frege (1902). (Reprint in: van Heijenoort, 1999)
144. Sacks, G.E.: A Minimal Degree Less than $0'$. Bulletin of the American Mathematical Society **67**, 416–419 (1961)
145. Sacks, G.E.: Degrees of Unsolvability, *Ann. Math. Stud.*, vol. 55, 2nd edn. Princeton University Press (1963)
146. Sacks, G.E.: The Recursively Enumerable Degrees are Dense. Annals of Mathematics **80**(2), 300–312 (1964)
147. Sainsbury, R.M.: Paradoxes, 3rd edn. Cambridge University Press (2009)
148. Shannon, C.E.: A Universal Turing Machine with Two Internal States. In: Automata Studies, *Ann. Math. Stud.*, vol. 34, pp. 157–165. Princeton University Press (1956)
149. Shen, A., Vereshchagin, N.K.: Computable Functions, *Student Mathematical Library*, vol. 19. American Mathematical Society (2003)
150. Shoenfield, J.R.: Mathematical Logic. Addison-Wesley (1967)
151. Shore, R.A.: The Structure of the Degrees of Unsolvability. In: A. Nerode, R.A. Shore (eds.) Recursion Theory, pp. 33–51. American Mathematical Society (1985). (Proceedings of Symposia in Pure Mathematics, vol. 42)
152. Shore, R.A.: The Recursively Enumerable Degrees. In: E.R. Griffor (ed.) Handbook of Computability Theory, pp. 169–197. Elsevier/North-Holland (1999)
153. Sieg, W.: Mechanical Procedures and Mathematical Experience. In: A. George (ed.) Mathematics and Mind, pp. 71–117. Oxford University Press (1994)
154. Sieg, W.: Hilbert's Proof Theory. In: D.M. Gabbay, John Woods (eds.) Handbook of the History of Logic, vol. 5. Logic from Russell to Church, pp. 321–384. Elsevier (2009)
155. Sieg, W.: Hilbert's Programs and Beyond. Oxford University Press (2013)
156. Simpson, S.G.: Degrees of Unsolvability: A Survey of Results. In: J. Barwise (ed.) Handbook of Mathematical Logic, Studies in Logic and the Foundations of Mathematics, pp. 631–652. Elsevier/North-Holland (1977)
157. Sipser, M.: Introduction to the Theory of Computation, 2nd edn. Thompson Course Technology (2006)
158. Slaman, T.A.: The Global Structure of the Turing Degrees. In: E.R. Griffor (ed.) Handbook of Computability Theory, pp. 155–168. Elsevier/North-Holland (1999)
159. Smith, P.: An Introduction to Gödel's Theorems. Cambridge Introductions to Philosophy. Cambridge University Press (2007)
160. Smullyan, R.M.: Gödel's Incompleteness Theorems. Oxford Logic Guides. Oxford University Press (1992)
161. Smullyan, R.M.: First-Order Logic. Dover Books on Mathematics. Dover Publications (1995)
162. Smullyan, R.M.: A Beginner's Guide to Mathematical Logic. Dover Books on Mathematics. Dover Publications (2014)
163. Smullyan, R.M., Fitting, M.: Set Theory and the Continuum Hypothesis. Dover Books on Mathematics. Dover Publications (2010)
164. Soare, R.I.: Recursively Enumerable Sets and Degrees. Perspectives in Mathematical Logic. Springer (1987)
165. Soare, R.I.: The History and Concept of Computability. In: E.R. Griffor (ed.) Handbook of Computability Theory, pp. 3–36. Elsevier/North-Holland (1999)
166. Soare, R.I.: An Overview of the Computably Enumerable Sets. In: E.R. Griffor (ed.) Handbook of Computability Theory, pp. 199–248. Elsevier/North-Holland (1999)

167. Soare, R.I.: Turing Oracle Machines, Online Computing, and Three Displacements in Computabiliy Theory. Annals of Pure and Applied Logic **160**, 368–399 (2009)
168. Soare, R.I.: Interactive Computing and Relativized Computability. In: B.J. Copeland, C.J. Posy, O. Shagrir (eds.) Computability: Turing, Gödel, Church, and Beyond, pp. 203–260. The MIT Press (2013)
169. Soare, R.I.: Computability Theory and Applications: The Art of Classical Computability. Theory and Applications of Computability. Springer (2015)
170. Spector, C.: On Degrees of Recursive Unsolvability. Annals of Mathematics **64**(2), 581–592 (1956)
171. Steprāns, J.: History of the Continuum in the 20th Century. In: D.M. Gabbay, Akihito Kanamori, John Woods (eds.) Handbook of the History of Logic, vol. 6. Sets and Extensions in the Twentieth Century, pp. 73–144. Elsevier (2012)
172. Stewart, I.: Concepts of Modern Mathematics. Dover Publications (1995)
173. Stillwell, J.: Classical Topology and Combinatorial Group Theory, 2nd edn. Graduate Texts in Mathematics. Springer (1993)
174. Stoll, R.R.: Set Theory and Logic. Dover Publications (1979)
175. Suppes, P.: Axiomatic Set Theory. Dover Publications (1972)
176. Tarski, A.: Introduction to Logic and the Methodology of Deductive Sciences. Dover Publications (1995)
177. Tent, K., Ziegler, M.: A Course in Model Theory. Lecture Notes in Logic. Cambridge Univeristy Press (2012)
178. Turing, A.M.: On Computable Numbers, with an Application to the Entscheidungsproblem. Proceedings of the London Mathematical Society **42**(2), 230–265 (1936). (A.M. Turing, A Correction, ibid. 43:544–546, 1937) (Reprint in: Davis, 1965)
179. Turing, A.M.: Computability and λ-definability. Journal of Symbolic Logic **2**, 153–163 (1937)
180. Turing, A.M.: Systems of Logic based on Ordinals. Proceedings of the London Mathematical Society **45**(2), 161–228 (1939). (Reprint in: Davis, 1965)
181. Urquhart, A.: Emil Post. In: D.M. Gabbay, John Woods (eds.) Handbook of the History of Logic, vol. 5. Logic from Russell to Church, pp. 617–666. Elsevier (2009)
182. van Atten, M., Kennedy, J.: Gödel's Logic. In: D.M. Gabbay, John Woods (eds.) Handbook of the History of Logic, vol. 5. Logic from Russell to Church, pp. 449–510. Elsevier (2009)
183. von Neumann, J.: First Draft of a Report on the EDVAC. (1945) (A modern interpretation in: Godfrey, M.D, Hendry, D.F.: The Computer as von Neumann Planned It, *IEEE Annals of the History of Computing*, 15(2):1993)
184. Wang, H.: Proving Theorems by Pattern Recognition, II. Bell System Technical Journal **40**, 1–41 (1961)
185. Wang, P.S.: The Undecidability of the Existence of Zeros of Real Elementary Functions. J. Assoc. Comput. Mach. **21**(4), 586–589 (1974)
186. Weber, R.: Computability Theory, *Student Mathematical Library*, vol. 62. American Mathematical Society (2012)
187. Whitehead, A.N., Russell, B.: Principia Mathematica. Cambridge University Press (1910-13). (2nd edn., 1925–27)
188. Winfried, J., Weese, M.: Discovering Modern Set Theory., vol. I. The Basics. American Mathematical Society (1996)
189. Yates, C.E.M.: Three Theorems on the Degrees of Recursively Enumerable Sets. Duke Mathematical Journal **32**(3), 461–468 (1965)
190. Yates, C.E.M.: A Minimal Pair of Recursively Enumerable Degrees. The Journal of Symbolic Logic **31**(2), 159–168 (1966)

Index

© Springer-Verlag Berlin Heidelberg 2015
B. Robič, *The Foundations of Computability Theory*,
DOI 10.1007/978-3-662-44808-3

Printed in the United States
By Bookmasters